A LEADER'S
GUIDE TO
SCIENCE
CURRICULUM
TOPIC STUDY

A LEADER'S GUIDE TO
SCIENCE CURRICULUM TOPIC STUDY

SUSAN MUNDRY
PAGE KEELEY
CAROLYN LANDEL

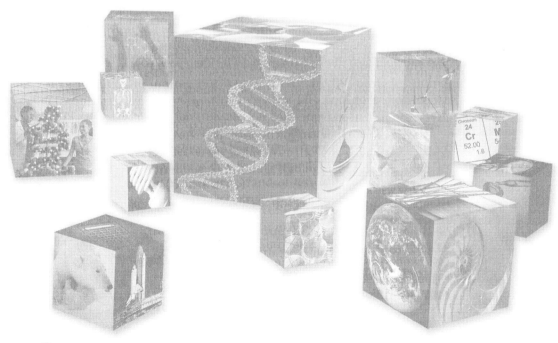

CORWIN
A SAGE Company

WestEd

Maine **MATHEMATICS** and **SCIENCE** Alliance

NSTApress®
NATIONAL SCIENCE TEACHERS ASSOCIATION

A JOINT PUBLICATION

This material is supported with funding from the National Science Foundation under a Teacher Professional Continuum, Category III, Grant no. ESI-0353315, "Curriculum Topic Study: A Systematic Approach to Utilizing National Standards and Cognitive Research," awarded to Page Keeley, PI, at the Maine Mathematics and Science Alliance. Any opinions, findings, conclusions, or recommendations expressed in this material are those of the authors and do not necessarily reflect the views of the National Science Foundation.

For information:

Corwin	SAGE India Pvt. Ltd.
A SAGE Company	B 1/I 1 Mohan Cooperative
2455 Teller Road	Industrial Area
Thousand Oaks, California 91320	Mathura Road,
(800) 233-9936	New Delhi 110 044
Fax: (800) 417-2466	India
www.corwinpress.com	
SAGE Ltd.	SAGE Asia-Pacific
1 Oliver's Yard	Pte. Ltd.
55 City Road	33 Pekin Street #02-01
London EC1Y 1SP	Far East Square
United Kingdom	Singapore 048763

Printed in the United States of America

Library of Congress Cataloging-in-Publication Data

Mundry, Susan.
A leader's guide to science curriculum topic study/Susan Mundry, Page Keeley, Carolyn Landel.
    p. cm.
"A Joint Publication with WestEd, Maine Mathematics and Science Alliance, and the National Science Teachers Association."
Includes bibliographical references and index.
ISBN 978-1-4129-7816-3 (cloth w/cd)
ISBN 978-1-4129-7817-0 (paper w/cd)

    1. Science—Study and teaching—United States. 2. Curriculum planning—United States. I. Keeley, Page. II. Landel, Carolyn. III. WestEd (Organization). IV. Maine Mathematics and Science Alliance. V. Title.

LB1585.3.M85 2010
507.1—dc22   2009024046

This book is printed on acid-free paper.

09 10 11 12 13 10 9 8 7 6 5 4 3 2 1

Acquisitions Editor:	Dan Alpert
Associate Editor:	Megan Bedell
Production Editor:	Jane Haenel
Copy Editor:	Claire Larson
Typesetter:	C&M Digitals (P) Ltd.
Proofreader:	Victoria Reed-Castro
Indexer:	Maria Sosnowski
Cover and Graphic Designer:	Michael Dubowe

Contents

Preface

OVERVIEW OF THE CTS PROJECT

Curriculum Topic Study (CTS) is a National Science Foundation–funded project that developed a process, guidelines, and materials for K–12 educators to deepen their understanding of the important science and mathematics topics they teach. CTS builds a bridge between state and national standards, research on students' ideas in science and mathematics, and opportunities for students to learn science and mathematics through improved teacher practice. Principal Investigator Page Keeley, of the Maine Mathematics and Science Alliance, directed the project, in collaboration with Co-Principal Investigator Susan Mundry, from WestEd.

The CTS process, tools, and materials engage educators in a systematic and scholarly method of using national and state standards documents and research summaries on student learning to study a curricular topic; analyze findings from the study; and apply the new knowledge gained about content, curriculum, instruction, and assessment to teaching and learning. Rather than providing the answers, CTS promotes inquiry among educators by guiding them in discovering new ideas about teaching and learning connected to the curricular topics they teach.

The four books resulting from the project include *Science Curriculum Topic Study: Bridging the Gap Between Standards and Practice* (Keeley, 2005); *Mathematics Curriculum Topic Study: Bridging the Gap Between Standards and Practice* (Keeley & Rose, 2006); this book, *A Leader's Guide to Science Curriculum Topic Study*; and *A Leader's Guide for Mathematics Curriculum Topic Study* (2010 anticipated release). In addition to the four books, there is a Web site that provides updates on the project and supplementary materials to support CTS. The URL for the Web site is www.curriculumtopicstudy.org.

THE KNOWLEDGE BASE THAT INFORMED CTS

There has never been a greater need for students to learn and excel in science. Our future society will increasingly be highly technological and scientific, yet recent reports warn that many U.S. students are not being adequately prepared to contribute to an increasingly scientific and technological workplace. As the Committee on Science Learning K–8th Grade recently wrote,

> We are underestimating what young children are capable of as students of science—the bar is almost always set too low. Moreover, the current organization of science curriculum and instruction does not provide the kind of support for science learning that results in deep understanding of scientific ideas and an ability to engage in the practices of science. (National Research Council [NRC], 2007, p. vii)

Recent results from an international assessment of students' understanding in science and the ability to use scientific knowledge to address questions and solve problems in daily life (Programme for International Student Assessment [PISA], 2007) shows U.S. students lagging behind many of their counterparts in other nations. Educators are increasingly being called on to use research-based practice and to adopt methods that support all students to reach challenging learning outcomes in science. Experienced teachers with strong backgrounds in science subject matter and extensive pedagogical content knowledge may be our very best hope for supporting student learning and interest in science. Pedagogical content knowledge (PCK) is the specialized knowledge about teaching and learning that helps teachers understand what makes the learning of specific topics easy or difficult for students and develop strategies for representing and formulating subject matter to make it accessible to learners (Shulman, 1986). Teachers with background knowledge in subject matter content and specialized knowledge of teaching science tend to produce higher achievement outcomes among their students (Darling-Hammond, 2000; Goldhaber & Brewer, 2000; Monk, 1994). Research on professional development programs in science and mathematics shows greater positive effects on student learning from programs that focus on building teachers' content knowledge and on understanding of how students learn subject matter (Brown, Smith, & Stein, 1996; Cohen & Hill, 2000; Kennedy, 1999; Weiss, Pasley, Smith, Banilower, & Heck, 2003; Wiley & Yoon, 1995). That is why effective teacher professional development in science must not only address the content teachers need to know in order to successfully teach developmentally and conceptually appropriate ideas and skills at their grade level; it must also be designed to help teachers understand how to best identify, organize, and teach important content.

The National Science Education Standards make a strong argument for why we need tools that help teachers develop PCK.

> Effective teaching requires that teachers know what students of certain ages are likely to know, understand, and be able to do; what they will learn quickly; and what will be a struggle. Teachers of science need to anticipate typical misunderstandings and to judge the appropriateness of concepts for the developmental level of their students. In addition, teachers of science must develop understanding of how students with different learning styles, abilities, and interests learn science. Teachers use all of that knowledge to make effective decisions about learning objectives, teaching strategies, assessment tasks, and curriculum materials. (NRC, 1996, p. 62)

CTS was developed to provide tools that help teachers develop this type of specialized science teacher knowledge.

PURPOSE OF THIS BOOK

All of the above describe why CTS is an important process for teachers to learn and use. In 2005, the CTS parent book, *Science Curriculum Topic Study: Bridging the Gap Between Standards and Practice* (Keeley, 2005), was published. It provides the introduction, process, and material teachers need to conduct a curriculum topic study. The purpose of this book is to support leaders in facilitating the CTS process and applications. It offers designs and suggestions using CTS in a variety of professional development configurations to improve teachers' content knowledge and various aspects of their curricular, instructional, and assessment work.

Furthermore, this guide is designed to strengthen the ways national standards and research on learning are embedded within effective professional development strategies. This *Leader's Guide* is designed to provide leaders with a standards- and research-based "tool box" filled with a variety of content-specific professional development designs, tools, and resources that will strengthen professional development and help educators become more effective teachers of science.

AUDIENCES

The primary audiences for this book include the many professionals who lead or support teacher professional development and preservice education in science in grades K–16. These include national, regional, and local science professional developers; science teacher leaders; coaches and teachers on special assignment; facilitators of professional learning communities (PLCs); state and local science specialists and supervisors; staff from school-university partnerships, including science education faculty in schools of education, science faculty in schools of arts and sciences, district-based science teachers, and faculty and student-teacher supervisors in teacher education programs; and university faculty who teach science methods courses in teacher preparation programs. All of these primary audiences can use this resource book to enhance their own teaching and provide courses, workshops, institutes, and other professional development experiences for current and future science teachers.

Secondary audiences include principals; district curriculum and assessment coordinators; informal science specialists who design and implement informal programs for K–12 students and adult consumers of science; and curriculum and assessment developers, who can all use the book as a resource to strengthen the science programs in their schools, improve the effectiveness of committee work, and increase teachers' understanding of the research on science teaching and the most important science concepts they need to teach in K–12 to improve learning results.

THE NEED FOR CTS IN PROFESSIONAL DEVELOPMENT

The CTS approach to professional development that is the subject of this book is designed to enhance teachers' understanding of the science content that is most important for all students to learn and how to improve students' opportunities to learn the content through effective curriculum, instruction, and assessment. CTS provides educators with processes, tools, and resources to link content standards and the research on learning to classroom practice. Classroom practice includes teachers' content knowledge, the curriculum or instructional materials they use, instructional contexts and strategies, and uses of assessment. CTS provides a powerful yet simple way for science educators to engage in professional development that will help them

- enhance their adult science literacy,
- explore implications for effective instruction,
- identify the key ideas and skills students need to progress through their K–12 learning,
- use research on students' ideas in science to inform teaching,
- recognize connections within and across topics in science, and
- be a better consumer of their state standards and district curriculum.

CTS is a valuable resource for leaders and designers of teacher learning in a variety of settings ranging from one- or multiday workshops, to weeklong institutes, to semester preservice and graduate courses, to PLCs that meet regularly over a year or more. Over the past decade or so, professional development in science has been undergoing a transformation from primarily "one size fits all" workshops and field experiences to more ongoing, subject- and need-focused programs, often situated in teachers' real work, such as through examining student work, reviewing and selecting instructional materials, developing exemplary lessons, and coaching and mentoring (Loucks-Horsley, Stiles, Mundry, Love, & Hewson, 2010; Sparks, 2002). These new forms of professional development have come about as the field has gained a deeper understanding of how people learn and begun to seek ways to embed teacher learning in their real work. However, a major challenge to making these forms of professional development work is ensuring that the teachers and facilitators have tools to focus this work on the appropriate K–12 content and how to teach and assess it effectively. For example, we have seen many groups coming together to examine student work with insufficient content knowledge, a lack of knowledge about students' ideas that provides a lens for identifying misconceptions and learning difficulties, and inconsistent interpretations of the meaning and intent of the learning goal being assessed. Furthermore, the protocols needed to support teachers to learn from their professional development experience and make productive decisions about using what they learn in the classroom were often missing or ineffective.

This *Leader's Guide* was developed to bring the content-specific knowledge of teaching and learning science into the center of the many "new" more building-based or job-embedded professional development strategies, such as looking at student work. It provides guidelines and structure for facilitators to engage teachers in evidence-based dialogue, supported by standards and research on learning. For example, a facilitator's use of CTS can enhance a group's collegial learning in contexts such as looking at student work by first engaging the teachers in answering key questions such as the following:

- What should the students in this grade be expected to know about this topic?
- What common misunderstandings do children of this age group tend to have?
- What prior knowledge is necessary to support the understanding of this topic?
- Is there content you are unsure about that you would like to learn more about in order to interpret student responses?

This information provides teachers with a stronger foundation for looking at their students' work and connecting it to key ideas in the standards and the research on how students think about the key ideas. Furthermore, it strengthens facilitators' ability to lead a group through the process by enhancing their knowledge.

BRIDGING THE GAP

The CTS project set out to increase the use and application of national standards and research in the classroom and in professional development for teachers. The four major national standards publications that guided the development of state standards and curriculum frameworks—*Science for All Americans* (American Association for the Advancement of Science [AAAS], 1989); *Benchmarks for Science Literacy* (AAAS, 1993), including published summaries of research on learning in Chapter 15; *National Science Education Standards* (NRC, 1996); and *Atlas of Science Literacy* (Vols. 1–2) (AAAS, 2001–2007)—along with the research compendium, *Making Sense of Secondary Science* (Driver, Squires,

Rushworth, & Wood-Robinson, 1994), collectively provide educators with a rich professional knowledge base. This knowledge base includes the key ideas and skills needed for science literacy, commonly held ideas students bring to their learning, contexts and implications for instruction, conceptual difficulties and developmental implications for learning, and the coherent growth of learning from kindergarten to high school graduation. These publications have been available to teachers since the start of standards-based reform in the mid-1990s. Yet through the hundreds of CTS workshops the authors have given and observed, the majority of teachers who are first introduced to CTS indicate they have never used these resources. Every introductory CTS workshop we piloted began with an overview of the national standards and research publications listed above, followed by a show of hands when the following questions were asked: "How many of you have heard of or own this book and use it? How many of you have heard of or own this book but have never used it? How many of you are hearing about and seeing this book for the first time?" Most surprising is the number of hands that still go up when the last question is asked. This lack of use of the professional publications that form the backbone of standards- and research-based teaching and learning is further amplified in studies such as the NRC's 2002 report, *Investigating the Influence of Standards*. A conclusion of this report was that although they have been out for almost a decade, standards have not made a significant impact where they matter most: in the classroom.

Furthermore, new teachers are now entering the profession who were middle school students at the time standards were introduced. Career changers are also entering the profession as new teachers. They missed the 1990s wave of learning about standards while they were working in noneducation fields. During the 1990s and early part of the twenty-first century, today's experienced teachers were introduced to standards and the basic principles of standards-based reform. For almost a decade, they were immersed in a flurry of activities such as alignment of curriculum, backwards planning for standards-based lessons, and development of aligned assessments. While the attention to standards has subsided somewhat as teachers developed familiarity and acceptance of standards, we still need to acknowledge that new teachers, both young and career changers, entering teaching today did not have the introduction to standards that our experienced teachers did. Today standards are almost taken for granted. However, that doesn't mean that experienced teachers are using them effectively. In addition, teachers outside of a university setting have not easily accessed the research. Clearly we still have many more years ahead of us to reach the level of implementation where all teachers are effectively focusing on standards and using research to inform their teaching.

Translating the standards and research into practical use in the classroom is a continuing challenge for science education reform. The materials developed by the CTS project are designed to support educators to meet this challenge. The developers of CTS realized that teachers needed both a reason to use the national standards and research on learning and an efficient process for working with them. Facilitators of teacher learning needed realistic designs and suggestions for how to engage teachers in new forms of professional development that lead them to identify and examine the various considerations that support science literacy learning for all students. CTS was developed to help teachers and professional developers methodically incorporate the standards and research into their work.

CTS began with identifying 147 key curricular topics in the standards and thoroughly vetting the common standards and research documents to identify the readings that could be used by teachers and professional developers to explore and study these key topics. Tools for applying the study results were developed, followed by designs for professional development that would embed CTS into a variety of teacher learning contexts.

IMPLICATIONS FOR PROFESSIONAL DEVELOPMENT

Finding time to stay abreast of the standards and research is difficult. A teacher's day is already jam-packed, and many legitimately ask, "Where can we find the time we need to learn all this?" The answer lies in making better use of existing professional development time and in engaging in learning about and using the standards and research in the regular school day. The authors believe every professional development experience—whether it is a content institute, an overview to new curriculum materials, a teacher-directed study group, grade-level team meetings to examine student work, or others—is a ripe opportunity to use CTS. The current trend to providing more substantive and ongoing professional development and forming learning communities is transforming what and how teachers learn, yet it is essential that these new forms of professional development focus on important content and how to teach it (Feger & Arruda, 2008; Hord & Sommers, 2008; Loucks-Horsley et al., 2010; Wei, Darling-Hammond, Andree, Richardson, & Orphanos, 2009).

The CTS process provides the resources that link the theoretical knowledge that comes from a careful examination of standards and research to the situations teachers face in their schools and classrooms. It is this situated use of CTS that supports such a wide variety of strategies for teacher learning.

Studies have also shown that the types of professional development closely linked to improved student learning provide opportunities for teachers to engage in professional dialogue and critical reflection (Birman, Desimone, Garet, & Porter, 2000; Cohen & Hill, 1998; Weiss et al., 1999). Increasingly, professional development is focused on breaking down the isolation of teaching and building a professional culture in schools characterized by groups of teachers examining student results, thinking and reflecting on practice, discussing research and what works, and developing teacher leadership (Schmoker, 2004). The advantage of teachers working collaboratively and learning from their own practice is that teaching practice, aligned with clear and explicit student learning goals, becomes the centerpiece of the professional development. Teachers examine and critique actual artifacts of teaching and learning, such as lessons, student work, and cases of teaching and learning. CTS can provide the focus and direction collaborative groups need to advance their learning. Teachers in these groups learn to engage in authentic dialogue about the specific teaching and learning ideas in the standards and how these ideas compare with their own curriculum, instruction, and assessment. They analyze research on children's ideas and compare those to their own students' ideas. They look for ways to incorporate the research into their own curriculum and reflect on how to use this knowledge to enhance learning. For science teacher collaboration to be successful, it needs a strong content focus and well-developed skills for dialogue and reflection. CTS provides those critical elements.

CTS IMPACT ON TEACHERS AND TEACHER EDUCATORS

This *Leader's Guide* is designed to enhance teachers' and professional developers' knowledge and performance (e.g., strengthen content knowledge and support the design of content-rich professional development). In the science disciplines, teachers and professional developers are particularly challenged because even if they are knowledgeable in one science discipline or grade span, they may not know all the science disciplines or grade spans they are working with. CTS meets the diverse needs of a wide range of teachers and leaders in science (e.g., the biologist who is teaching astronomy, the high school teacher leader who is working with elementary teachers on instructional strategies for first graders, or the university scientist codeveloping a middle school curriculum).

In working with teachers around the country, the CTS developers have documented the substantial impact the process has had on what teachers understand about science learning and how they approach their curriculum, instruction, and assessment. (A summative evaluation report of the CTS project will be available in spring 2010 at www .curriculumtopicstudy.org.) In the evaluation surveys and interviews, teachers who use CTS have reported that they

- deepen their understanding of the science content they need to teach K–12 curricular topics effectively;
- translate science content for adult learning (such as through university courses) to content appropriate for K–12 settings;
- identify and clarify core knowledge and skills in science including unifying themes, big ideas, concepts, skills and procedures, specific ideas, terminology, and formulas embedded in the content standards and curricular objectives;
- improve coherency of scientific ideas as they develop over multiple grade levels;
- make effective use of the research base on student learning to identify potential learning difficulties, developmental considerations, and misconceptions associated with a science curricular topic;
- apply effective content-specific pedagogical strategies and identify useful contexts for teaching scientific ideas as they relate to a particular topic;
- improve their ability to make connections within and across science topics; and
- engage in substantive evidence-based discourse with their colleagues about goals for student learning, modifications needed in instructional materials, and methods for enhancing student understanding.

Science leaders (teacher leaders, university faculty, professional developers, and science specialists) also benefit personally from using CTS. Leaders have commented that they have been able to increase the focus of the professional development on content and PCK and achieve greater results. For example, as part of engaging teachers in a science lesson, some professional developers now have teachers do a CTS on the topic of the lesson before they experience the lesson. This both enhances teachers' understanding of the important content and builds awareness of common preconceptions students might bring to their learning. By embedding CTS into their content institutes, scientists increase the opportunity for teachers to translate their content-learning experiences into developmentally and conceptually appropriate content and activities for the grade level they teach. This bridge between adult learning, often the "science of scientists," and K–12 student learning or "school science" has been missing from many of the professional content learning experiences designed for teachers by experts in a scientific field. These experts know their content well but are unfamiliar with PCK and "school science." CTS is the tool that leaders can use to help teachers situate their learning in the classrooms or other contexts in which they teach.

Leaders have also commented on the versatility of the materials and how they have been able to use them in a variety of situations including one-on-one coaching, PLCs, small informal teacher meetings, and large content institutes. While the materials may be used differently depending on where a teacher is on the teacher professional continuum, CTS adds value to every level of teaching from preservice to novice teachers to experienced teachers to teacher leaders and to those who leave the teaching profession to support teachers. The process and the quality of the materials are helping all types of leaders in science education strengthen their work with teachers by focusing more precisely on standards and the research on students' ideas in science.

Ultimately the impact of CTS is demonstrated through the improved learning of students taught by teachers who regularly use the process to inform their teaching. These students benefit from higher levels of engagement due to teachers' understanding of how to make content accessible to all students. They also benefit from increased coherence in the use of curriculum materials and design of instruction as their teachers use CTS to identify gaps in the curriculum, select appropriate phenomena and representations, strengthen the connections among concepts, and use a variety of formative assessment techniques to elicit students' ideas and monitor learning throughout the course of instruction.

ORGANIZATION OF THIS BOOK

This *Leader's Guide* is organized into seven chapters. Chapter 1 provides the overall rationale for CTS, addresses the question, "Why should educators use CTS?" and introduces leaders to the language of CTS. Chapter 2 provides the leader with an understanding of what the CTS process is and discusses how it supports the development of science literacy. Chapter 3 is written especially for professional developers planning to use CTS in their work with teachers. It summarizes key information about research on effective professional development for science teachers and provides the overall tips and strategies for using CTS in professional development. Chapter 4 includes the designs to lead introductory sessions for CTS. If you are just getting started with CTS, you can use these designs to help your participants experience CTS by doing partial or full guided topic studies on several topics that will acquaint them with the process as well as the resources. Chapter 5 provides the designs and guidelines for leading full topic studies on particular topics (e.g., atoms and molecules), guidelines for designing your own full topic study, and suggestions for combining topics. Chapter 6 leads you through ways to use the applications of CTS in a content, curricular, instructional, or assessment context. Finally Chapter 7 discusses examples of how to embed CTS in ongoing professional development strategies, such as how to use CTS within lesson study, study groups, and mentoring. The CD-ROM contained in the back of this *Leader's Guide* provides masters for all handouts and PowerPoint presentations to accompany the introductory material in Chapters 1 through 3 as well as the designs and suggestions in Chapters 4 through 7.

Acknowledgments

This *Leader's Guide* was informed by the work of many dedicated professionals from across the nation who are working to improve the quality of science education for all students. The Maine, New Hampshire, and Vermont teacher leaders participating in Maine's Governor's Academy and the Northern New England Co-Mentoring Network, and the staff at the Maine Mathematics and Science Alliance were the CTS pioneers, trying out the earliest versions of the CTS process and applying it to their roles as teacher leaders, mentors in their districts, and professional developers. They generously gave of their time and suggestions for improving the process and embedding it into ongoing professional development.

Our many national field testers and pilot sites contributed significant suggestions for the development and revisions of the tools and designs included in this *Leader's Guide*. Our pilot and field testers included science specialists, university faculty from colleges of science and education, Mathematics and Science Partnerships program project directors, fellows from WestEd's National Academy for Science and Mathematics Education Leadership, regional professional development providers, state and district science coordinators and specialists, and professional developers. They generously took time to try out our designs, provide us with feedback, and gather and provide workshop comments from their participants. The authors wish to thank all of the hundreds of science facilitators and workshop participants who generously agreed to provide us with valuable feedback.

We also wish to thank our project design team, who gave their time and expertise to advise us and guide the development process. Special thanks go to Cathy Carroll and Karen Cerwin from WestEd; Joëlle Clark, Northern Arizona University; Francis Eberle, National Science Teachers Association; Jeanne Harmon, Center for Strengthening the Teaching Profession; Mark Kaufman, principal in Lincoln, MA; Cheryl Rose and Fred Gross, Education Development Center; Nancy Kellogg, science consultant, Boulder, CO; Pam Pelletier, Boston Public Schools and adjunct instructor, Harvard University; and Joyce Tugel, Maine Mathematics and Science Alliance. We also wish to thank our evaluator, Bill Nave, for the insights he shared with us that informed development of this project and helped us understand the impact CTS has on teacher learning.

We thank all the authors and organizations that developed the resource books used for the CTS process. Without the vision and ideas you put into these resources, the CTS process would not exist. These include Dr. James Trefil and Dr. Robert Hazen, coauthors of *Science Matters*; the American Association for the Advancement of Science and Project 2061, *Science for All Americans, Benchmarks for Science Literacy,* and *Atlas of Science Literacy* (Vols. 1–2); the National Research Council, *National Science Education Standards,* and Drs. Rosalind Driver, Ann Squires, Peter Rushworth, and Valerie Wood-Robinson, coauthors of *Making Sense of Secondary Science: Research Into Children's Ideas.*

We are grateful to the National Science Foundation for funding this project. We especially wish to thank our program officer, Dr. Michael Haney, for his support and encouragement.

Our very special thanks go to our colleagues, Kathy DiRanna, Karen Cerwin, and Jo Topps at WestEd, and Joyce Tugel, Lynn Farrin, and Nancy Chesley at Maine Mathematics and Science Alliance, who have provided input, examples, and materials for this book. Andrew Boudreaux, Mark Emmet, Steven Gammon, Mary Janda, and Adrienne Somera from the North Cascades and Olympic Science Partnership contributed to the development of the CTS Lesson Study design and tools included in Chapter 7. We are indebted to them for their work. Alexis MacNevin provided the student work that is used in the Lesson Study session. Thank you for your generous contribution. Our deep appreciation also goes to Deanna Maier, production coordinator at WestEd, who managed the final production and created a polished manuscript we could be proud to submit, and to Laurie Mitchell, CTS administrative assistant at the Maine Mathematics and Science Alliance, whose organizational skills were invaluable in assembling materials for the national review and keeping the hundreds of files for the CD-ROM straight. We also wish to thank graphic artist, Jonathan Somers, who designed our CTS icons, and Nathan Keeley, who designed and maintains the CTS Web site.

Our deep gratitude to Dan Alpert, our acquisitions editor at Corwin, who never gave up on us, despite many new ideas, revisions, and delays. Thank you for your patience.

We especially give our thanks to our wonderful family members David, Nate, Chris, and Mary Keeley, Richard and Katherine Maines, and Hans and Jordan Landel, who give us tons of support and put up with our long hours of work on weekends and holidays and love us anyway.

We wish to acknowledge the manuscript reviewers:

Dr. Robert Barkman, Springfield College, MA

Karen Cerwin, WestEd, CA

Dr. Joëlle Clark, Northern Arizona University

Linda Frame, Juneau School District, AK

Dr. Nancy Kellogg, Metropolitan State College of Denver, CO

Dr. Susan B. Koba, Retired, Omaha Public Schools, NE

Dr. Jim McDonald, Central Michigan University

Theresa Roelofsen Moody, New Jersey Astronomy Center at Raritan Valley Community College

Dr. Christine Anne Royce, Shippensburg University of Pennsylvania

Joyce Tugel, Maine Mathematics and Science Alliance

Dr. Wil van der Veen, NJACE Science Education Institute, NJ

Amy Wilson, Mary K. Goode Elementary School, MA

About the Authors

Susan Mundry is currently deputy director of Learning Innovations at WestEd and the associate director of WestEd's Mathematics, Science, and Technology program. She directs several national or regional projects focused on improving educational practice and oversees the research and evaluation projects of Learning Innovations. She is codirector of a research study examining the distribution of highly qualified teachers in New York and Maine for the Northeast & Islands Regional Education Laboratory and is the co–project director for the evaluation of the Intel Mathematics Initiative, a professional development program for elementary and middle grade teachers aimed at increasing student outcomes in mathematics. She is also a principal investigator for two National Science Foundation projects that are developing products to promote the use of research-based practice in science and mathematics. Since 2000, Ms. Mundry has codirected the National Academy for Science and Mathematics Education Leadership, which provides educational leaders with training and technical assistance on professional development design, leading educational change, group facilitation, data analysis and use, and general educational leadership, as well as access to research-based information to improve teaching and learning. Building on this work, she provides technical assistance to several large urban schools districts engaged in enhancing leadership and improving math and science programs.

As a senior research associate for the National Institute for Science Education (1997–2000), Ms. Mundry conducted research on attributes of effective professional development. She served on the national evaluation team for the study of the Eisenhower Professional Development program led by the American Institutes for Research, where she worked on the development of national survey instruments and the protocols for case studies. From 1982 to 1997, Ms. Mundry served in many roles from staff developer to associate director at The NETWORK, Inc., a research and development organization focused on organizational change and dissemination of promising education practice. There she managed the work of the National Center for Improving Science Education and the Center for Effective Communication, provided technical assistance to schools on issues of equity and desegregation, oversaw national dissemination programs, and codeveloped the "Change Game" (*Making Change for School Improvement*), a simulation game that enhances leaders' abilities to lead change efforts in schools and districts.

Ms. Mundry has written several books, chapters, and articles based on her work. She is coauthor of the best-selling book *Designing Effective Professional Development for Teachers of Science and Mathematics* (2nd ed.), as well as *Leading Every Day: 125 Actions for Effective Leadership,* which was named a National Staff Development Council Book of the Year in 2003, and *The Data Coach's Guide to Improving Learning for All Students* (2008). She holds a bachelor's degree from the University of Massachusetts–Amherst and a master's degree in education from Boston University.

Page Keeley is the senior science program director at the Maine Mathematics and Science Alliance. She directs projects in the areas of leadership, professional development, standards and research on learning, formative assessment, and mentoring and coaching and consults with school districts and organizations nationally. She has been the principal investigator on three National Science Foundation–funded projects: the *Northern New England Co-Mentoring Network*, a school-based mentoring program that supported science and mathematics professional learning communities for middle and high school mentors and new teachers, www.nnecn.org; *Curriculum Topic Study: A Systematic Approach to Utilizing National Standards and Cognitive Research*, www.curriculumtopicstudy.org; and *PRISMS: Phenomena and Representations for Instruction of Science in Middle School*, prisms .mmsa.org, a National Digital Library collection of Web resources aligned to standards and reviewed for instructional quality. In addition she is a co–principal investigator on two statewide projects: Science Content, Conceptual Change, and Collaboration, a state mathematics and science partnership focused on conceptual change teaching in the physical sciences for K–8 teachers; and a National SemiConductor Foundation grant on Linking Science, Inquiry, and Language Literacy. Ms. Keeley is the author of ten nationally published books, including four books in the *Curriculum Topic Study* series (Corwin), four volumes in the *Uncovering Student Ideas in Science: 25 Formative Assessment Probes* series (NSTA Press), *Science Formative Assessment: 75 Practical Strategies for Linking Assessment, Instruction, and Learning* (Corwin and NSTA Press), and *Mathematics Formative Assessment: 50 Practical Strategies for Linking Assessment, Instruction, and Learning* (in press).

Ms. Keeley taught middle and high school science for 15 years, after working as a research assistant in immunogenetics at the Jackson Laboratory of Mammalian Genetics in Bar Harbor, Maine. As a teacher, she was an active leader at the state and national level. She received the Presidential Award for Excellence in Secondary Science Teaching in 1992 and the Milken National Educator Award in 1993. She has served as an adjunct instructor at the University of Maine, is a Cohort 1 Fellow in the National Academy for Science and Mathematics Education Leadership, served as a science literacy leader for the American Association for the Advancement of Science/Project 2061 Professional Development Program, and has served on several national advisory boards. She is a frequent speaker at national conferences and served as the 63rd president of the National Science Teachers Association for the 2008–2009 term. Ms. Keeley received her bachelor's degree in life sciences from the University of New Hampshire and her master's degree in secondary science education at the University of Maine.

Carolyn Landel received her doctoral degree in biochemistry and molecular biology from the University of Chicago and pursued postdoctoral studies at University of Massachusetts Medical School and the Fred Hutchinson Cancer Research Center. While maintaining an active research program, Dr. Landel brought together her strong scientific training and her commitment to education by supporting state and national science education reform efforts. In 2002, Dr. Landel joined the Science, Mathematics and Technology Education program at Western Washington University (WWU), home to the state's largest teacher preparation program. Here she serves as project director of a National Science Foundation–funded Mathematics and Science Partnership that unites scientists from WWU and four community colleges with twenty-eight small and rural school districts to improve science education K–16. Dr. Landel directs the day-to-day work of all aspects of the project, including participation in science education research studies in collaboration with Westat, Horizon Research, the Education Development Center, and Georgia Tech University.

Dr. Landel was a fellow in the National Academy for Science and Mathematics Education Leadership and principal investigator of a state-funded initiative to develop and pilot a science education leadership program at WWU for K–12 teachers. This successful pilot was scaled initially through the National Science Foundation–funded Mathematics and Science Partnership program and is currently being sustained through continued funding from the state department of education.

Dr. Landel led the early dissemination efforts of CTS in Washington State in both inservice and preservice contexts. Based on her experiences with CTS, she joined a panel of national experts assembled to inform the development of print and Web-based publications to help teachers and professional developers utilize CTS to improve content-focused professional development and teacher practice.

Dr. Landel's current collaborations include working with WestEd on the development of a simulation to help science education leaders understand the elements of effective professional development and how to design meaningful content-rich experiences for science teachers. She is a lead consultant with the Education Development Center and Vulcan Production to produce an innovative print, Web, and video resource to support school-based leadership teams improve student achievement. She is also actively involved with Horizon Research and Project 2061 in the creation of instruments to assess teacher opportunities to learn and that measure changes in teacher science content knowledge, teacher pedagogical content knowledge, classroom practice, as well as changes in student achievement.

Dr. Landel has authored numerous publications in scientific journals and, more recently, in the education literature. Her most recent publications can be found in NSTA Press and *Education Leadership*. Her current research interests include understanding how partnerships between higher education scientists and teachers can improve teacher, faculty, and student learning and what attributes of professional development lead to improved science content and pedagogical content knowledge required of teachers to increase student learning.

Introduction to
A Leader's Guide

This book, *A Leader's Guide to Science Curriculum Topic Study*, is the natural companion to the parent book, *Science Curriculum Topic Study: Bridging the Gap Between Standards and Practice* (Keeley, 2005). The CTS parent book provides an introduction to the process of curriculum topic study (CTS); the resources used to engage in CTS; various ways to use CTS to support content knowledge, curriculum, instruction, and assessment; and the 147 CTS guides that contain the prevetted readings used in CTS. It has become an essential resource used by science educators to improve their practice. This *Leader's Guide* offers practical suggestions for using the CTS parent book, including tools, designs, and additional resources for incorporating CTS into the work science leaders do to support teacher learning. The *Leader's Guide* was developed to assist teacher educators and leaders, such as preservice faculty, scientists who provide content support to teachers, science specialists, teacher leaders, and professional development providers, in developing the professional knowledge base science teachers need to be effective in the classroom. It provides tested strategies for introducing the CTS process that builds preservice and inservice teachers' knowledge of the research on learning science and the national standards and benchmarks that are the bedrock for ensuring quality teaching and science literacy for all. Furthermore, it supports forms of teacher learning in collegial groups and professional learning communities (PLCs) that are guided by a common knowledge base as teachers work together to plan lessons, examine student work, develop assessments, select curriculum, and go about the daily business of educating our nation's youth.

ADVICE FOR USING THIS *LEADER'S GUIDE*

Users of this *Leader's Guide* may wonder how they should begin using this book. You may be asking, *Where do I start? In what order do I use it? What else do I need to effectively use this book?* There is no single answer to these questions. It depends on your familiarity with CTS and your purpose for using it. We do encourage all leaders who use this *Leader's Guide* to have a copy of the CTS parent book, as well as the resources listed in Table 1.1. Often, throughout this *Leader's Guide*, we will be referring you to sections and pages in the CTS parent book, and it will help you to have it to consult.

Before you begin using this *Leader's Guide*, it is important to become familiar with the CTS parent book, CTS resources, and experience a CTS. If you have never conducted a CTS on your own, pull out your CTS parent book. From it, select one of the 147 CTS topics of interest to you and follow the process described in Chapter 3 of that book, "Engaging in Curriculum Topic Study." Wear two hats as you conduct your own CTS: (1) As a learner, reflect on what knowledge you gained as you did the CTS, and (2) as a professional developer, consider what you need to do to facilitate this type of learning with others. Compare what learners do with what a facilitator would do throughout the process on pages 48–49 of the CTS parent book to get a sense of what the teachers with whom you work will be doing and what you will be doing as a CTS facilitator.

Table 1.1 Essential Resources for Leaders of CTS

CTS Resources	Available Through
Science Curriculum Topic Study (Keeley, 2005)	Corwin: http://www.corwinpress.com NSTA Press: http://www.nsta.org/store/
Science Matters (Hazen & Trefil, 1991, 2009)	Major bookstores and online sellers
Science for All Americans (AAAS, 1989)	Oxford University Press: http://www.us.oup.com/us/ NSTA Press: http://www.nsta.org/store/ Available to read online at: http://www.project2061.org/publications/sfaa/default.htm
Benchmarks for Science Literacy (AAAS, 1993)	Oxford University Press: http://www.us.oup.com/us/ NSTA Press: http://www.nsta.org/store/ Available to read online at: http://www.project2061.org/publications/bsl/default.htm
National Science Education Standards (NRC, 1996)	National Academy Press: www.nap.edu NSTA Press: http://www.nsta.org/store/ Available to read online at: http://www.nap.edu/catalog.php?record_id=4962
Making Sense of Secondary Science (Driver et al., 1994)	NSTA Press: http://www.nsta.org/store/ Online booksellers
Atlas of Science Literacy (Vols. 1–2) (AAAS, 2001, 2007)	Project 2061: http://www.project2061.org/publications/atlas/default.htm NSTA Press: http://www.nsta.org/store/

CTS is a versatile professional development tool with multiple uses and purposes. We do not prescribe a linear, step-by-step process for using CTS in your work. Where you start, the sequence you use, the designs you select, the tools you use, and the supplementary resources you include will be as varied as the diverse types of leaders who are using CTS. Each of the chapters in this book will begin by describing what is in the chapter and how leaders might use it. While step-by-step scripts are provided for many of the CTS designs, we encourage you to adapt the materials to the needs of your audience and to your own facilitation style.

Before implementing the designs, tools, and suggestions in this book, it is important to have a deep understanding of how CTS enhances professional development, the different purposes it achieves for teacher learning, the variety of ways to embed it into your own teacher learning contexts, and the language used throughout CTS and this book. This groundwork should be done first if you plan to regularly use CTS in your work. This chapter addresses the question, "Why use CTS?" It will provide you with the rationale for using CTS and lay the groundwork for you to use the material provided in the subsequent chapters.

UNDERLYING BELIEFS

As suggested in the preface to this book, the education field has undergone a tremendous transformation in beliefs about what constitutes effective learning for both children and adults and what it takes to be a quality teacher. There is a growing recognition of the complexity of teaching and the vast array of knowledge a teacher must possess to meet the needs of a wide range of students. We know more now than ever before about conceptual learning in science and the skills necessary to do science, and we are learning more all the time. As the education field, and in fact our entire culture, becomes one that is knowledge-using and knowledge-producing, teachers are increasingly using and contributing to the education knowledge base. These developments have provoked the following two strong beliefs that undergird the CTS work:

1. Teachers, like other professionals, must possess and continue to build their own *specialized knowledge base*. For teachers, this consists of content knowledge and knowledge about teaching in a specific content area, including an understanding of how children of a certain age learn, called pedagogical content knowledge (PCK). Teachers continue throughout their careers to develop and actively use their specialized knowledge base to guide their educational practice.

2. Teachers, like other professionals, should be engaged in *collegial learning communities* that are guided by strategic and enduring goals and focused on enhanced learning and ongoing improvement. These communities should be knowledge-using and knowledge-producing and be guided by two very basic ideas: Use the knowledge generated from standards and research to provide evidence and justification for your ideas, and learn from the expertise of others shared through peer-reviewed literature, conference presentations, and the wisdom of thoughtful practitioners.

Each of these beliefs has changed how we think about the content and purpose for professional development for science teachers and has had a major impact on how teachers are engaged in professional learning. CTS supports teachers to build their specialized teaching knowledge and participate in productive collegial communities focused squarely on putting research and standards to work in the classroom. Throughout this book, you will see examples of how teachers can use CTS to enhance their content and pedagogical knowledge in collegial, collaborative learning environments. You will see how teachers use knowledge gained from CTS as well as contribute new knowledge about teaching and learning in their own unique contexts through professional development strategies described in Chapter 7, such as action research, lesson study, and video demonstration lessons.

THE NEED FOR A COMMON PROFESSIONAL KNOWLEDGE BASE

Over the last few decades the education field has learned more about what it takes to develop qualified teachers in science and the knowledge, skills, and mind-sets that support teaching and learning in this discipline. New ideas have been shaped and influenced by the growing research base that provides educators with insights into how students develop their understanding of specific ideas in science and how preconceptions may impede learning if they are not surfaced and taken into account when designing instruction. There is greater awareness and use of recommended practices such as establishing a clear and coherent curriculum, focusing on an explicit set of standards-based learning goals, using instructional strategies that support and deepen student learning of key ideas in science, and embedding standards- and research-based assessments throughout instruction that inform teaching and provide information on the extent to which students are achieving a learning goal. Increasingly educators are asked by administrators and others to justify requests for new programs or practices with objective evidence of success. Stakeholders want to know what works and are looking to professional educators to identify and explain effective practices that can lead to increased student achievement.

These developments have given way to new ideas about teacher professional development. We know that science teaching involves much more than hands-on activities, teaching tips, and general pedagogical techniques. Professional teachers must possess both content and PCK, the specialized knowledge of content and how children learn it. This knowledge enables teachers to focus on important learning goals and provide developmentally appropriate instruction and assessments. Quality professional development programs are increasingly focused on enhancing teachers' understanding of their content and how to teach it. Teachers are learning to review and revise their instructional materials and methods to better reflect alignment with standards and research on learning and are taking collective responsibility for knowing not only their content, but also how children think about ideas in science and what types of experiences, phenomena, and representations can best support learning.

One of the characteristics of professional teachers is their belief in the importance of acquiring their own professional libraries or having access to professional resources to regularly inform their teaching and expand their knowledge base. Knowledge of effective science teaching does not end after graduation from a teacher preparation program or graduate program. Teachers are constantly seeking the wisdom and knowledge shared by researchers and expert practitioners that help them grow and develop as professional science teachers. There is a plethora of professional literature to support science teacher learning. However, the vast collection of literature can be narrowed down into six major publications that best support standards- and research-based teaching and learning across all the science disciplines, grade levels, and teacher expertise. These are the common and collective resources identified by the CTS Project that can be used with the 147 science curriculum topics identified in the CTS parent book. These professional resources should be in the library of every science teacher and teacher educator, whether they are part of their own collection or shared within a school or organization. These resources are listed in Table 1.1 and provide a common knowledge base that all teachers can refer to and use. The fact that these books were authored by highly respected scientists, researchers, and science educators and some, such as the national standards documents, went through an extensive national review process that involved consensus from the science education community at large, makes them credible and relevant to all science educators striving to develop shared understandings of content, teaching, and learning. Having access to these books is like having an expert at your fingertips 24/7!

CTS MAKES THE KNOWLEDGE BASE ACCESSIBLE

As described in the preface, many teachers, and even some teacher educators, have never used or even heard of some of these resources, even though they have been out for more than a decade. As the use of CTS grows, these resources are becoming better known and more frequently used in the science education community. As a facilitator, one of the changes you will see firsthand as you use CTS with teachers is the renewed emphasis and embrace of the national standards and research literature, even though states have their own standards.

Prior to CTS, getting to know and use national standards and research on learning posed several difficulties. The focus on state standards shifted teachers' attention away from the more detailed source documents on which many of these state standards were based. As we discussed in the preface, we found that many teachers had no knowledge of publications like *Science for All Americans* and *Making Sense of Secondary Science.* Some had heard of the *National Science Education Standards* and the *Benchmarks for Science Literacy* but had never opened a copy or even realized that these publications contained much more than a list of what students were expected to know and be able to do in science. They didn't know enough about the publications to know how useful they could be in informing teaching and learning.

For others, the standards and research publications were available but a process for using them was missing. Some teachers found navigating through the publications to be difficult and unwieldy. They consulted the standards documents, sifting through the hundreds of pages of text to find what was relevant to their curriculum, their students, or their teaching and often struggled and became frustrated because the answers they were seeking were so hard to find. They didn't know how to use the essays or why the learning goals were written a certain way. They struggled with figuring out how to sequence and connect standards coherently. Many never realized there was a chapter in the back of the *Benchmarks* that contained summaries of research connected to the chapters describing what all students should know. They didn't realize how *Science for All Americans* is the seminal, enduring document that lays out the vision for the standards documents. And few knew of an obscure little book published in England, *Making Sense of Secondary Science: Research Into Children's Ideas* (Driver, Squires, Rushworth, & Wood-Robinson, 1994), that contains a vast compendium of summarized research on learning across the disciplines of life science, physical science, and Earth and space science.

For those who persevered, their efforts paid off in gaining clarity about the standards and how they relate to teaching and learning, but it took a substantial time commitment and their searches often ended when they identified learning goals for their particular grade span. As advocated by Project 2061 of the American Association for the Advancement of Science (AAAS), all teachers need a broad and deep understanding of all science topics. They should know what every twelfth-grade graduate is expected to know and the level of schooling in which students are expected to learn certain ideas in science. They should understand not only the learning goals at each grade span, but also the research suggesting what is difficult or easy for students to learn, the contexts and strategies that support learning, the connections within and across science topics, and how a coherent understanding grows over the K–12 sequence. But how do teachers develop this knowledge? Where can they find the tools and the time? CTS provides the means and the organized process to help education professionals use these professional publications efficiently and effectively. Most important, it has gotten the books off the shelf and into the hands of teachers so they could use them. The CTS Project identified 147 relevant curriculum topics and prescreened and identified all the readings from the resource books that would contribute to a teacher's understanding of the professional knowledge described above. These readings are combined in study guides and facilitated through a process that engages teachers in a deep and thoughtful study of teaching and learning connected to a curricular topic they teach.

In our experience introducing CTS to teachers, familiarity with and access to the CTS resources tended to be more on the side of the standards documents than the research. Teachers seemed to have less familiarity with and access to the research base on student learning. We know from cognitive research that students often have strong preconceptions about the world around them that may support or interfere with their learning (Bransford, Brown, & Cocking, 2000; Donovan & Bransford, 2005; National Research Council, 2007). For example, Driver et al. (1994) reported that several researchers have studied student ideas about "gas and found that, initially, [students] do not appear to be aware that air and other gases possess material character. For example, although young children said that air and smoke exist, they regard such material as having transient characters similar to that of thoughts" (p. 80). Even when students develop an awareness of the material character of gases, they still may not think of gases as having weight or mass. Until students understand that gases have mass, they are "unlikely to conserve mass when describing chemical changes that involve gases" (p. 77). In the past, teachers were either not aware of such misconceptions or were not prepared to identify and address them in their instruction. Teachers must know that their students may have these difficulties and understand the importance of helping young children see that "air is there." Brooks and Driver (1989) pointed out that by understanding the research, teachers of different grade levels would know how to focus their instruction on building understanding for children of different age groups. These teachers would know that five- to seven-year-olds can develop an understanding of air existing all around them and moving naturally by observing windy days, effects of moving air on objects such as streamers and parachutes and can begin to see that we can make air move by sucking and blowing and that air takes up space through balloons or can be "squashed" by observing squeeze bottles. Seven- through eleven-year-olds learn that air occupies space, has mass and therefore weight, and pushes out in all directions by observing bubbles in water, balloons, and inflating tires. This paves the way for eleven- through fourteen-year-olds to learn that air can be compressed and expanded and to develop understanding of atmospheric pressure. All of these ideas are encompassed either explicitly or implicitly in the state standards teachers and students are held accountable for achieving. By using CTS, teachers can identify key ideas in their standards and then refer to research summaries to know what may make the learning difficult or comprehensible to students. They can use this information to plan instruction at their own grade level or comprehensively across grade levels. They now have a way to access and link the research to K–12 student learning goals.

BUILDING PROFESSIONAL COMMUNITY

The other underlying belief that is changing teacher professional development for the better is the growing commitment to building PLCs among teachers. After more than a century of schools that operated like multiple one-room schoolhouses under one roof, the idea of a PLC and teamwork in schools is finally taking hold. In the recent past, teachers across the hall or just next door may have been struggling with the same questions and problems with no reason or way to collaborate to find solutions. Increasingly, teaching is being deprivatized by the growing number of PLCs in schools that examine practice and results on a regular basis and pursue solutions to the problem of poor student performance. However, like other innovations, building a PLC doesn't happen by magic, and there are many pitfalls that must be addressed. In our work, we have focused on putting the "professional" into the PLC. We have asked, "What are the tools teachers need to make sure they reflect the knowledge of the profession in their learning communities?" Our conclusions are that PLCs must be research-based and standards-driven to be "professional." Historically, isolation among teachers led to very little sharing of what works among educators. Recent technologies and new organizational structures are helping to change that. Yet teachers' days are still highly structured and scheduled and they need efficient and effective ways to work together and put their professional knowledge to work. Through CTS, once teachers learn the process, they can quickly and efficiently explore the readings on any given science topic to address questions of practice and inform deliberations and decisions for their own practice and to share with others in their PLCs. Whether they are involved in a formal PLC that meets to examine results and pinpoint areas for improved student learning, or a grade level team monitoring how new curriculum materials are working, the CTS process will support and enhance these collegial groups of teachers to use the research and the standards to inform their work. Through the use of CTS, we have seen the conversations in these groups shift from the autobiographical stories that emanate from, "What are you doing in your classroom?" to scholarly discussions that pertain to all teachers such as "What do the national standards and research say, and how might we apply that in our classrooms and to the implementation of our state standards?" Examinations of curricular or instructional strategies are enriched because teachers base their analyses on whether the materials and strategies reflect important and challenging key ideas and research on how children learn as opposed to focusing only on their own opinions, biases, or the materials' style, layout, or reading level.

As schools and school districts support new organizational arrangements that reduce hierarchy and promote collaboration, CTS can help at every juncture. As Ann Jolly, a former middle school science teacher and an Alabama Teacher of the Year reports,

> PLCs involve teams of teachers in working together to study, learn, and support one another as they make changes in classroom practice. The process is collaborative rather than isolated. Ongoing learning and support continue throughout the school year. This professional development occurs at the school site and focuses on needs of the specific students in that school. Teachers work as interdependent colleagues, and a culture of collaboration and collective responsibility takes root. When teachers work together in PLCs to implement new teaching practices, over 90% of teachers do so successfully. Teamwork and collaboration work! (Jolly, 2007, ¶ 8)

PLCs are usually organized as collaborative teacher groups focused on learning and achieving desired results. Eaker, DuFour, and Burnette (2002) suggested that PLCs systematically address four key learning questions:

1. What do we want students to learn?

2. How will we know if they have learned it?

3. What do student-learning data reveal?

4. What are we going to do if students are not learning?

Too often, however, these collaborative groups lack a systematic focus on disciplinary content or lack the knowledge base on learning science to adequately address these questions. Table 1.2 shows how the CTS process and specific sections of the study guides in the CTS parent book can be used to address these questions. Beginning with the first question, CTS can guide the school community to ensure that the science learning objectives the group chooses are enduring and that important ideas reflected in the national standards are clarified so that key ideas are clear and explicit and supported developmentally and conceptually by research. In addressing the second and third questions, CTS can also help the community use assessments that probe for understanding by using the CTS process to develop and use ongoing formative assessments that link key ideas in the standards to common misconceptions and reveal whether students have similar ideas to those identified in the research. Teachers examine students' results on assessments that reveal their thinking to decide what is needed next. The fourth question may be answered by examining the K–12 articulation of learning goals to determine whether gaps exist that may pose barriers to learning, by analyzing curriculum materials to see the extent to which they promote learning of the key ideas, by identifying instructional contexts or phenomena that have proven effective in supporting learning, or even by examining teachers' own content knowledge to determine whether they are making the right connections. Table 1.2 shows how the different sections of the CTS process can support key questions for PLCs. (For a refresher on the six different sections of a CTS guide and the resources that are used with the sections, turn to page 22 in the CTS parent book or refer to Handout A1.6: Anatomy of a Study Guide in the Chapter 4 folder of the CD-ROM for this *Leader's Guide*.)

Table 1.2   Key Questions for PLCs and How CTS Can Help

Key Question	*PLC Use of CTS*
What do we want students to learn about a particular science topic?	• *CTS Section III, V, and VI*: Identify the learning goals that align with the topic; unpack the concepts, ideas, or skills within the learning goals for the topic. • *CTS Section V*: Identify the connections among related concepts; examine how key ideas build. • *CTS Section I*: Examine the culmination of K–12 science literacy ideas for enduring understanding and use into adulthood.
How will we know if they have learned it?	• *CTS Section IV*: Identify common misconceptions that may be revealed through instruction and assessment. • *CTS Sections III, IV, and VI*: Develop and use formative assessments and culminating performance tasks to check for understanding before, throughout, and at the end of instruction.

Key Question	PLC Use of CTS
What does the student learning data reveal?	• *CTS Section III*: Identify the extent to which students' ideas match the key ideas in the standards. • *CTS Section IV*: Identify common misconceptions and barriers that impede learning.
What are we going to do if they do not learn?	• *CTS Section II*: Examine instructional contexts and suggestions to determine if curricular or instructional changes are needed. • *CTS Section IV*: Examine ways to address student misconceptions. • *CTS Section V*: Examine the K–12 articulation of learning goals to see if there are gaps that need to be filled; look for ways to make stronger connections among a coherent set of learning goals.

OBSERVATIONS AND VOICES FROM THE FIELD

The results of using CTS have been impressive. (*Note:* The final summative evaluation of the CTS Project will be available on the CTS Web site in 2010.) As one teacher leader who used CTS said, "The process is an essential tool to bridge the gap between research- and standards-based practice in teaching science." A mentor teacher reported that CTS is especially useful in situations where mentors work with novice teachers and that the novice teachers are not the only beneficiaries. She described how mentors also show tremendous growth in skills and understandings when they use CTS as part of an induction program for beginning teachers.

Many users have commented on the ease and versatility of the CTS materials. For example, one participant said, "The [materials] allowed for a directed view of where to look in the CTS guides for the information we needed." Another pointed out that "developing our own [assessment] probes helped to give insight into what it takes to develop a good formative assessment item that can uncover the misconceptions our students have."

One of the greatest results we have seen comes from the teacher "aha" moments. For example, one leader reported this insight:

> I was sitting with some 8th grade science teachers when one of them turned to me and said, "I can't believe it, I'm not teaching what I am supposed to be teaching!" That went on in various forms throughout the two days with the math and science teachers. CTS really had the teachers looking intently at what their students needed to know on a topic. They were clearly dissecting the standards. Not only did they dig in but their vocabulary changed.

As another teacher commented, "The most valuable aspect of CTS was finding out what I need to know to be scientifically literate in order to teach content."

Teachers have pointed out that another valuable aspect of CTS for them is the review and discussion of misconceptions and misunderstanding based on research. As one teacher said, "It's a 'wake-up call' to all teachers that instruction strategies/techniques that address misconceptions are keys to learning."

One professional developer we worked with summed up the value of CTS this way:

> CTS is a systematic procedure anyone can use; it provides synthesized information on specific topics and thus saves time in looking for answers; it helps users work from a common understanding of a particular topic to answer a specific question; and it helps users develop the habits of good research strategies.

After experiencing CTS, leaders of professional development for science teachers immediately saw the significance of using CTS to enrich and invigorate teacher-learning programs. Many professional developers in science are not experts in every science topic area. Some with excellent backgrounds in one science area, such as Earth science, may be responsible for designing teacher learning programs in all other science domains including physical, life, and space science. They need an easily accessible process for gaining a clear vision of the important science for the K–12 classrooms and how to translate that science into grade level–appropriate curriculum and instruction. CTS provides such a tool.

PROFESSIONAL DEVELOPMENT DESIGNS ARE ENHANCED THROUGH CTS

Professional development for science teachers comes in a variety of forms and structures ranging from half-day workshops to weeklong institutes, to ongoing collaborative structures like PLCs and lesson study. Regardless of the type or length of the professional development experience, CTS increases the focus on the content by connecting it to the key ideas in the standards and the research on learning. This increased focus ultimately translates into improved student achievement. For example, leaders who are instructional coaches should routinely do a CTS on the topics they are addressing in their coaching. Sometimes these leaders use this information for their own planning, but more often they incorporate it into the work they do with teachers, such as collaboratively planning and providing feedback on a lesson together. In Chapter 7, you will find examples of professional development where the leader uses CTS with teachers in the context of particular professional development strategies. CTS is so valuable for leaders, we believe that no professional development leader should plan content-focused workshops, study groups, lesson study, or any science professional development without first doing a CTS on the topics they will address in their professional development session. This book contains all the tools and resources to support them not only to do that, but also to build in rich CTS experiences for the teachers with whom they work to improve science education.

CTS is a versatile resource designed to address multiple needs, audiences, and contexts. Likewise, this *Leader's Guide* addresses the multifaceted nature of professional learning and the different types of leaders who may design and support teacher learning by using CTS. To give you a sense of the versatility of CTS when it is used within different contexts, the following are just a few examples of the various ways leaders use CTS. They also show how the CTS underlying beliefs of building specialized knowledge of science teaching and learning, as well as supporting collaborative group learning, are manifested through these examples.

CTS USE BY COLLABORATIVE SCHOOL TEAMS

Student learning is at the heart of what teachers do. When teachers encounter students having learning difficulties, one reaction has been to simply teach the content over again in the same or slightly different way. Another approach is to gather data to find out what the students are having difficulty understanding and if other teachers at the same grade level are experiencing similar results. In this second approach, CTS resources help to pinpoint how to make the content more accessible to students. Improving opportunities for students to learn science content by first examining teachers' own content knowledge, identifying key ideas they want their students to learn, identifying instructional contexts that can enhance learning, becoming aware of the research on students' ideas in science and how they impact learning, and understanding how learning progresses from one idea to the next in a coherent sequence of ideas improve both the quality of teaching and subsequently the depth and endurance of learning.

For example, a middle school team might observe that their students seem to perform poorly on assessment items measuring inquiry skills on the state test. The teachers wonder what they can do to improve students' abilities to design and carry out a scientific experiment. Although their students have had opportunities to experience science in a hands-on learning environment, the students always score low on the inquiry section, especially experimental design that involves controlling variables. The team decides to use the Experimental Design CTS module (see Chapter 5) to investigate the topic. During the CTS process, they define for themselves what an experiment is and how it differs from other forms of investigation. Several of the teachers on the team had thought all investigations were experiments and the CTS helps them understand the various ways scientists investigate the natural world, including experimentation. Furthermore the team had not been aware that the research indicated how difficult it is for students to identify and control variables.

After using the instructional scaffold activity in the Experimental Design CTS module, the teachers understand how important it is to explicitly scaffold and teach students how to identify and control variables. As they use the *Atlas of Science Literacy* (AAAS, 2000, 2007), they become more aware of the precursor skills and knowledge that build up to students' being able to design their own experiments and know why identifying and controlling variables are important to scientists. Overall, the CTS helps the team understand why their instruction hadn't been working for their students and what they need to do so that their students can achieve the targeted learning goals. As a group, they revisit their curriculum and instructional materials and strengthen them. The conversation shifts in the team from an activity focus to a learning focus, guided by the common knowledge base they now have as a team as a result of doing the CTS together.

CTS USE IN PRESERVICE EDUCATION

Preservice teachers in science education courses benefit tremendously from using CTS at the beginning of their careers. Not only does it establish a habit of practice that will be useful to them throughout all stages of their careers, but it also helps them link their preservice experiences to the current emphasis in many schools on standards and research-informed instruction. For example, most preservice teachers are asked to design at least one lesson as part of their methods course requirements. CTS provides the information they need up front to ensure their lesson appropriately addresses important key ideas in science and anticipates the commonly held ideas students might bring to their learning. CTS creates the awareness needed for preservice teachers to design effective lessons that address standards and are informed by research on learning. In the process, many preservice teachers, particularly those who have limited science backgrounds, find they are gaining new knowledge about content, teaching, and learning and realize areas where they would like to continue their science content learning. It also brings a deep appreciation early on in their careers of the need to have and use a professional library of resources that will help them teach in a system that increasingly focuses on accountability to standards and instructional decisions based on research.

CTS ENHANCES SCIENCE EDUCATION LEADERSHIP

CTS has many benefits for people in science education leadership roles, such as school and district administrators, teacher leaders, and coaches. In our CTS work, we encourage all leaders working in these roles to use CTS to increase their familiarity with science standards and learning research and how they are used to inform curriculum, instruction, and assessment. Depending on one's role, these leaders may need more in-depth understanding of the key ideas in the learning goals and the commonly held ideas noted in the research on learning in any particular grade spans.

District Science Coordinators

For example, district science coordinators are often responsible for curriculum adoption and development committees, overseeing and supporting coaches, arranging and approving professional development in science, making classroom observations, requesting resources and lab materials and related tasks that require them to have a very broad and deep understanding of K–12 science. Very often they may be a specialist in one particular area of science (e.g., high school chemistry), but may not have first-hand knowledge about what learning strategies are effective for teaching basic life science topics in Grade 4 or the common difficulties students have with key ideas in astronomy. They can benefit from engaging in full curriculum topic studies on topics that are the focus of new curriculum, to inform the selection of professional development programs and to make a research-based case for using certain materials and labs. Every science coordinator should own a copy of the CTS parent book and all of its accompanying resource books listed in Table 1.1. They should actively share these books with the people with whom they work to encourage others to use CTS to inform teaching actions and decisions.

Principals

Programs for principals can use CTS to help prospective school administrators develop an understanding of what they should look for when teachers are teaching certain science topics. For example, one principal said he was wondering why a teacher he observed had been asking her students questions about content that was not directly addressed in the lesson being taught, and he noted it in the evaluation as a concern. After the lesson, he asked the teacher about it. She shared with him that she had a hunch that students were missing some of the prior knowledge needed to understand the topic she was teaching. She pulled out a map from the *Atlas of Science Literacy* (AAAS, 2000, 2007) and showed the principal the concepts and specific ideas that students are expected to develop at each grade level and pointed out the areas in which she was probing students to see if they had the prior knowledge that served as a precursor to the ideas she was trying to develop in her lesson. She showed him the CTS guide that indicated which *Atlas* map to use to examine the topic she was teaching. The principal developed a greater appreciation for what this teacher was doing to assess prior knowledge and quickly saw how using CTS himself would inform his classroom observations. While principals do not have to know every bit of information on the 147 science topics included in the CTS parent book, it is important for them to experience CTS enough so that they know its purpose and can suggest teachers use it as they are planning lessons, developing assessments, and reviewing the essential content students should learn. In addition, administrators should encourage teachers to use CTS to justify decisions they make regarding teaching and learning.

Teacher Leaders and Instructional Coaches

Teacher leaders and coaches are other key leadership groups that can strengthen and enhance their leadership capacity through CTS. Teacher leaders and coaches are often chosen because they stand out among peers and are successful with their own students. When they begin to work with a variety of teachers, they are challenged to know the goals for science learning across many grade levels, and they must be a resource for teachers who may be teaching topics at a grade level with which the coach or teacher leader is not familiar firsthand. CTS helps in both areas. When leaders use CTS themselves, they can quickly and efficiently review the standards and research to inform what they do to support their teachers. They can also introduce CTS tools to the teachers with whom they work so they have access to the information they need any time. It is like having a virtual expert on call any time the coach, mentor, or teacher leader is not readily available to provide assistance.

THE LANGUAGE OF CTS

Like any new tool or resource, CTS comes with its own language, specialized terminology, and operational definitions. For the purpose of clarity, Table 1.3 lists words and terms frequently used throughout this *Leader's Guide*. Descriptions of how this terminology is used in the context of CTS, as well as operational definitions for words that may have different meanings in other contexts, is provided for leaders to ensure consistency when using CTS in your professional development contexts. In addition, this chart is provided as Handout 1.1 in the Chapter 1 folder on the CD-ROM at the back of this book if you choose to share it with teachers, adapt it, or add additional terminology you use in your CTS professional development.

Chapters 1 and 2 in the CTS parent book introduce the user to CTS and the tools and collective resources used with the process. In Chapter 2 of this *Leader's Guide*, we will expand upon these two chapters by describing what leaders need to know to introduce CTS effectively in their work with science educators.

Table 1.3 CTS Specialized Terminology

CTS Terminology	Clarification of Terminology Used in the CTS Context
Commonly held idea	A pervasive notion about a scientific idea that has been studied in groups of students with results published in the research literature and is likely to be held by students outside of the study
Concept	A mental construct used to conceptualize a scientific idea (e.g., gravity, space between molecules, interaction in a food web)
Content knowledge	Knowledge of disciplinary subject matter (e.g., physical science, astronomy, nature of science, inquiry, etc.)
CTS	An acronym that stands for curriculum topic study
CTS Learning Cycle	An instructional model for CTS adult learning
CTS parent book	A shorthand way of referring to the book, *Science Curriculum Topic Study: Bridging the Gap Between Research and Practice* (Keeley, 2005)
Grain size	How broad or specific a topic is (e.g., *properties of matter* is a large grain size topic whereas *density* is a small grain size topic)
Key idea	An important idea unpacked from a learning goal—sometimes there are several key ideas embedded in one learning goal
Leader's Guide	A shorthand way of referring to this book, *A Leader's Guide to Science Curriculum Topic Study*
Learning goal	A teaching and learning target that specifically describes what students should know or be able to do
Misconception	A catch-all term for ideas students have that are not entirely scientifically correct (e.g., naïve ideas, partial understandings, preconceptions, misunderstandings, alternative frameworks)
National standards	Both the *Benchmarks for Science Literacy* and the *National Science Education Standards*
PCK	An acronym that stands for pedagogical content knowledge—this is the specialized knowledge about science teaching and learning that teachers need to understand in order to make content accessible to students
Preconception	An idea formed, often early on, before students formally encounter the content. Preconceptions can form outside of school or during previous curricular contexts
Professional learning community (PLC)	A group of team members who regularly collaborate to make improvements in meeting student learning needs through a shared vision and focus on curriculum, instruction, and assessment
Research	Although there are many kinds of research in education, in CTS, this term refers specifically to cognitive research (research on learning)
Science literacy	The understandings and ways of thinking that are essential for all citizens in a world shaped by science and technology (AAAS, 1989)
Skill	A process skill used in science (e.g., measurement, explanation, controlling variables) or a habit of mind (e.g., critical reasoning, skepticism, identifying bias)

CTS Terminology	Clarification of Terminology Used in the CTS Context
Sophistication	The complexity of an idea at a given grade level; for example, the Grades 3–5 gravity key idea that "The Earth's gravity pulls any object toward it without touching it" is at a lower level of sophistication than the Grades 6–8 idea that introduces the notion of a gravitational force between objects that depends on the mass of the objects and how far apart they are from each other
Standards	Common goals established nationally, statewide, or locally that are widely accepted by the science community and provide a focus for teaching and learning
Study guide	One of the 147 CTS study guides
Teacher educator	Anyone who facilitates teacher learning such as preservice faculty, scientists working with teachers, staff developers, coaches, etc.
Topic	A conceptual organizer or category for related learning goals that can be taught in a variety of contexts or themes (e.g., *life cycles* is a CTS topic, but *life cycle of the butterfly* is a contextual theme)
Topic study	A shorthand way of referring to a curriculum topic study

2

Introduction to Curriculum Topic Study for Leaders

The purpose of this chapter is to provide the CTS leader with basic background on curriculum topic study (CTS) and guidance about using the CTS parent book, *Science Curriculum Topic Study: Bridging the Gap Between Standards and Practice* (Keeley, 2005), to prepare to lead CTS sessions. The chapter reviews what CTS is and what it is not. It provides an overview to the resource books that are used with CTS, defines what is meant by science literacy, and demonstrates the value of knowing and using standards and research to guide practice. In this *Leader's Guide*, the authors provide substantial background and guidance on using CTS based on our experiences, but we think of new ways to apply CTS often. We therefore encourage CTS users to visit our Web site regularly for updated ideas and suggestions and to use the "contact the CTS Project" option on our Web site to share ideas and materials you have developed using CTS.

WHAT IS CTS?

CTS is a rigorous, methodical study process designed to help science educators deeply examine a common curricular topic. The CTS process utilizes a common set of books that includes standards and cognitive research resources, 147 different science topic study guides, and a variety of tools and processes to systematically examine a curricular topic. In addition to the science CTS described in this *Leader's Guide*, there is also a mathematics version of CTS. CTS is best described as follows:

- A *process* that incorporates a systematic study of standards and research
- A set of *tools* and collective *resources* for improving curriculum, instruction, and assessment

- An intellectually engaging, collaborative *professional development* experience where teachers come together to develop a common knowledge base about teaching and student learning

The CTS approach was adapted from the American Association for the Advancement of Science's (AAAS) Project 2061 study of a benchmark. This detailed study procedure involved clarifying the meaning and intent of a benchmark learning goal from the *Benchmarks for Science Literacy* (AAAS, 1993) using several tools from Project 2061's collection of science literacy resources. The premise behind this study procedure is that in order for educators to help students achieve the learning described in a benchmark, they must first understand what the benchmark statement intends students to be able to know or do. Taken at face value, a benchmark statement, a standard, a performance objective, or any learning goal regardless of the language we use to label it, can be easily misinterpreted as to its meaning, the boundaries that delineate the extent of knowledge for science literacy, experiences that contribute to learning the key idea(s) and skills, and the level of sophistication expected by the learning goal. Educators who have experienced the Project 2061 systematic study procedure note a significant difference in their understanding of a learning goal after completing a benchmark study.

While studying a single learning goal is important and worthwhile, the CTS Project developed a procedure similar to Project 2061's study of a benchmark to examine teaching and learning at the larger grain size of a curricular topic, eventually drilling down to specific learning goals. Since many teachers target multiple learning goals within a curricular unit or cluster of lessons, similar tools and resources for studying a curricular unit are needed. While standards may differ from state to state and instructional materials may include different curricular objectives, curricular topics are similar in schools across states. CTS expands the resources for a topic study beyond the Project 2061 tools to also include the *National Science Education Standards* (National Research Council [NRC], 1996) as well as an adult science trade book, *Science Matters* (Hazen & Trefil, 1991, 2009), and a compendium of research on student learning in *Making Sense of Secondary Science* (Driver, Squires, Rushworth, & Wood-Robinson, 1994). In addition, there is a Web site, www.curriculumtopicstudy.org, that provides optional suggestions for supplementary readings and materials that are topic specific. The CTS resources are listed in Table 2.1 and discussed further in this chapter.

We recommend that professional developers and leaders of CTS become thoroughly familiar with what CTS is and the underlying knowledge and research base that supports it. Take the time to read Chapter 1, "Introduction to Curriculum Topic Study," in the CTS parent book. It will give you the background you need to

Facilitator Note

Making Sense of Secondary Science is a publication from the United Kingdom where secondary education refers to the education of eleven- through sixteen-year-olds, as compared with the United States, where secondary education usually refers only to high school students. The book also addresses the science ideas of young children ages five to ten. In essence, it is a K–12 resource.

Table 2.1 CTS Resources

- *Science for All Americans* (AAAS, 1989)
- *Science Matters* (Hazen & Trefil, 1991, 2009)
- *Benchmarks for Science Literacy* (AAAS, 1993, 2008)
- *National Science Education Standards* (NRC, 1996)
- *Making Sense of Secondary Science* (Driver et al., 1994)
- *Atlas of Science Literacy* (Vols. 1–2) (AAAS, 2001–2007)
- State standards or frameworks and curriculum guides
- Optional: Additional supplementary resources (videos, journal articles, Web sites, trade books)

be prepared to answer questions about what CTS is, what CTS means by a "curricular topic," how CTS addresses content and pedagogy at the topic level, and what CTS can do to deepen science teachers' understanding of content, teaching, and learning.

WHAT CTS IS NOT

While it is important to know what CTS is, it is also important to know what CTS is not. CTS helps educators seek solutions or answers to problems or questions related to standards-based content, curriculum, instruction, or assessment, but it does not provide *the* solution. CTS provides the analytic lens and tools needed to make sense of these problems or questions and helps educators think through what they need to better understand in order to be effective teachers (e.g., how their students learn, curriculum alignment, and state standards interpretation). CTS does not provide the answers unique to their own contexts. Instead it engages educators in a process that helps them uncover what they need to know and helps them think about how to apply that knowledge to their own situations. The tool draws on and enriches professional dialogue among teachers as they use their professional knowledge to apply their CTS discoveries to their own situation. The following describes limitations to be aware of when using CTS:

- CTS is not the complete remedy for weak content knowledge (CTS is used to enhance and support content learning and to raise awareness of content that teachers need to learn).
- CTS is not a collection of teaching activities (CTS describes considerations one must take into account when planning or selecting teaching activities and contexts).
- CTS is not a description of *how to*'s (CTS helps you think through effective teaching based on knowledge of learning goals and how students learn, but does not prescribe a particular pedagogical approach).
- CTS is not a quick fix (CTS takes serious, dedicated time to read, analyze, and reflect on teaching and learning).
- CTS is not the end-all for professional development (CTS helps teachers identify the need for additional learning experiences).

CTS AND SCIENCE LITERACY

When educators use CTS, they develop a deeper understanding of science literacy and the implications for curriculum, instruction, and assessment. When leading a CTS session, it is important to share with educators what is meant by *science literacy*. There are as many definitions of science literacy as there are of *inquiry*. In CTS, we use the description in *Science for All Americans:* "Science literacy includes the understandings and ways of thinking that are essential for *all* citizens in a world shaped by science and technology" (AAAS, 1989, p. xiii). This broad, multifaceted description of science literacy includes

- being familiar with the natural and human-designed world;
- recognizing and respecting unity in the natural world;

- being aware of the ways science, mathematics, engineering, and technology depend on one another;
- understanding key concepts and principles of science;
- being able to use scientific ways of thinking and communicating;
- knowing that science is a human enterprise and works in a certain way that may differ from other disciplines and ways of knowing;
- knowing the strengths and limitations of science; and
- being able to use scientific ways of thinking for uses other than scientific work, including personal and social purposes.

Furthermore, the *National Science Education Standards* document goes on to say that "individuals will display their scientific literacy in different ways, such as appropriately using technical terms, or applying scientific concepts and processes. And individuals will often have differences in literacy in different domains, such as more understanding of life-science concepts and words, and less understanding of physical-science concepts and words" (NRC, 1996, p. 22).

Science literacy develops during the school years and continues to evolve in adulthood. For example, the ideas and skills important to teaching science continue to deepen over the course of teachers' continual professional learning. "Scientific literacy has different degrees and forms; it expands and deepens over a lifetime, not just during the years in school. But the attitudes and values established toward science in the early years will shape a person's development of scientific literacy as an adult" (NRC, 1996, p. 22).

When educators use CTS, they are identifying the science knowledge and skills that *all* students are expected to know and use, regardless of whether they will go on to pursue higher education and scientific careers. Because the K–12 content of CTS is rooted in the standards, CTS describes the threshold of science teaching and learning, not the ceiling. It does not describe the science needed for advanced coursework or concepts and skills that exceed science literacy at a given grade level nor does it spell out all the science taught in the K–12 curriculum. It focuses on the core knowledge and skills described in the standards that every student must attain in order to progress to the next stage of increasingly sophisticated ideas or understand the important connections between concepts. To learn more about adult and K–12 science literacy, CTS leaders are encouraged to read the introduction to *Science for All Americans*, the "About Benchmarks" section in the front of *Benchmarks for Science Literacy*, and the overview to *National Science Education Standards*.

In addition, Chapter 13 of *Science for All Americans* describes several principles of teaching and learning that characterize the approach of effective teaching that focuses on science literacy. The following are some of these important principles:

- Learning is not necessarily an outcome of teaching.
- What students learn is influenced by their existing ideas.
- Progression in learning is usually from the concrete to the abstract.
- People learn to do well only what they practice doing.
- Teaching should be consistent with the nature of scientific inquiry.
- Science teaching should reflect scientific values.
- Science teaching should aim to counteract learning anxieties.
- Teaching should take its time.

There are six overall categories of science standards found in the *National Science Education Standards* (NRC, 1996). One of these categories, the science content standards, is most commonly referred to and used by educators. The content standards include eight specific content areas, including unifying concepts and processes in science, science as inquiry, physical science, life science, Earth and space science, science and technology, science in personal and social perspectives, and the history and nature of science. Within each of these areas, there is specific content identified as essential for learning within three grade bands (K–4, 5–8, and 9–12). For example, in the content area of physical science, Grade K–4 students will develop an understanding of properties of objects and materials; position and motion of objects; and light, heat, electricity, and magnetism. The other five categories of standards in the *National Science Education Standards* are equally important but often not given the same attention as the content standards. These include science teaching standards, professional development standards, assessment standards, science education program standards, and system standards. A careful examination of these standards will show how CTS easily fits into and supports these comprehensive standards as shown by the examples in Table 2.2.

Table 2.2 CTS and the National Science Education Standards (Noncontent Standards)

National Science Education Standards (NRC, 1996)	*Example Standard (NRC, 1996)*	*Example of How CTS Supports the Standard*
Science teaching standards	Teachers select teaching and assessment strategies that support the development of student understanding and nurture a community of science learners.	CTS helps teachers examine content-specific instructional implications that inform effective teaching. Examining the research on learning informs assessment strategies that can be used to elicit preconceptions.
Standards for professional development for teachers of science	Professional development activities must provide opportunities to know and have access to existing research and experiential knowledge.	CTS connects teachers to learning research that helps them understand what makes the learning of specific concepts difficult. Teachers link the findings from research to their own instructional contexts and knowledge of their students.
Assessment in science education	Assessments are deliberately designed and have explicitly stated purposes.	CTS provides a process to develop diagnostic assessments that explicitly link key ideas to common misconceptions for formative purposes.
Science education program standards	The science program should be coordinated with the mathematics program to enhance student use and understanding of mathematics in the study of science and to improve student understanding of mathematics.	There are several CTS study guides related to mathematics (e.g., data collection and analysis, graphs and graphing, correlation, etc.) that can build a common understanding among science and mathematics teachers and promote greater consistency across the disciplines.
Science education system standards	Policies must be supported with resources.	CTS fuels and sustains science education reform. Policy makers should ensure that all teachers have access to the materials used in CTS and opportunities to learn how to use them.

Facilitators of CTS will benefit from becoming familiar with all of the sections of *Science for All Americans, Benchmarks for Science Literacy,* and *National Science Education Standards* when questions about science education or issues that are related to CTS but not specifically addressed in the CTS process arise. A knowledgeable facilitator can make the link between the broader teaching and learning recommendations in *Science for All Americans, Benchmarks for Science Literacy,* and *National Science Education Standards* and the specific recommendations that are revealed through the study of a specific curricular topic. As we point out in Chapter 1, there is a professional knowledge base that all of us working in science education should know and be able to apply. These resources are the foundation of that knowledge.

KEY POINTS ABOUT STANDARDS AND RESEARCH ON LEARNING

CTS is described as a "standards- and research-based" study of a curricular topic. In order to understand what this means, beliefs about standards and research need to be addressed in order to make the best use of the CTS process and resources. These beliefs have surfaced during the CTS Project's work with many educators and represent some of the misconceptions that still linger since the introduction of standards. In addition, many teachers have entered the teaching profession post-standards movement and may not be as aware of what standards really are and how they can be used as their colleagues who experienced the wave of standards-based reform. CTS session leaders may wish to precede the introductory CTS sessions with an opportunity to surface and discuss participants' beliefs about standards and research. Handout 2.1: Ten Common Beliefs about Standards and Research on Learning, found in the Chapter 2 folder on the CD-ROM at the back of this *Leader's Guide,* can be used to surface teachers' beliefs and initiate dialogue about standards and research prior to learning about the specific resources used in the CTS process. You might use this as a kickoff discussion or prior to beginning your CTS work. If you are working with a group of teachers over months or years, you can help them revisit their beliefs from time to time and reflect on how they are changing.

Key points to make about standards and research that can be surfaced and discussed using Handout 2.1 during any of the CTS sessions include the following:

1. *Standards are for all students.* Science standards describe the important skills and knowledge that all students should be expected to learn, regardless of their academic placement. This does not mean that students who struggle with learning cannot start with simpler concepts and gradually work their way to achieving a standard. Whether students are in honor classes or needing remedial services, the standards are written for all students. Although there are circumstances that may make it difficult for some students to achieve particular standards, it is important that they have every opportunity to learn the ideas and skills in the standards.

2. *Standards are the threshold, not the ceiling.* A common myth about standards is that they "dumb down" content and do not allow students to learn more advanced ideas. The standards examined in Sections III, V, and VI when using CTS guides are not intended to limit students' knowledge. They describe what *all* students are expected to know in order to be science literate. They are the thresholds for building science literacy. However, that does not mean students can't go further. Depending

on the readiness of students and the context for learning, students can engage in learning beyond the standards—if they have demonstrated they have met the standards that are prerequisite to building more sophisticated knowledge.

3. *Standards are not intended to limit what is taught.* Standards provide clear direction and focus for teaching and learning. While teachers must focus on helping students achieve the standards, this does not mean that other topics of science interest cannot be taught. Since there is limited time in the curriculum to teach all the important science ideas, teachers must ensure that standards are being met. However, other ideas in science can sometimes be integrated with the standards, or special topics of interest can be taught as long as they do not replace students' opportunities to learn the ideas essential to achieving science literacy.

4. *National standards can and should be used with state standards.* Many educators believe that state standards replace national standards and that there is no use for the latter when they are being held accountable to state standards. National standards (*Benchmarks for Science Literacy* and *National Science Education Standards*) are just as useful today as they were when they were released before states crafted their own standards. Using state and national standards should not be an either/or decision, for several reasons. First, instructional materials are not explicitly developed for a particular set of state standards, and their alignments to state standards are, at best, topical. Materials are developed to align with the national standards and often to the standards used in large states that have a formal textbook adoption process. Understanding the connection to the national standards and their relation to state standards can strengthen the alignment and use of instructional materials. Second, there is wide inconsistency across state standards both in coherence, interpretation, and specificity. Using national standards to interpret the meaning and intent of a state standard improves interpretation within a state as well as promotes consistency from one state to another. This is particularly important as our society has become increasingly mobile and students may attend school in more than one state. Third, for assessment purposes, most state standards are written with a performance verb that precedes a knowledge or skill statement. Often the very addition of a verb, particularly if it was arbitrarily selected to encompass various levels of cognitive demand, can change the meaning of a learning goal.

For example, a state standard that says "Demonstrate the law of the conservation of matter" is very different from the AAAS Project 2061 benchmark that describes that students will know the following:

> No matter how substances within a closed system interact with one another, or how they combine or break apart, the total mass of the system remains the same. The idea of atoms explains the conservation of matter: If the number of atoms stays the same no matter how the same atoms are rearranged, then their total mass stays the same. (AAAS, 1993, p. 79)

One can demonstrate the law of the conservation of matter with a particular interaction yet fail to recognize the importance of the interaction taking place in a closed system, that there are a variety of different types of interactions that can be used to demonstrate this principle, and that atoms are used to explain why mass is conserved. The benchmark provides much greater specificity to help teachers develop the knowledge their students need to successfully meet a state standard.

The national standards are written to describe the knowledge and skills students should attain, not how they will demonstrate knowledge on an assessment item or how it should be taught. Taking the time to study the national standards, particularly the language and specificity used, can help users of state standards be clearer about what the knowledge is in a performance-based learning goal and helps ensure that key ideas are not masked by the verbs used for assessment purposes.

5. *Standards are not a curriculum.* Standards inform curriculum but by themselves, they are not *the* curriculum. Curriculum is the way content is organized for teaching and learning. Content standards are the learning goals that are targeted by the curriculum. Often textbooks or other instructional materials have been aligned to the standards, but one must keep in mind that most instructional materials try to cover everything. Teachers must still have opportunities to examine standards and organize the teaching of their content in ways that are developmentally and conceptually sound.

6. *Standards can be repeated in the curriculum.* Repeatedly teaching the same content, especially after students have learned the content is counterproductive. However, research on learning indicates that some concepts must be encountered several times in different contexts and at different levels of complexity before students can attain an enduring understanding. Encountering a key idea from a standard in different contexts ensures that students can transfer their knowledge to different situations. For example, if one taught that "substances react chemically in characteristic ways with other substances to form new substances with different characteristic properties" (NRC, 1996) only in a physical science context without revisiting the key idea in a life science context, students may fail to recognize that this key idea also applies to chemical reactions in living systems.

7. *Rigor does not mean difficult.* Some educators believe that standards undermine efforts to provide a rigorous and challenging curriculum. Rigor actually refers to high levels of intellectual engagement, not difficulty. If students are deeply immersed in higher-level thinking and applying content, then they are engaged in rigorous learning. Too often students are taught advanced topics that they lack the necessary prerequisites to learn for the sake of "rigor" and end up memorizing facts at the expense of conceptual understanding.

8. *Standards do not prescribe a pedagogical approach.* Standards describe the goals for learning. Some of these goals are best achieved through inquiry or conceptual change instruction while others may be learned through other more direct approaches. There is no one method of instruction prescribed in the standards.

9. *Knowledge and skill standards are not intended to be taught separately.* Standards are not designed to be a "checklist" curriculum, with each standard taught separately from the others. This approach not only makes it seem as if there are too many standards to cover, but it also leads to disjointed learning and a curriculum that lacks coherence. Most standards can and should be taught as a cluster of interrelated ideas and skills connected in a coherent way. This includes clusters of knowledge standards as well as clusters of skills standards. In addition, the skills standards should be taught and applied in the context of knowledge acquisition of disciplinary content.

10. *Cognitive research informs teaching and learning but does not prescribe it.* Research is very helpful in informing educators about difficulties students are likely to encounter, commonly held ideas they may have, and the developmental appropriateness of content and context. However, students are unique individuals, and their readiness depends on several factors including their prior knowledge and experiences, the quality of the school's curriculum, the skill of the teacher, and other factors. Because research may point out that some students struggle with a concept at a certain age, it does not necessarily mean one must not teach it. Teachers should use their best judgment about their own students and assess their readiness before teaching ideas that may be conceptually challenging to students. Research should be used to inform, not dictate what is taught.

USING CTS TO LEAD PROFESSIONAL LEARNING

CTS provides a methodical way to ensure that professional development is focused on the K–12 content and opportunities to learn necessary for science literacy. It provides leaders with a standards- and research-based focus for their professional development designs and decisions. Whether leaders are involved at the school, district, or state level, working in a university or science museum, leading staff development, facilitating science curriculum or assessment committees, or engaged in other types of science reform efforts that require knowledgeable and skilled leadership, CTS is a tool that brings increased content alignment, coherency, and consistency to leaders' work.

Leaders of teacher learning come from a variety of backgrounds where they may have strengths in one area but not others. For example, university scientists working with teachers may have strong expertise in an area of science, but little understanding of K–12 curriculum or developmental readiness of students at different grade levels. CTS helps them build a bridge from the university content they teach to what teachers need to take from that experience to shape curriculum and instruction at their specific grade levels. Teacher leaders may know the science and instructional implications of their own grade level, but lack knowledge of the other grade levels. A professional developer may be a masterful facilitator of adult learning, but lack specific knowledge of content and misconceptions students and teachers might have. A high school teacher in one discipline, such as physics, may have little knowledge of teaching and learning in a different discipline such as biology. A preservice educator may have particular expertise in instructional design and teaching methodologies but lack experience in using standards and cognitive research to inform these designs and methodologies. All of these are examples of how CTS can strengthen the professional development work that leaders do to improve teachers' practice.

CTS is a versatile tool that meets the diverse needs of a wide range of leaders in science. It not only informs leaders and raises their level of knowledge, but most important, it provides them with an intellectually stimulating process for developing a common core of content, teaching, and learning understanding within the various groups and types of teachers they work with, from preservice to novice teachers to experienced veteran teachers to teacher leaders.

CTS RESOURCES AND TOOLS

The primary resource used for CTS is the parent book, *Science Curriculum Topic Study: Bridging the Gap Between Standards and Practice* (Keeley, 2005). It provides several tools for using CTS. The primary tool is the CTS guide. In addition, there are several other tools described in Chapter 4 of the CTS parent book that are used with the CTS guides and study process. The CTS Project also maintains a Web site at www.curriculumtopic study.org that provides additional material for CTS users and leaders of CTS. Below is an at-a-glance review of these resources.

The Parent CTS Book

To use this *Leader's Guide* to CTS, it is necessary for leaders to have the CTS parent book. It contains the background material for teachers, a variety of tools, authentic vignettes of using CTS in a variety of contexts, and the 147 study guides that are at the heart of the CTS process.

The Collective Set of CTS Resources

What makes CTS unique as a professional learning strategy is its systematic use of a collective set of professional resources that have been analyzed and vetted to match the content of the 147 curricular topics. The vetting process saves time for the busy teacher by selecting the readings and page numbers that are most relevant to studying a topic. These resources include the primary national standards resources (*Benchmarks for Science Literacy* and the *National Science Education Standards*); two resources that describe adult science literacy (*Science for All Americans* and *Science Matters*); two visual depictions of the K–12 growth of understanding and coherent sequences and connections of learning goals (*Atlas of Science Literacy* [Vols. 1–2]); and three compendia of cognitive research summaries (Chapter 15 of *Benchmarks for Science Literacy*, *Making Sense of Secondary Science: Research Into Children's Ideas*, and the more recent research summaries in Volume 2 of the *Atlas of Science Literacy*). (*Note:* When the first edition of the CTS parent book came out in 2005, Volume 2 of the *Atlas of Science Literacy* was not yet published and therefore is not referenced in the study guides. If you are using the 2005 edition of the CTS parent book with this *Leader's Guide*, you can download a crosswalk between the CTS guides and the new *Atlas* maps from the CTS Web site, www.curriculumtopicstudy.org, or print a copy of Handout 2.2: CTS Crosswalk to *Atlas of Science Literacy* from the Chapter 2 folder on the CD-ROM at the back of this book.)

This suite of CTS resources is described further in the CTS parent book. Leaders of CTS are encouraged to take the time to become thoroughly familiar with these books by reading the descriptions of each resource and how it is used in Chapter 2 of the CTS parent book. In addition, examine each book for its

- layout and organization of content,
- introduction and overview of the resource,
- unique features that distinguish it from the other CTS resources, and
- background material on science literacy or children's ideas.

> **Facilitator Note**
>
> A new updated edition of *Science Matters* (Hazen & Trefil, 1991) was published in 2009. If you are using the 2009 edition of *Science Matters* with CTS, please check the CTS Web site for a crosswalk to the 147 CTS guides that includes the new sections and page numbers of the 2009 edition of *Science Matters*.
>
> It is anticipated that there will be a second edition of the CTS parent book released sometime after 2010 that will update the study guides to include Volume 2 of the *Atlas of Science Literacy* (AAAS, 2007), the 2009 version of *Science Matters* (Hazen & Trefil, 2009), as well as include additional recent material. When the second edition of the CTS parent book comes out, the CTS Web site will reflect the changes. Every effort will be made to make sure subsequent editions are matched to this *Leader's Guide*.

Facilitator Note

Of all the resources used for the CTS process, the *Atlas of Science Literacy* is probably the most complex to use and understand. It is essential for leaders to read the introduction to the *Atlas* as well as the supporting material that describes ways the maps are used, map keys, the nature of connections, and so forth. The Project 2061 Web site at www.project2061.org includes useful information about the *Atlas* as well as an introductory PowerPoint presentation explaining the resource.

One question that often comes up when leading a CTS session is "Science changes so rapidly, aren't these resources out of date?" It is true that science has advanced considerably in the past decade, but it is also true that the big ideas in science are still pretty much the same. The ideas necessary for science literacy as described in *Science for All Americans* and in the standards still hold true today. *Science Matters* is slightly different from *Science for All Americans*, as it describes science ideas that may go beyond what is learned in K–12 education but are explained to help everyone, including nonscience majors, understand the science that is presented in the media or that may come up in conversations or decisions that depend on some knowledge of science. In this respect, there are some sections in *Science Matters* where our current knowledge has advanced, such as stem cell research, which is not mentioned in the 1991 book, but is included in the revised 2009 edition. However, it still describes most of the important, big ideas that are central to the secondary science curriculum. It is up to the facilitator to decide whether to use it or not, depending on the context of your professional development session. The research described in Chapter 15 of *Benchmarks for Science Literacy*, Volume 1 of *Atlas of Science Literacy*, and *Making Sense of Secondary Science* was conducted in the decade up to and preceding the standards. However, much of this research on learning is still relevant today. The second volume of *Atlas of Science Literacy* contains research summaries from research that was available after the standards and Volume 1 was published. These summaries can be used to update the research base in Section IV of the CTS parent book and will be referenced in the second edition.

Another question that may come up is how "dated" the standards resources are. At this time, there is no scheduled effort to revise *Benchmarks for Science Literacy* or *National Science Education Standards* and create a new set of standards. However, you will notice in Volume 2 of *Atlas of Science Literacy* that some benchmarks have been revised or added that were not in the original version of *Benchmarks for Science Literacy*. As we become more familiar with standards and their use in the classroom, changes have been necessary to reflect current research, our use of words, complexity of ideas, or key ideas that were left out and are necessary for developing a coherent understanding of science. These are indicated on the *Atlas* maps by * (for edited benchmark) or ** (for new benchmarks). You can also see a comparison between the original benchmarks and the updated benchmarks by visiting the Project 2061 Web site at www.project2061.org and accessing the online version of *Benchmarks for Science Literacy*. These updates reflect necessary revisions after a decade of standards implementation but do not discard the original standards documents in favor of a completely new set of standards. This would be counterproductive to the hard work states, schools, and districts have done over many years of standards-based reform. While we know a lot more about standards today than we did in the past, one can think of the standards documents as dynamic tools that are continually updated as needed without starting all over again and developing new sets of standards.

At this time, the National Science Teachers Association (NSTA) is working on the development of *Science Anchors*. These "anchors" will not replace the national standards but rather focus on a smaller set of core ideas common across both sets of national standards as well as important, common ideas in states' standards. The NSTA *Science Anchors* will be similar to *Curriculum Focal Points: For Prekindergarten Through Grade 8 Mathematics* (National Council of Teachers of Mathematics, 2006) as far as their impact on standards development and curriculum. As the *Science Anchors* develop, or other initiatives that aim to develop a smaller set of core national standards take hold, there will be updates on the CTS Web site to describe how they can be used with CTS.

The CTS Guides

The heart of the CTS process is the CTS guide. What makes these guides so useful to educators who have been using, or tried to use, national standards and research documents is that the relevant readings for particular science topics have already been identified and vetted. For more information about the study guides and the six sections of each, review pages 19–21 in the CTS parent book and refer to "Anatomy of a Study Guide," which you can find in the Module A1 handouts folder for Chapter 4 on the CD-ROM at the back of this *Leader's Guide.*

The page numbers in each CTS guide reference the editions of each resource book that are described in the CTS parent book. If you use a different edition in which the page numbers do not coincide with the study guides, it should not pose a problem, as each study guide specifically lists the section titles and subtitles where the readings can be found. Likewise, for those resources that have online versions that are not numbered by pages, you can find the section for each reading by using the chapter or section titles.

It is important to note that there is overlap in the 147 study guides in terms of their grain size. The overlap is purposefully designed to allow the CTS user to study a topic using a broad sweep or a narrow one. For example, the topic "Properties of Matter" is much broader than the topic of "Density," which is subsumed in "Properties of Matter." One can even select a grain size that is in between broad and highly specific, such as "Physical Properties and Change." Grain sizes range from broad to average to narrow and are identified on Handout 2.3: Science Curriculum Topic Study Guides—Grain Size Indicator in the Chapter 2 folder of the CD-ROM at the back of this *Leader's Guide.* Generally, the larger the grain size, the more sections there are to read and synthesize. Before designing a CTS session, leaders should carefully choose the appropriate grain size for their audience and the content of their professional development.

> ### Facilitator Note
>
> One thing to be aware of is that each study guide identifies the related readings. However, it is not always necessary to read all of the material listed in each study guide. As the facilitator, there will be times when you will select certain readings and skip others. However, you will need to examine these carefully in order to ensure that important information needed for the purpose of your professional development session is not left out. For example, some study guides contain the historical episodes relevant to a topic that you may choose to eliminate to save time or because they do not fit with the primary purpose of your session. It is important to do the CTS first yourself and then decide which readings will best fit your audience and focus of your professional development.

CTS Application Tools

Chapter 4 in the CTS parent book provides a description of some of the tools used with CTS such as tools for curriculum selection, K–12 scope and sequence articulation, hierarchy of content knowledge, instructional design, and assessment probes. Take the time to read the background on each of these applications and how CTS findings are applied in these various contexts. In Chapter 6 of this *Leader's Guide* are suggestions for integrating these application tools into your CTS professional development designs.

The CTS Web Site: www.curriculumtopicstudy.org

CTS is a dynamic resource that will continually be updated with new information, tools, and resources. Because new research on learning continues to be published, the CTS Project maintains a searchable database with updated research findings linked to specific CTS guides. While CTS uses a collection of common resources across all topics, the Web site lists topic-specific resources such as videos of teaching and learning, assessment items, and content articles that can be used with specific topic guides. CTS leaders should check the Web site periodically for additional materials to supplement the CTS parent book as well as this *Leader's Guide*. Examples of ways to use the supplementary resources on the Web site with the CTS sections are shown in Table 2.3. When you visit the Web site, you have the option of e-mailing the supplemental resources to yourself or to colleagues, making them easy to access later for your own work.

FROM GENERIC TOOLS TO CONTENT-SPECIFIC TOOLS

Leaders have a variety of excellent, general tools at their fingertips to improve and enhance teaching and learning. Many leaders incorporate these tools into their professional

Table 2.3 Examples of CTS Supplementary Web Site Resources and Their Use

CTS Study Guide Section	Examples of Supplement Type and Use
Section I: Identify Adult Content Knowledge	• Content trade books in science—provide information and visuals to support science content learning. • Web site—provides content information specific to a topic. • Videos—content experts explaining science concepts.
Section II: Consider Instructional Implications	• Instructional resource—model lesson to illustrate effective instruction of a specific topic. • Journal article—connecting research to practice in the classroom or information about specific teaching techniques. • DVD or Web videos—video footage of a teacher teaching a CTS topic.
Section III: Identify Concepts and Specific Ideas	• Science education books—provide specific information about the science standards such as "unpacking" a learning goal. • Web site—clarifications of learning goals.
Section IV: Examining Research on Student Learning	• Journal article or Web site—link to recent research papers or articles about students' ideas in science. • Assessment task—example assessment probes designed to elicit research-identified misconceptions.
Section V: Examine Coherency and Articulation	• *Atlas* strand maps—links to specific online *Atlas* maps. • Web sites—links to strand map services such as the Harvard Smithsonian Center for Astrophysics Digital Video Library (www.hsdvl.org).

development. However, these tools are designed to be used by teachers of all disciplines. As a result, the unique features of science as a discipline are often not explicitly addressed by many of these tools. It is assumed that science educators can "fill in" their own content. However, in many cases, this perpetuates the notion of "practice as usual" if teachers have not been confronted with new ways of thinking about teaching and learning as they apply to specific content. For example, a teacher who is not aware of why the word *evaporation* is intentionally not used in elementary science standards may design an assessment that uses the word *evaporation* and shows the upward arrow pointing toward a cloud, yet fail to develop the bigger idea that when water leaves an open container, it goes into the air in a form we cannot see called water vapor. The bigger idea is that the water is in the air around us even though we cannot see it. Without CTS, a teacher's approach may perpetuate the misconception that water immediately goes up to the clouds, rather than exists mostly in the surrounding air. Students can use the word *evaporation* by saying the water evaporates, yet have no idea that it is in the air in a form they cannot see. CTS helps bring content specificity to many of the generic tools schools and professional developers are using to develop curriculum, instruction, and assessment, such as the following:

- *Understanding by Design* (Wiggins & McTighe, 2005)
- *Classroom Instruction That Works* (Marzano, Pickering, & Pollock, 2001)
- *Concept-Based Curriculum and Instruction* (Erickson, 1998)
- *Getting Results With Curriculum Mapping* (Jacobs, 2004)
- *Enhancing Professional Practice: A Framework for Teaching* (Danielson, 1996)
- *Professional Learning Communities at Work* (DuFour & Eaker, 1998)
- *Whole-Faculty Study Groups: Creating Student-Based Professional Development* (Murphy & Lick, 2001)

If you are using these tools with teachers in your professional setting, consider ways to use CTS to bring more content specificity to these tools. Many of the designs and suggestions provided in this *Leader's Guide* can be used with the generic tools. For example, when using *Understanding by Design* to design a performance task, it makes sense to conduct a CTS first in order to be well grounded in the research on learning and clarify what knowledge and skills are desirable according to the standards. Study groups often focus on implementing specific instructional strategies such as infusing reading and literacy strategies across the subject areas. The teachers could benefit from doing a CTS first to identify what important topics students could be writing and reading about and how to avoid introducing vocabulary that may confuse students or perpetuate misconceptions.

What Do Leaders Need to Know Up Front to Use CTS Optimally?

In order to help others to use CTS effectively, leaders should first become thoroughly familiar with the CTS parent book. The following guiding questions based on material in the CTS parent book that is used with this *Leader's Guide* will prepare you to be an effective leader of the designs included in this book or to design your own. In the spirit of constructivist learning, if you can answer each of these questions on your own, you will be well prepared to facilitate the designs in this book as well as develop your own and field any questions about CTS. Consider teaming up with a colleague from your school or organization who is using CTS and go through these questions together as a study group. Thoroughly knowing the CTS parent book that this *Leader's Guide* is designed to support will enable you to be a knowledgeable, effective, and confident leader of CTS.

Questions you can answer from the CTS parent book, Chapter 1:

1. Can you describe what CTS is and what educators gain from it?

2. Can you explain why standards and research are used in CTS?

3. Why is CTS called a "systematic study process"?

4. Why does CTS focus on topics? What does "topic" mean in CTS?

5. Where did the CTS process originate? What informed development of CTS?

6. How does CTS address teachers' needs at all levels of the professional continuum?

Questions for Chapter 2 of the CTS parent book:

1. Can you describe the components of a CTS guide and what the guide is used for?

2. What are the resource books used in CTS?

3. Why is CTS described as "having experts at your fingertips 24/7"?

4. Can you describe why CTS is a versatile and flexible tool for teachers?

Questions for Chapter 3 of the CTS parent book:

1. What are some examples of ways to process information from CTS readings?

2. Where would you find guiding questions for each of Sections I through VI of the CTS parent book? How would you use these questions?

3. What is the CTS learning cycle? How would you describe each of its components in terms of what the facilitator does?

4. How is a student inquiry using a learning cycle similar to adult CTS inquiry using the CTS learning cycle?

5. What does a CTS summary sheet look like and what is its purpose?

Questions for Chapter 4 of the CTS parent book:

1. What are some ways to use CTS to improve content knowledge?

2. What are some ways to use CTS in a curriculum context?

3. What are some ways to use CTS in an instructional context?

4. What are some ways to use CTS in an assessment context?

Questions for Chapter 5 of the CTS parent book:

1. Which vignettes connect most to your work and the goals of teachers with whom you are working?

2. How do the vignettes illustrate the use of CTS for curriculum, content learning, instruction, and assessment?

3. Which vignettes address assessment, and how do they illustrate the use of CTS?

4. How do the vignettes illustrate how the different grades or levels engage in the CTS process?

5. How can vignettes be used to help educators see the value in using CTS?

Questions for Chapter 6 of the CTS parent book:

1. How are the CTS guides organized?

2. How many study guides are there in all?

3. Besides the life, Earth, space, and physical categories, what other categories are included in the CTS guides?

4. How would you describe the differences in grain size of the study guides?

Questions for Resources A and B (pp. 273–281) of the CTS parent book:

1. What types of resources are found here?

2. How might you use these resources?

Once you become thoroughly familiar with the CTS parent book and the various resources and tools it uses or contains, it is time to consider ways to design and facilitate effective learning using CTS. The next chapter connects CTS to the research on effective professional development and provides facilitation strategies for getting started, ideas for managing and accessing resources and choosing reading and processing strategies, and general tips to make your use of CTS successful. Subsequent chapters offer the specific designs for partial and full topic studies, applications that can be embedded in CTS, and designs and suggestions for embedding CTS within many different types of professional development strategies.

3

Considerations for Designing and Leading Curriculum Topic Study

In this chapter we review research-based principles on effective professional development and demonstrate how using the curriculum topic study (CTS) process as a resource for professional development planning and as a component of any professional development strategy can lead to more effective teacher learning programs. In later sections in this chapter, we describe the CTS learning cycle that is used for all CTS sessions and provide many facilitation strategies and tips that can be used to ensure your CTS sessions fully engage participants in active and thoughtful learning and reflection.

CTS AND THE PRINCIPLES OF EFFECTIVE PROFESSIONAL DEVELOPMENT

In the seminal work describing new ways of thinking about and designing professional development for science and mathematics teachers, Loucks-Horsley, Hewson, Love, and Stiles (1998) and Loucks-Horsley, Stiles, Mundry, Love, and Hewson (2010) suggested that professional development designers must pay attention to several key principles for quality teacher learning. In this chapter, we discuss how using CTS strengthens teacher learning programs by addressing many of these research-based principles and provide a strong rationale for how the CTS tools can lead to more effective teacher learning programs.

Education leaders, teacher educators, and professional developers in science endeavor to create professional development programs that have the necessary ingredients and characteristics for success. In our work, we have found that programs designed with several research-based principles for effective professional development first identified by Loucks-Horsley et al. (1998) by synthesizing national standards from the National Research Council (1996), National Council of Teachers of Mathematics (2000), and the National Staff Development Council (2001) can serve as guideposts for the design of quality teacher development programs in science. We have applied these principles directly to our designs for leading teachers through CTS and find that building the CTS strategy into any teacher professional development from preservice to inservice to teacher leadership development supports these programs to embody the important principles for effective professional development. When these principles are present in professional development, teachers have a greater chance of transforming their practice and enhancing student learning. The principles of effective professional development that we find most connected to the work of CTS are as follows:

1. The professional development design is based on a well-defined image of effective classroom learning and teaching and immerses teachers in developing a clear vision of standards- and research-based practice that ensures learning for all students.

2. The professional development experiences provide opportunities for teachers to build their science content and pedagogical content knowledge (PCK) and examine practice to make direct connections to the classroom.

3. The professional development engages teachers as adult learners using effective means of gaining new knowledge and applying their insights to learning approaches they can use with their students.

4. Effective professional development builds a learning community based on collaboration that is focused on enhancing practice among science teachers.

5. The professional development supports teachers to serve in leadership roles by developing their expertise in their subject matter and providing ample opportunities for them to share knowledge with their colleagues through many different leadership roles as mentors, coaches, facilitators, members of professional learning communities (PLCs) and study groups, and often just by opening their classroom doors to share practice (Loucks-Horsley et al., 2010).

In the next section, we discuss these five principles and provide examples of how CTS brings them to life in professional development initiatives.

Professional Development Is Driven by a Vision of Effective Learning and Teaching

What does it mean for professional development to be driven by a well-defined image of effective classroom teaching and learning, and how can CTS help you get there? As novelist Lewis Carroll once said, "If you don't know where you are going, any road will get you there." Unfortunately, since too often professional development is planned and carried out without a clear vision of where it is headed or what purpose it will accomplish, educators set off on any path. Using CTS, professional developers can better define where they are going and design the right learning paths to ensure their teachers will reach their learning destinations using navigation tools that will keep them from veering off on side trips, traveling unmarked roads and confusing highways, ending up in dead ends, and embarking on inefficient journeys that cover far more territory than needed.

The image of effective classroom learning and teaching reflected in this principle calls for all students to have an equitable opportunity to learn science deeply and in ways recommended by national standards, for example, by engaging in inquiry and investigation. It envisions that students (and teachers) will be active members of a rigorous and challenging learning community. Students will have access to knowledge that is appropriate for them based on national and state standards for their grade span, is developmentally appropriate, and builds upon what they already know and understand. The effective learning community will provide meaningful ways for students to assess what they know, to correct misunderstandings, and be challenged to move to the next level (National Research Council [NRC], 1996).

Some of the first questions professional developers, teacher educators, and other leaders need to ask as they aim to provide teacher learning programs that are aligned to this principle are, *What do we want to see in the classroom? What would students be doing? What would teachers be doing?* Once they are clear about these, they are better prepared to decide, *What professional development will get us there?* But these questions are not always easy to answer. It is sometimes difficult to identify concepts and clarify the key ideas important for students to master, as well as recognize the instructional implications that impact mastery of that content. Many teacher-learning programs allow the instructional materials to dictate the answers to these questions. The problem with that, however, is that instructional materials are not always research and standards based. Textbooks and other materials may be written too broadly to appeal to many different standards. They may include activities that contextualize learning to the extent that students have difficulty transferring key ideas to other contexts. They often lack examples of relevant phenomena that can provide explanatory power for key ideas or use imprecise or inappropriate language that can develop and further reinforce students' misconceptions (Garnett & Treagust, 1990; Sanger & Greenbowe, 1997). Analyses of U.S. science textbooks have shown that they cover too many topics, use difficult vocabulary, make little contact with students' background knowledge, and do not address commonly held misconceptions (Anderson, 1995; Roseman, Kesidou, Stern, & Caldwell, 1999; Van den Akker, 1998). They also lack logical structures that systematically develop concepts and relate topics to one another in principled and meaningful ways, often presenting content in the absence of the important explanatory function of science (Ajewole, 1991; De Posada, 1999; Shymansky, Youre, & Good, 1991; Strube & Lynch, 1984).

To ensure that teacher education and professional development is based on a well-defined image of effective teaching and learning, we recommend that all teacher educators and professional developers consult the standards and research by engaging in targeted curriculum topic studies. As discussed in Chapter 2 of this *Leader's Guide*, CTS affords access to key information to inform the design of professional learning for teachers. In each guide, Section I reveals the enduring understandings all adults, including teachers, should know about the topic; Section II uncovers suggestions for providing effective instruction on the topic and the student learning difficulties of which a teacher should be aware; Section III provides insights into the concepts, specific ideas, or skills that make up the learning goals for the topic and helps clarify what is most important to teach; Section IV points out the specific misconceptions students might have about the topic and raises ideas about what might contribute to the misconceptions; Section V illustrates the progression of concept development from its most basic beginning in the early grades to a culminating, sophisticated scientific idea in the later grades; and Section VI provides the opportunity to better understand how state and local standards and curriculum align with national standards and research. This is critical information educators need to create professional development that is based on a well-defined image of effective teaching and learning.

For example, if you are planning professional development focused on climate change for middle and high school teachers, turn to the study guides in the "Earth" category of the CTS parent book and choose to do a few topic studies on the most relevant topics to enhance your understanding of the content, curricular, and instructional considerations in that subject area to inform your design. You might pull out the topic guides "Weather and Climate," "Models," and "Human Impact on the Environment" to identify and understand what is important for teachers to know to be well versed in Grades 7 to 12 content necessary for a science literate person to understand climate change as well as the teaching and learning implications for the grade levels. Summarize your CTS findings to help you consider the key ideas to incorporate into your professional development design. Review your summary and use it to consider what materials and experiences make sense for the teachers or prospective teachers you work with and what experiences you can provide them that would best reflect the standards. You might focus your CTS work on either the specific grade levels of the teachers with whom you work or on all of Grades K–12 if your purpose is to work across grade levels or help teachers build a vision of how their grade level fits into the entire K–12 learning experience. Conducting a topic study on the content for your work with teachers as part of your design and planning process will lead to a more focused and effective program, driven by a vision of effective teaching and learning.

Professional Development Builds Content and Pedagogical Content Knowledge and Examines Practice

A primary purpose of professional development is for teachers to gain content and pedagogical content knowledge and reflect on their practice for the purpose of enhancing student learning. In recent years, there has been a growing research base pointing to the need for all teachers to have strong content and pedagogical content knowledge. Research studies that examined the relationship between teacher qualifications and background and student achievement in mathematics and science found certified high school math and science teachers (usually indicating coursework in both subject matter and education methods) had higher-achieving students than teachers without certification in their subject area (Darling-Hammond, 2000; Goldhaber & Brewer, 2000; Monk, 1994). Several professional development programs that focus on building teachers' content and pedagogical content knowledge in science and mathematics show greater positive effects on student learning (Brown, Smith, & Stein, 1996; Cohen & Hill, 2000; Kennedy, 1999; Weiss et al., 2003; Wiley & Yoon, 1995). While many professional development programs focus on teaching content, Weiss et al. (2003) found it is necessary, but not sufficient, for teachers to have content knowledge:

> [Teachers] also must be skilled in helping students develop an understanding of the content, meaning that they need to know how students typically think about particular concepts, how to determine what a particular student or group of students thinks about those ideas, and how to help students deepen their understanding. (p. 28)

The CTS process clarifies the content necessary for science literacy (i.e., what twelfth-grade graduates should know and be able to use throughout their adulthood) and specifically, what students need to learn at a particular grade span. By engaging teachers in a CTS inquiry, they may discover that some of the activities they are using to teach certain science topics are not aligned with the important content that students need to become science literate or that they are not addressing concepts in enough depth or level of sophistication to promote real understanding. The CTS process also raises teachers' awareness of the content they need to understand better to teach more effectively.

Deepening PCK is equally essential for effective teaching. This involves helping teachers learn

- what children of certain ages are capable of learning and doing;
- how to anticipate ideas that are likely to be difficult for students to learn and have strategies to help them;
- how to assess students' prior knowledge and help bridge the gap between where they are and where they need to go; and
- strategies for representing and formulating subject matter to make it comprehensible to different learners with varying styles, abilities, and interests. (NRC, 2007; Shulman, 1986)

Furthermore, we know from the work of Bransford, Brown, and Cocking (2000) that expert teachers

- know the structure of knowledge in their disciplines,
- know the conceptual barriers that are likely to hinder learning, and
- have a well-organized knowledge of concepts and inquiry procedures and problem-solving strategies.

CTS, by virtue of its focus on the structure of scientific content, research into students' conceptions, and pedagogical suggestions linked to key ideas and skills, supports teachers to develop this essential expertise. Because teacher knowledge and expertise has such a profound impact on student learning, processes that develop and strengthen science teachers' knowledge and skills, such as CTS, are a sound investment toward improving professional development and ultimately student achievement in science.

All professional development leaders are encouraged to do a CTS on the topics for their program so they can investigate the concepts and key ideas from the content that teachers should know (see Sections I and III of a study guide in the CTS parent book) and the connections between ideas (see Section V). They can use what they learn to fine-tune their professional development designs and teacher education programs to focus directly on the key ideas that are found in the standards. With respect to enhancing PCK, professional developers can turn to the readings in Sections II and IV to examine teaching and learning considerations. Section II helps teachers identify appropriate contexts for learning, relevant phenomena that provide explanatory power for key ideas, and effective strategies for teaching the content. Section IV is used to explore the research on student learning for their topics. The research on learning reveals the commonly held ideas and misconceptions students bring to their learning, and it is also likely that some teachers, especially those who lack preparation in science, may hold similar ideas that need to be surfaced and challenged. Based on what professional developers glean from their topic studies, they may modify their plans to include activities that better reflect the suggestions for teaching the topic and build in formative assessments that help teachers identify and build on prior knowledge of the key ideas in the standards.

In addition to building content and pedagogical content knowledge, this principle calls for teachers to have opportunities to make direct connections to their teaching practice. Increasingly teachers are involved in teams and PLCs that afford the opportunity to examine practice for the purpose of increasing student learning. Early forays into collaborative work suffered from a lack of focus on the content and how to teach it (Loucks-Horsley et al., 2010). Groups lacked processes for drawing upon trusted science education resources and a strong knowledge base to guide their discussions. With few resources to guide them, teachers often relied solely on their own opinions or how things worked in their classrooms. Teachers gathered to examine student work and often used wonderful protocols, but needed help to make sound professional judgments about the quality of the content learning and to learn to look for evidence that students were learning important content, which involved knowing and being clear about the key ideas in a learning goal and the appropriate level of sophistication expected by it. They also needed to be aware of commonly held ideas and misconceptions that might surface during an examination of the student work. Examining student work often focused on what the students knew or did not know, but often failed to link the student results back to teacher practice and what they might do differently to improve students' opportunities to learn and demonstrate their learning. CTS directly supports teachers to deepen their content and pedagogical content knowledge and to use their new insights to examine practice in an accountable, collegial, and content-based way. Here is where engaging the teachers themselves in doing a CTS as part of their professional development really pays off. The authors have observed and participated in countless institutes, workshops, lesson studies, study groups, teachers examining student work, coaching sessions, and other forms of professional development. What stands out for us is how using CTS to examine practice changes the conversation teachers have. For example, consider the following dialogue reconstructed from notes from observing groups of teachers examining student work.

Example 1: Teachers Examine Student Work Without the Benefit of Using CTS

Facilitator: I am distributing four different examples of student work that come from fourth-grade student notebooks. The work was completed after the electricity unit was complete. Take a few minutes to look over the work and see what you notice.

Teacher 1: One thing that stands out is that only one of these students answered the question. One of the others just listed the parts, battery, bulb, wire and the other two I can't make out what they said, but their drawings look interesting, like they might know how this works.

Teacher 2: At least that one boy could spell "battery."

Teacher 3: Yes, all these children obviously need more work on writing and literacy.

Example 2: Teachers Who Use CTS Before Examining Student Work

Teacher 1: I did the CTS readings last night and learned a lot about the main ideas our fourth graders should know about electricity. [Other teachers agree.]

Facilitator: Was there anything that surprised you in the readings?

Teacher 3: Well, as I was reading, I started thinking that I had not really been looking for that understanding in my students' work. Before I had focused more on whether they knew the parts of a circuit rather than the idea of a complete circuit. [Other teachers agree.]

Facilitator: Let's make a list of the key ideas from the standards we read last night that we will use as we look at examples of student work on the electricity unit.

[Teachers reported that they would want to see evidence of student understanding that electrical circuits require a complete loop through which an electrical current can pass. They also listed the common misconceptions they read about related to electrical circuits and agreed to look at the student work for evidence of them in their students.]

Facilitator: Now let's look at the student work. What do you notice that reflects an understanding or lack of understanding of the key ideas in the standards or shows evidence of commonly held ideas noted in the research?

As the example illustrates, when teachers use a standards-and-research base to examine student work, they focus more on what matters, the content students know, and the difficulties they may have in learning the content. This opens the door for rethinking lessons and intervening when students do not learn the content. Collaborative groups involved in lesson study, examining student work, setting standards, and choosing and implementing curriculum have enriched their abilities to examine practice by using CTS first to develop a shared vision as to what to look for in student work or whether a lesson includes instructional strategies likely to address misconceptions or if curriculum reflects appropriate content for certain grade spans. Once teachers have CTS experience, they are able to discuss how the object of their inquiry matches the standards and research. There is less talk about how things work in "my classroom," and more professional discourse based in evidence and standards.

While CTS is not a substitute for in-depth learning on content, it is a helpful tool for professional developers to use to focus what content they will immerse teachers in learning and a great resource for teachers to use to discover what content and pedagogical content they need to know.

Professional Development Is Research Based and Engages Teachers as Adult Learners in the Learning Approaches They Will Use With Their Students

This principle of effective professional development suggests that teachers need to have time to engage as learners in some of the same lessons that they will use with their students. For example, like classroom instruction, the professional development should have some means of assessing prior knowledge and using what teachers already know to inform the professional development instruction. Teacher educators and professional development can use CTS to develop assessment probes (see discussion and guidelines for doing this in Chapter 6, p. 202) or use existing formative assessment probes (e.g., Keeley, Eberle, & Tugel, 2007) to help surface teachers' existing ideas. A crosswalk between the 147 topics in the CTS parent book and the CTS-developed formative assessment probes from the *Uncovering Student Ideas in Science* series (Keeley, Eberle, & Dorsey, 2008; Keeley, Eberle, & Farrin, 2005; Keeley & Tugel, 2009; Keeley et al., 2007) can be found on the CTS Web site at www.curriculumtopicstudy.org.

From there, you can design learning experiences for teachers that immerse the teachers in the content they need to learn. Teachers need the opportunity to experience the activities, inquiry, and tools they will use with their own students. Their learning is facilitated by guidance and probing questions from the professional development facilitator. The teachers are guided to explain their understanding and to revisit how that changes or adds to their own prior knowledge. Their experience is "book-ended" by a reflection on their experience as adults and in their role as teachers. The facilitator helps teachers identify what they did as learners, what the facilitator did in terms of the lesson design, assessment of prior knowledge, use of probing questions, and providing feedback on whether the explanations teachers gave were correct. The teachers develop a view of how they would do these same steps with their own students and develop a plan to translate them to their own classrooms.

Once again, CTS is a valuable tool for designing and immersing teachers in this type of learning. Conducting the readings in Section I of each study guide in the CTS parent book gives a picture of what the science literate adult should know about the science topic. Teacher educators can use this to begin to think about what teachers should know and understand and to develop or choose an assessment to measure prior knowledge. Focusing on the grade level of the materials being used, the professional developer would also want to review Sections II, III, and IV of the study guide. The findings from this review would help them to ask probing questions during the lesson and address persistent misconceptions. After teachers experience the topics as learners, they might be interested in doing a CTS themselves to explore what their students should know and understand and use that to plan their instruction. Professional developers need to design experiences for teachers that mirror what their students will learn, but that honor them as teachers and adults. CTS supports this by helping them get a view of both what they as a scientifically literate adult should know and what goals they need to work toward for their students.

Effective Professional Development Builds a Learning Community Among Science Teachers

In recent years, we have learned more and more about what makes professional development have an impact on teachers' thinking and their practice. Programs that provide opportunities for teachers to collaborate with their colleagues and other experts to improve their practice, take risks, and try out new ideas have a greater chance of having such an impact in the classroom where it counts (Hord, 1997; Vescio, Ross, & Adams, 2008). Hord and Sommers (2008, p. 9) pointed out that there are five components of PLCs. These are as follows:

1. Shared beliefs, values, and vision

2. Shared and supportive leadership

3. Collective learning and its application

4. Supportive conditions

5. Shared personal practice

As introduced in Chapter 1, the importance of developing PLCs among teachers is an underlying belief that has guided the CTS work. PLCs by their very nature are bringing professional discourse into the regular school day. Teachers are committing to goals for student learning and reflecting and discussing how their actions will lead to those goals. The success of such groups demands that they use existing research and resources so they can "stand on the shoulders of giants," which is a primary goal of CTS, rather than continuously reinvent the wheel. The CTS resources and tools provide the help PLCs need to effectively conduct their work in a collaborative, content-focused setting.

CTS Develops Teacher Leaders

To support and sustain professional development, there needs to be a focus on building teacher leadership. Teacher leaders are encouraged to contribute to the profession in many ways, by shaping and providing professional development, mentoring and coaching, facilitating PLCs, developing and supporting ongoing improvements in their buildings, and leading others to use effective practice. In the Northern New England Co-Mentoring Network, an NSF-supported program that developed teacher leaders to serve as mentors and leaders of PLCs in schools in Maine, New Hampshire, and Vermont, teachers identified CTS as one of the essential tools in their leadership toolbox. (See the example in Chapter 7 of how CTS can be used to support mentor teacher leaders.) Whether they were guiding teachers to examine demonstration lessons, helping to select curriculum, planning teacher workshops, or mentoring a beginning teacher, the mentor teachers started by doing a CTS on the topic they were using. By using CTS and demonstrating the various ways CTS can be used in teachers' work, teachers became stronger leaders with a deep knowledge of the standards and research on learning. In their work with peer teachers, they shared their own opinions less; instead, they engaged others in examining the standards and research to inform and justify decisions. CTS is a tool that not only guides and informs leaders' work, but also provides them with a process to develop a common core of understanding within the various groups they work with and a shared language for discussing teaching and learning.

CTS LEARNING CYCLE

CTS is similar to inquiry in the classroom—both involve a systematic investigation designed to address an important question. In classroom inquiry, the focus is on investigating the natural world. In CTS inquiry, the focus is on investigating a curricular topic. Both processes result in the construction of new knowledge and shared understandings. Classroom inquiry is often guided by an instructional model. In the early 1960s, J. Myron Atkin and Robert Karplus (1962) formulated a constructivist model of guided discovery designed to be similar to the way scientists invent and use new concepts to explain the natural world. This instructional model, called the learning cycle, was designed to allow students an opportunity to surface and examine their prior conceptions. Once ideas are revealed, students have an opportunity to explore their ideas, arguing about, seeking new information, and making meaning during the process. When students see that their prior ideas do not fully match their new findings, a disequilibrium results that opens the door to the construction of new ideas. When students reach the stage where they develop the formal scientific understandings and ways of reasoning that promote conceptual understanding, they are encouraged to extend their learning and apply their understanding of science concepts to new contexts (Keeley et al., 2008).

The stages through which leaders guide teachers in investigating a curriculum topic with CTS are very similar to the learning cycle model used in classroom instruction. A number of studies have shown that the learning cycle has many advantages when compared with other approaches to instruction, specifically the transmission model (Bybee, 1997). For this reason, we developed a learning cycle model to use in CTS with adult learners called the CTS learning cycle of inquiry, study, and reflection. Figure 3.1 is a graphic representation of the CTS learning cycle.

The original learning cycle model has evolved and undergone a variety of adaptations including two of the most popular instructional models used in science classrooms—the

| Figure 3.1 | The CTS Learning Cycle of Inquiry, Study, and Reflection |

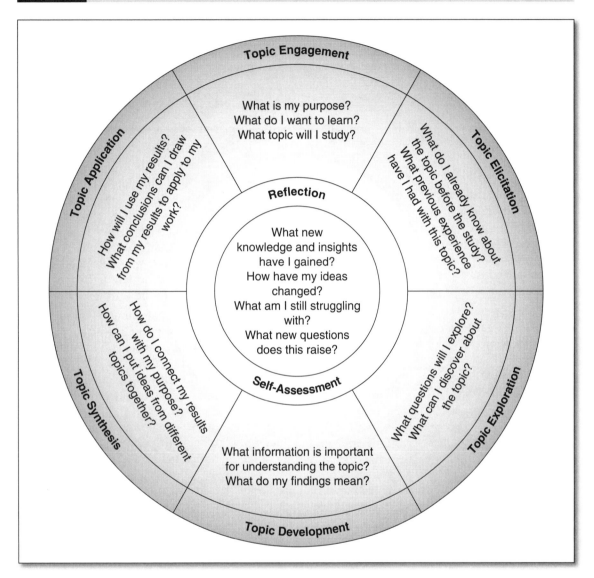

5E model (Bybee, 1997) and the conceptual change model (CCM; Posner, Strike, Hewson, & Gertzog, 1982). Many leaders use one of these models to design their teacher professional development so that it mirrors the way instruction unfolds in the classroom. Whether you decide to use the CTS Learning Cycle of Inquiry, Study, and Reflection, or adapt your use of the 5E or CCM to design your session, the outcome is still the same—teachers go through a carefully designed process of learning to construct new understandings that build upon or challenge their prior knowledge and lead to deeper content and pedagogical knowledge of the topics they teach. Table 3.1 shows the similarities between two commonly used instructional models and the CTS learning cycle.

Leaders are encouraged to read and become familiar with the CTS learning cycle described on pages 36–49 in the CTS parent book so that you will know and understand how participant learning evolves from one stage to the next. Figure 3.13 on pages 48–49 of the CTS parent book provides examples of what CTS participants and CTS facilitators do during each stage in the CTS learning cycle. Many of the CTS designs included in

Table 3.1   Three Learning Cycle Models

5E Learning Cycle (Bybee)	Conceptual Change Model (CCM; Posner et al.)	CTS Learning Cycle of Inquiry, Study, and Reflection (Keeley)
Engage—provides an opportunity to capture interest in the topic and identify students' existing conceptions (including misconceptions). **Explore**—provides an opportunity to test their ideas, gather information, and begin constructing concepts and skills before formal introduction of concepts. **Explain**—provides an opportunity for resolving misconceptions, sense-making, and developing formal concept understanding and terminology. **Elaborate**—provides an opportunity to apply or extend the development of concepts to new contexts, form generalizations, and work further to resolve conceptual difficulties. **Evaluate**—provides an opportunity to determine how well students understand a concept or can apply a skill. Also includes an opportunity for reflection and self-assessment.	**Commit to an Outcome**—provides an opportunity for students to become aware of their own preconceptions by making predictions about the result of an activity. **Expose Beliefs**—provides an opportunity to surface and share ideas in small groups and then with the whole class. **Confront Beliefs**—provides an opportunity to test ideas, make observations, and discuss them in small and large groups. **Accommodate the Concept**—provides an opportunity to resolve any discrepancies between their existing ideas and their observations or new information and to develop formal understanding of the concept. **Extend the Concept**—provides an opportunity to make a connection between their formal understanding of the concept from their class experience and new situations. **Go Beyond**—provides an opportunity to pursue additional questions or problems related to the concept.	**Topic Engagement**—provides an opportunity to generate interest in the topic and identify the purpose of the study. **Topic Elicitation**—provides an opportunity to activate and identify current knowledge and beliefs about the topic. **Topic Exploration**—provides an opportunity to explore the readings, raise questions, record ideas, and begin to get a sense of how the information fits the purpose of the CTS section. **Topic Development**—provides an opportunity to read more in depth and analyze information, focusing on what is relevant to the topic. Findings are shared with colleagues, meaning is made from findings, and discussions provide clarification for constructing a full understanding of the topic. **Topic Synthesis**—findings are synthesized across CTS sections or multiple topic studies to address the purpose of the study. **Topic Application**—information from CTS is applied to a context relevant to teachers' work. **Reflection and Self-Assessment**—provides a metacognitive opportunity to think about new knowledge and insights gained, unresolved issues, and new questions to pursue.

Chapters 4 through 7 of this *Leader's Guide* are designed using the CTS learning cycle with specific strategies for each of the seven stages in the learning model, including engagement, elicitation, exploration, development, synthesis, application, and reflection. You may wish to vary these strategies to address the specific goals of your groups. Table 3.2 provides suggestions for facilitating each stage.

Table 3.2 CTS Learning Cycles and Suggestions for Facilitation

Stage in the CTS Learning Cycle	Icon	Suggestions for Facilitating Each Stage
TOPIC ENGAGEMENT: Establish a purpose for doing CTS; create interest in studying the topic.		• Use a hook showing student difficulty with a topic (e.g., video, student work, performance data, experience a phenomenon). • Ask why the topic is important to K–12 science. • Ask participants to record their personal goals for the session (what do they hope to learn more about?).
TOPIC ELICITATION: Activate prior knowledge related to the topic (content, instructional considerations, learning goals, misconceptions, connections).		• Brainstorm ideas about the topic prior to doing CTS. • Use Frayer Model for terminology and operational definitions. • Provide 4-square template (CTS parent book, p. 42). • Generate list of concepts that make up the topic. • Use assessment probes to predict student difficulties. • Generate beliefs about learning related to the topic. • Administer pre-CTS questionnaire.
TOPIC EXPLORATION: Explore the topic by reading assigned sections of the text listed on the study guide with clear purpose.		• Choose grouping strategy and assign readings. • Review study guide sections with the group or create a reading assignment guide. • Provide time for individuals to read and process their assigned section. • Provide materials for recording information.
TOPIC DEVELOPMENT: Engage in group discussion to clarify and make meaning of the readings.		• Provide time for individuals in each small group to share and discuss each of the readings. • Select a large group debriefing strategy. • Raise and clarify major points with the large group. • Encourage groups to create a summary sheet of their findings (this is usually done later, using the group's study results).
TOPIC SYNTHESIS: Look across all sections of the CTS—combine CTS findings to support purpose for doing the CTS; combine findings when studying two or more topics.		• After each participant reports out on their reading section, have group discuss the overall significance of the study. • Encourage groups to share three to five major insights or key points gained from the CTS. • Compare and contrast "old" ideas with "new." • When two different topic studies are done simultaneously, combine similar and related findings from each topic.
TOPIC APPLICATION: Apply findings to a specific context related to participants' work.		• Choose a context application from Chapter 6 of the CTS parent book to apply CTS results. • Revisit elicitation questions and respond by applying new information gained from the CTS.
TOPIC REFLECTION: Reflect on how CTS changed one's understanding of the topic; identify unanswered questions.		• Choose a strategy for individual reflection. • Encourage large group reflection. • Encourage participants to revisit their own goals for the CTS and reflect on the knowledge gained. • Provide time to raise unanswered questions.

As you examine the full topic study design modules included in Chapter 5 of this *Leader's Guide*, notice how the stages of the learning cycle have been incorporated throughout the design. To print out a poster of the CTS Learning Cycle, see the file on the CD-ROM under Chapter 3. The icons in Table 3.2 are included to indicate the learning cycle stage that each of the module activities is designed to address.

FACILITATION TIPS AND STRATEGIES

As a professional learning strategy, CTS differs from other types of science professional development, which often involve teachers in "doing science," because it relies heavily on text-based readings and sense-making discussions based on text. While inquiry is at the heart of CTS professional learning, the inquiry is directed at formulating questions about teaching and learning specific to a curricular topic, gathering data from text, analyzing the information, drawing conclusions from the information and sense-making discussions, and communicating findings to others. In many ways, CTS also parallels the technological design process in that a teaching or learning problem is identified, information is gathered from text readings to address the problem, the constraints of the classroom and context are considered, the findings from CTS are applied to propose a solution or develop a product (e.g., improve curriculum coherency, develop assessment probes that can be used to inform instruction, and enhance content knowledge), the applied results of CTS are evaluated for their impact on teaching and learning, and CTS results and products are communicated and shared with others.

A major difference between facilitating professional learning in a hands-on inquiry or technological design setting compared with facilitating learning in CTS is that the CTS leader must incorporate a variety of facilitation skills to guide teachers in managing the text materials and constructing new learning from the readings. It involves having a repertoire of facilitation tips and strategies to guide adult learners and manage group processes, including establishing productive norms for collaboration, reading, sense-making discussions, and teacher reflection. In this section, we share some of our tips and strategies for facilitating the type of learning that results from the CTS process of reading and discussing text. The first step facilitators must consider is how to acquire and manage the resource materials used for CTS.

ACQUIRING AND MANAGING MATERIALS

When designing professional development that incorporates CTS, facilitators must first acquire their own copies of the CTS parent book and all the CTS resources used to conduct a CTS. (See a complete list of resource books on p. 2.) Second, facilitators must decide how to provide access to the CTS resources when they use CTS in their professional development designs. If the purpose of the professional development is to train teachers in the use of CTS with the expectation that they will continue to use it, then it is important to provide all teachers with their own copies of the CTS parent book. Whether or not you provide copies of the resource books depends on your budget, whether teachers have access to the resources in their schools or personally own copies, and the availability of an Internet connection during your sessions so that teachers with laptops can electronically access *Science for All Americans, Benchmarks for Science Literacy, National Science Education Standards,* and selected maps from the *Atlas of Science Literacy* (Vols. 1–2). The following are a variety of options for acquiring and managing materials. Choose an option that best fits your audience and professional development context.

1. *Provide a set of resources to each participant.* This is obviously the most expensive and resource intensive approach. Professional developers who choose this option often fund this option with grant funds for participant materials.

2. *Provide a set of resources to each school team.* If participants attend your professional development as members of school teams and your funding source can support it, consider providing a set of materials to each team of up to five teachers from the same school. These materials can then continue to be used by the school team and their school colleagues for their CTS-related work.

3. *Obtain and provide a session set of resources.* If you regularly use CTS in your work with teachers, consider acquiring your own set of materials to provide for use during your CTS professional development. For each group of six teachers provide the following:

   One *Science for All Americans*

   One *Science Matters*

   Two *Benchmarks for Science Literacy*

   Two *National Science Education Standards*

   One *Making Sense of Secondary Science*

   One copy of each volume of the *Atlas of Science Literacy*

4. *Obtain a partial session set and borrow or have participants bring their resources.* You may wish to purchase a few sets of resources and find an organization or school that can lend you additional session copies. You might also contact teachers in advance who may already have copies of some of the resources and ask them to bring them. Between your set, borrowed copies, and ones teachers own, you may have enough copies for your audience to use.

5. *Use paired readings.* If you have session copies but not enough for everyone in your audience, consider having teachers read the CTS sections in pairs.

6. *Alternate readings.* If all participants do not have their own books and there is a set of shared books provided on each table, alternate readings when you choose the option where every person reads all the sections. For example, while one person is reading *Science for All Americans,* have others at the table select a different book and section. Once that section reading is complete, the book is returned to the stack so others can use it when they finish with their books.

7. *Provide handouts of the readings.* If resource book sets are not available, consider making copies of each of the readings and distributing the copies to groups. If you do this, be sure to put your full set of the resources on a "library display" so that teachers can look at the actual books during the breaks. Alternatively, if you have some session sets, but not enough to go around, or you have copies of some books but not others, you might combine providing copies of the actual books with copies of readings from the books.

8. *Combine books with electronic access.* If you have a wireless Internet connection at your setting and some teachers bring their own laptop computers, provide access to the online versions of the resources to the teachers with laptops and provide the books to those who do not have laptops.

9. *Purchase used copies for sessions.* If your budget is limited, consider acquiring used copies of the books to make up your session sets. Amazon.com and other booksellers often sell used copies of each of these resources; you can assemble a set for a small percentage of the cost of buying all the books new.

One of the laments leaders frequently hear after introducing CTS to teachers is that they love the process but won't be able to use it after the session because their schools do not have the funds to buy the materials. From our experience, we have found that teachers who are excited about using CTS go back to their schools and share its value with their administrators. As a result, most of the teachers we have worked with have convinced their administrators to provide the resources for their continued professional development, a set for school-based PLCs or grade-level teams, or a school set to keep in their school's professional library. What they originally perceived as a roadblock to using CTS was overcome when their administrators saw the value of investing in these resources to build the capacity of their science teachers to grow and collaborate as professionals for the purpose of improving student learning. One of our teachers remarked how her principal shifted from bringing in the "one-shot wonder" or motivational speaker that cost the school $3,500 for one day to using those same funds to provide eighteen sets of CTS materials for teachers to use. Additionally, many of the teachers we have worked with have gone on to buy their own sets of materials to be part of their professional library that they will own and keep throughout their professional career. Even though they have access to some of the resources online, they still purchase their own copies of the books. Many teachers recognize having a set of functional resources is part of the professionalism of teaching and recognize the value of investing in their own future (and they like that it is tax deductible). In summary, don't let remarks about not having the books discourage you after providing the professional development. You will be delighted to learn later when you see many of your participants again that this has not been an obstacle for their continued use of CTS.

> **Facilitator Note**
>
> NSTA Press regional distributors offer complete "course packs" of the books used in CTS at a discounted price. Contact NSTA Press at nsta.org for further information on ordering the complete set of CTS resource books.

DEVELOPING NORMS FOR COLLABORATIVE WORK

Collaborative teacher groups are guided by group norms that set parameters and expectations for how they will work together. In all of our professional development work, we have drawn upon the work of Garmston and Wellman (2008) to establish and set norms for collaboration and productive work. Their work has focused in part on ensuring that all members of a group develop the norms and skills to support collaborative work. They suggest groups adopt "seven norms of collaboration" that are driven by three conditions (p. 31):

1. Intention to support thinking, problem solving, and group development

2. Attention of the group members to attend fully to the other members

3. Linguistic skills of the listener or responder

The seven norms are universal and support all group development. They include the following:

Pausing: This includes four types of pauses: (1) when a question is asked to allow processing time; (2) when someone speaks, which allows others to hear what is being said before jumping into the conversation; (3) personal reflection time, when someone asks the group to give them a moment to think about a question or idea that has been raised; and (4) the collective pause, when the group stops and reflects or may write to collect their thoughts.

Paraphrasing: This norm ensures that the information shared has been understood. Garmston and Wellman suggest several "paraphrase stems" such as *you're suggesting . . . ; you're proposing . . . ; so what you are wondering is . . . ;* and *so you are thinking that . . ."* (p. 33). Group members check for understanding before launching into discussions of their own interpretations.

Putting inquiry at the center: The basic idea of this norm is that group members "explore the perceptions, assumptions, and interpretations of others before advocating one's own ideas" (p. 34).

Probing for specificity: This norm helps guard against the miscommunication that can result when group members use vague terms, or use words like *best, worst, valuable, great,* and so forth without grounding them in clear criteria.

Placing ideas on the table: Effective groups signal when they are putting an idea on the table for consideration and when they are taking the idea off when it seems to have little support or is bogging down the group.

Paying attention to self and others: Everyone in the group needs to stay aware of how they are communicating with others and how their messages are being received. Group members consciously monitor their own verbal and nonverbal behavior and that of their group members.

Presuming positive intentions: Group members hold on to the idea that each member is a "committed professional who wants to solve a real problem" (p. 39).

Establishing these norms and supporting our groups to develop the skills needed to use them has been a major part of the success of CTS groups. In addition, there are several norms that we have used that are specific to supporting the CTS work. We encourage you to start every CTS session by establishing your own norms, drawing from those above and the following norms specific to CTS work.

CTS GROUP NORMS

Honor Times

It will be very difficult to accomplish the goals of CTS if people are allowed to be off task or to use group time to take long "bird walks" sharing autobiographical stories, even when they are somewhat connected to the group task. Encourage your group to adopt a norm of honoring times and staying on task.

Cite the Evidence

This norm encourages all group members to share what they learned from the CTS readings without inserting their own opinions. Ask participants to stick to the facts and data from the readings, including citing the page and passage being cited. When it comes to CTS, we like the old line from TV's Jack Webb in *Dragnet*, "just the facts, ma'am!"

Contribute to the Learning of Others

This norm reflects Garmston and Wellman's norm of paying attention to self and others. It reminds group members to be responsible for their own and their group members' learning by pausing, paraphrasing, and probing for specificity and making sure time is managed to allow everyone to contribute.

Share Resources

In most CTS sessions, people are sharing the many resource books used for the reading. Encourage everyone to complete their reading as quickly as possible and take notes they can refer to later during discussions, so they can make books available to others as soon as possible.

Check for Understanding

During the report out phase of most CTS sessions, it is essential for people to learn from what others have read. Encourage groups to manage their time to allow a few minutes for report out and then a pause to process and then clarify any confusing questions. For information that is still unclear, encourage group members to jot a question on a sticky note and post it on a chart labeled *CTS Content Questions* that is always posted during CTS sessions.

Maintain a Safe Learning Environment

The goal of all CTS sessions is learning. If teachers are fearful, they will be unable to reveal what they do not know and this is counterproductive to the goals. Set the norm that it is safe to raise any question about something that is confusing or not understood. Make it safe by inviting everyone to post questions on the *CTS Content Questions* chart for those who do not want to ask a question publicly. Encourage small group members to ask each other for help with something they read that they do not understand. Since we often have a broad range of participants, from those who rarely teach science to those who may be university professors in science, it is important to establish that we are all responsible for one another's learning as listed above and ask all participants to recognize that they can each share their particular expertise with each other. Encourage participants to write any unresolved questions about content on a sticky note and post it on the *CTS Content Questions* chart.

OPTIONS FOR ORGANIZING AND DISCUSSING READINGS

As the authors and field testers worked with CTS, we discovered the need to have structures and activities for participants to use to organize and discuss the readings in small or large groups. The following are recommended strategies we have used in our CTS sessions. Choose the strategy that best fits your audience and facilitation style. Facilitators can select any of these strategies to use in any of the facilitation designs included in Chapters 4 through 7.

Jigsaw Strategies

A commonly used reading strategy, a jigsaw builds on the idea that we learn best when we have to teach others. It is also a way to cut back on the amount of reading one individual would have to do by having the readings distributed among a group, with each reading summarized and shared by the group member assigned to that CTS section. Participants are placed in cooperative learning groups of five to six people; group size is usually based on the number of CTS sections used, number of CTS resources used, or length or sections of a reading. The following are suggested ways to jigsaw the readings:

Option 1: Large Group Jigsaw With Expert Groups

In this option, sections are assigned by tables, counting off, or passing out cards with the assigned reading. Each table group is an "expert" for their assigned reading or section of a CTS guide. After the experts have read their section, they discuss their own reading. For example, all the people who do the reading for CTS Section IA meet and discuss the IA section, or all the people who read Section II meet and discuss their section. After expert groups have discussed the findings from their section, they summarize their results for the whole group on chart paper and briefly (one minute or less) point out their major findings.

Option 2: Small Group Jigsaw With Expert Groups

In this option, each person in a small group is assigned to be the "expert" for a particular reading or section of a CTS guide. Everyone in the group or at the table is assigned one of the topic guide readings. After the experts have read their section, they meet with all the other people in the room who read the same section for a discussion. For example, all the people who read Section IA meet and discuss their section, or all the Section IIs meet and discuss their section. After expert groups have discussed the findings from their section, they return to their small groups and each expert presents a summary of the discussion to their table group.

Option 3: Regular Small Group Jigsaw

This option is similar to the above except experts do not meet in expert groups first. Each person in a small group volunteers to take a section to read and then summarize for the rest of the small group.

Option 4: Large Group Fishbowl Jigsaw

This option works best when you want to engage the whole group in processing the readings but need to parcel them out for the sake of time. Count off by numbers or pass out cards with the assigned reading. The readings will be spread out among the participants in the room. Bring each group up to the front of the room to sit in chairs as a panel, or in a circle in the center of the room, where everyone can see and hear them (in "the fishbowl"). Facilitate a discussion of the assigned reading to the group in the fishbowl while others listen and take notes. Provide time for those outside the fishbowl to ask questions, then switch to the next group, and so on.

Assigning Jigsaw Readings

There are a variety of ways to assign the jigsaw readings. The jigsaw breakdown you choose depends upon the number of resource books you have available (or you can make copies and use handouts) and the grade levels of your participants. Individuals can be assigned a section, or sections can be assigned to pairs. The latter is particularly helpful when using collaborative reading strategies (discussed later in this section) to extract meaning from the CTS assigned text sections. The following describes options for breaking down the jigsaw readings:

Assign by sections: IA, IB, IIA, IIB, IIIA, IIIB, IVA, IVB, V, and VIA or VIB (where appropriate) or I, II, III, IV, V, and VI (where appropriate). Review the readings before the session, and if the lengths of some readings are short, consider combining some sections.

Assign by book: Science for All Americans, Science Matters, Benchmarks for Science Literacy, National Science Education Standards, Atlas of Science Literacy, state standards document (where appropriate).

Assign by grade span: K–2, 3–5, 6–8, 9–12 assigned to read only their grade span sections of all of the following: IIA, IIB, IIIA, IIIB, and V. Everyone reads IA and, if *Science Matters* is used, include IB. *Note:* Both K–2 and 3–4 teachers read K–4 in the *National Science Education Standards.*

Other Strategies

In field-testing our designs, we found that some groups prefer jigsaws, and others prefer not to be restricted in their readings and discussions. Jigsaws can be overused. We recommend varying strategies if your professional development includes multiple opportunities to do CTS. Whether you choose to jigsaw or not depends on your audience and the topic chosen. The following are other strategies you can use to read and discuss CTS findings with others.

Option 5: Large Group Discussion, All Readings

This option works best when participants have access to all of the resource books (note: several of the books are online) or copies of the readings are provided so participants are able to do all or some of the readings in advance of the session. Each participant reads all the sections of the CTS guide selected by the facilitator for a particular topic or module and makes notes that will be brought to the CTS session discussion. This option can also be used on-site without prereading if the time dedicated for reading during the session is considerably increased.

Option 6: Small Group Discussion, Assigned Expert

In this option, everyone has an opportunity to read and discuss any or all of the sections, depending on their pace and the number of books available. Prior to reading, the group assigns an "expert" for each selected reading. The "expert" starts with that reading to ensure that every selected section has been read by at least one person who can discuss it with the group, but can read other sections as well if time allows. This option allows the group flexibility in reading what they are most interested in, provides an opportunity to sample the different books, and gives the participants a more complete picture of the different findings for each section. Many participants prefer this option.

CTS RECOMMENDED READING AND SUMMARIZING STRATEGIES

To ensure learning for participants, it is important to use structures and activities to process and summarize readings. The following are recommended reading and summarizing strategies. Facilitators can select any of these strategies to embed into any of the facilitation designs in Chapters 4 through 7 or their own designs for CTS professional development.

Say Something

This strategy was introduced to us by Bruce Wellman (adapted from Lipton & Wellman, 2004), who provided the facilitation skills training for both the Northern New England Co-Mentoring Network and the National Academy for Science and Mathematics Education Leadership, both of which worked with the CTS Project to try out CTS materials. It provides a structure for quick and ongoing processing of text materials. Ask members of your group to find partners. They will individually read a section of text and stop when they get to a designated point. When partners are ready, they stop and "say something"; for example, you might ask them to say something about a significant idea or a connection to their work or something the passage made them think about. Remind them to keep the "something" they say short. Continue the process, stopping at several designated places to "say some-thing" (such as at the end of a subsection or every few paragraphs) until they reach the end.

Question Stems

Providing a list of question stems for participants to use as they read focuses their reading and helps them make connections to their own context. Examples of question stems include the following:

- What are the key points made in this section?
- Give an example of how this section relates to your practice.
- What might you infer from reading this section?
- What evidence from this reading can be used to support our work?
- How does this compare to your prior knowledge?

Read, Write, Pair, Share

This strategy is similar to think-pair-share, but with the emphasis on the reading shaping the discussion. Participants read a section, record important ideas to share, pair up with a partner, and discuss the reading. You can combine this with the use of question stems to focus the sharing conversations.

Paired Reading

In this strategy, pairs of teachers help each other explore and make sense of the readings by reading the text aloud to each other, each taking turns, and summarizing the main ideas. The first partner reads a paragraph aloud, and the second partner summarizes the paragraph. Roles are reversed for the next paragraph. Continue alternating roles until the reading is completed. Once the entire reading is completed, both partners cooperatively summarize the main points of that section and identify any questions they have.

Problem Scenarios

This reading strategy provides motivation to read the text and helps teachers focus on the main ideas as they read. To use this strategy, a problem is posed for each CTS section. As teachers read the section, they record information that would address the person's problem. Examples of problem scenarios to use with each CTS reading section include the following:

Section I

Jenna doesn't have a science background and is afraid she doesn't know enough about this topic to teach it. What does she need to know as a science literate adult?

Section II

Jim's instruction just doesn't seem to be working with his students. What are some things he should consider when teaching this topic?

Section III

Ricia knows she should focus on the most important ideas when she is teaching this topic but her state standards are too broad to precisely determine what the key ideas are. Furthermore, there is so much material in her textbook, she doesn't know what to focus on and what to leave out. How can you help Ricia focus on clear and meaningful learning goals?

Section IV

Andre is aware of the importance of considering students' prior knowledge. He knows there are common misconceptions his students are likely to have but doesn't know what they might be. How can you help Andre become more aware of the misconceptions students have related to this topic?

Section V

Florio's curriculum contains a list of learning goals he needs to teach this year. The curriculum provides no guidance on which ideas are connected and should be taught together, what precursor ideas should come first, and connections he can make to other content areas. How can you help him see the connections among important ideas in science and how one idea can inform understanding of another?

Section VI

Anita uses her state's standards to guide her teaching but isn't always sure about what the standards intend as far as the experiences she should provide students, where to draw the boundaries in teaching content that may not be necessary or exceed the standards, and exactly what the key idea is in a standard. How can you help Anita clarify her state standards related to this topic?

You can use these scenarios or write your own. Post the scenarios or give each section group a card or sheet of paper to write the problem they are addressing.

One-Word Summary

In table groups, have participants craft a one-word summary of the section they read (you can also use this as an end-of-the-day wrap-up). Each table chooses a spokesperson who is prepared to share the word and explain why the group chose it.

OSQ

As participants read a section, have them record and share the following:

- *One major new idea for me*
- *Something I already knew*
- *Questions I have*

Focused Reading

To use this strategy, introduce participants to three "focused reading" symbols (Lipton & Wellman, 2004).

1. √ —Got it. I know or understand this.

2. ! —This is really important or interesting.

3. ? —Something I do not understand.

As participants go through their CTS readings, have them mark the text using these symbols (or use sticky notes). After reading in groups of three to four people, the participants review the items they marked with each of the symbols. They can start with the checks and review what they learned, then move to the points that were important, and finish with questions. Invite participants to help each other to clarify the places where they had questions if possible.

REPORTING OUT STRATEGIES

It is important for groups to have an opportunity to report out their CTS findings to others, both to reinforce their own learning and to share new insights with the whole group. However, be aware that when one strategy is used too often or the reporting out takes too long, people may tune out. Participants' attention can wane if groups are not reminded to keep their report outs brief and to the point. We often use a timer to keep reports out to the designated time. It is also important for leaders to remember to encourage groups to speak from the CTS evidence, not their own personal opinions. The designs included in Chapters 4 and 5 show how various strategies can be used to report out CTS findings. Examples you can embed into any CTS design include the following:

Gallery Walks

In this strategy, groups record their CTS findings on chart paper in a form that others can understand (group agrees on main categories for what will be posted ahead of time). After all the charts have been posted, participants do a "gallery walk," visiting each poster and learning from what others did. During the gallery walk, you might encourage participants to jot down notes of things they want to remember or questions they might have for the groups.

Gallery Walk With Docents

This is a modified version of the gallery walk described above, to be used with groups of thirty or fewer participants. After groups post their CTS findings charts around the room, the entire room stands and gathers around the first chart while a group member acts as a "docent," giving a brief report of the findings described on group's chart. When the report is finished, the whole group travels to the next chart and the process is repeated. This keeps people on their feet and attentive during the report out process.

Group Presentations

This typical form of reporting out involves small groups giving a brief presentation to the other groups on their CTS task. Encourage the group to focus on the major "aha's" and "just the facts" they gained from studying their section(s).

Key Points

Each participant crafts four to five key points or take-away messages from their CTS findings. Each group member shares key points with others at the table. When all the small groups are done, ask each table to share just one key point from their table with the whole group.

Give Me Ten

After small groups have had time to discuss their findings and clarify understandings, ask the whole group to list ten new things they learned from doing the CTS. Hold ten fingers up and count down one at a time as each participant volunteers one learning. Wrap up after the tenth volunteer and provide a brief synthesis of the group's learning and add any essential points that were not raised.

Partner Speaks

This strategy works well in small groups where CTS participants work primarily in pairs rather than table groups. As pairs discuss their CTS findings, each person listens carefully to the partner and records key ideas and insights gained. After the discussion is finished, each person reports out a few of the key findings the partner shared. This strategy encourages careful listening and allows all voices to be heard.

REFLECTION ON CTS

Reflection is a vital part of any CTS professional development design. The process of personal and group reflection allows CTS users to connect to their own practice the theoretical basis of the standards and the research base on learning that contributes to a deeper understanding of a curricular topic. In the process of reflection, teachers think about what the CTS process means to them and how they and their students will benefit from it. The process of reflection promotes metacognition and helps teachers see how their beliefs, ideas, knowledge, and skills as a teacher may have changed as a result of CTS. The reflection process also surfaces issues that may need further attention or understandings that may need further development before teachers can apply them effectively. Group reflection provides an opportunity to see what others have learned and relate that to one's own learning. It also provides a formative assessment opportunity for the facilitator to gauge what participants gained from the CTS process. The following are some of the strategies we have used and field tested in CTS professional development designs. Any of these strategies can be embedded into the CTS designs that you use.

Bumper Sticker Statements

This engaging reflection strategy is used after participants are introduced to CTS and are reflecting on the value of the CTS tools and processes. Provide participants with sentence strips or twenty- to twenty-four-inch strips of paper (*tip:* you can use a sheet of chart paper to make the strips). Ask them to work in groups to create bumper stickers that capture their views of CTS in short, catchy expressions they would want to share with other science educators. When participants are done, ask each group to stand and read their bumper sticker and then post it on the wall for all to see.

Paired Verbal Fluency (PVF)

This technique is used as a partner reflection and involves partners taking turns in timed rounds, talking about their CTS (Lipton & Wellman, 2004). The activity only takes seven minutes yet the reflectors can surface quite a bit in this short time. Start by asking teachers to stand up and make eye contact with the first person they see and then moving to stand with that person. Once all the pairs are matched, ask each partner to decide who will be partner A and who will be partner B (after they select partners announce that partner B will go first). Provide a discussion prompt such as "What was most valuable to you during this CTS session and what do you plan to do next?" Announce that when you give the signal, one partner will talk for exactly two minutes while the other partner only listens. After two minutes announce, "switch" and partners trade roles and repeat. At the next "switch," the first partner talks for one minute, followed by "switch" and the other partner talks for one minute. The last round is just thirty seconds each. Have them thank their partners, return to their seats, and ask for three to four people to share with the whole group something they heard.

Idea Exchange

Ask participants to reflect back on their experience and jot down three to five new ideas of things they will go back and use as a result of their CTS experience. Have participants get up and exchange one idea with five different people and then return to their seats. After everyone has shared, ask the whole group to share three to four ideas.

I Used to Think . . . , But Now I Know . . .

On a PowerPoint slide or a handout, write the phrase "I used to think _____, but now I know _____." Ask participants to fill in the blanks based on their CTS experience. When they are finished, ask all the people at a table to read their statements to their small group. When finished, ask for three to five examples to share with the large group. If the group is small (twenty or fewer participants), you can quickly go around the room and have each person read their statement.

Three-Two-One

Three-Two-One is a technique that scaffolds participants' reflections (Lipton & Wellman, 2004). It provides participants with an opportunity to reflect on their success in using CTS as well as recognize what was challenging for them. Participants are provided with a Three-Two-One handout and given time to fill it out. In pairs or small groups, participants share their responses. An example of three prompts to use with this technique is shown in Table 3.3.

Table 3.3	Three-Two-One CTS Reflection

Three key ideas I will remember from CTS:
  1.
  2.
  3.

Two things I am still struggling with or wondering about:
  1.
  2.

One thing that I will change in my practice:
  1.

The [W]hole Picture

This strategy uses a metaphor to help participants reflect on CTS as a process for developing a complete "picture" of a curriculum topic. To prepare the materials for this strategy, take manila index folders and cut out a small hole or several holes (dime-sized or smaller) in the folder. Place a picture from a magazine or other source inside the folder and glue it down. Make sure the part of the picture seen through the hole is representative of the picture but does not give it completely away. Put paper clips on the folder so participants won't open it. Pass out the folders to groups of two or three or table groups (depending on size of group) and have them look at the view through the hole. Have them share ideas as to what they think the picture inside is and why. After they have shared ideas, have them open the folder and compare their responses to what the actual picture is. Ask them, "How do these two different views of the picture represent the CTS process? How does this activity serve as a metaphor for the CTS process?" Ask for volunteers to share their ideas.

SUMMARIZING FINDINGS

Before or after participants leave a CTS session, they should be encouraged to develop a summary of their findings that can be archived or shared with others and that serves as a record of the information gained from doing a CTS. (*Note:* Many people like to have their laptops available so they have their summary typed and easily accessible.) Participants can also be encouraged to create their own summaries as they go back and use CTS in their work. The following describe some of the ways to have participants create summaries during the professional development session or after they go back to their own settings.

Poster Boards or Poster Sessions

Scientists and other groups often display their work on posters as well as present it at poster sessions. As well as orally presenting the summary of their CTS findings at a poster session, CTS users can also present their findings on posters that can be displayed in schools, organizations, and other settings. Participants can work in groups or individually during a professional development institute to create a visual poster of their CTS summary. Alternatively, participants in an ongoing professional development program can be sent home with a tri-fold poster board to create a CTS summary poster of a topic of their choice to bring back and share at a later professional development setting. The posters provide a colorful, visual way to summarize and communicate CTS findings to other audiences. In the Chapter 3 folder on the CD-ROM at the back of this book are photographs of tri-fold posters created by participants in the Maine Governor's Academy for Science Education Leadership. These teacher leaders presented their CTS posters at a poster session during one of the Academy meetings. They also shared them at their schools, displaying them in the teachers' room and other public places for other science teachers to view and learn from.

Summary Sheets

When there is time in a CTS session and computers are available, encourage participants to create a summary sheet of findings that can be saved and shared with their group. Additionally, there are times when CTS facilitators might want to prepare summary sheets to either refer to during their CTS sessions or share with leaders they are working with who will be leading CTS sessions. There is no one format for developing a summary sheet. Several examples are included in Chapter 3 on the CD-ROM. There is also an example of a middle and high school CTS summary on energy transformation on pages 44–45 of the CTS parent book.

Digital Charts

The wall charts created by small groups during the CTS summarize the key findings. Leaders can use a digital camera to photograph the charts and post them on a Web site or e-mail them electronically to participants who wish to have a record of the wall charts.

Take-Home Messages

Encourage participants to work in small groups to develop five to six take-home messages that summarize their CTS experience and that they would want to remember when they leave the session. Have them record their take-home messages on chart paper and post them around the room. When groups are finished, have all participants do a gallery walk and note any additional take-home messages they want to add to their own lists.

STRATEGY FOR ADDRESSING THE ISSUE OF TIME AND DIFFICULTY

The facilitation tips and strategies described in this chapter should help you in managing materials, effectively forming groups, choosing processing strategies that promote learning, and facilitating learning during the different stages of a CTS professional development session. Before we get into the designs in Chapters 4 and 5 and the various professional development strategies CTS can be used with, we want to leave you with a strategy for addressing one of the common questions and concerns that comes up with CTS first-timers. This is the issue of time and difficulty of doing a CTS. Many teachers who have never previously experienced this type of in-depth, rigorous professional development that is not intended to be a one-time event are unsure about whether they would have the time or inclination to use CTS. Initially it may seem daunting to them and take inordinate amounts of time. One way of addressing this up front during an introductory CTS session is to use a strategy called "backwards spelling."

The strategy works as follows:

What

Take a word or phrase related to CTS, the name of the topic you will be using, or use a general CTS word like *curriculum* or *standards* and ask participants to spell it backwards.

When

Use backwards spelling at the beginning of CTS, especially for first-timers, when you are describing the additional backwards design step of starting with a thorough study of the topic before making decisions about curriculum, instruction, and assessment.

Why

Backwards spelling is a good brain activator, and in addition, it serves as a metaphor for explaining that science teachers don't typically start with a study of the curriculum topic or learning goal, and how it might feel uncomfortable, time-consuming, difficult, or awkward at first.

How

Ask participants to spell the word, for example, *photosynthesis*, backwards. When everyone finishes, talk about how difficult that was and why and connect it to the first experience with the CTS process. Now ask them to spell it backwards five times in their head. Ask them if it got easier and why. Make the link to why repeated practice with CTS makes the process easier, faster, and allows them to come up with their own strategies for using it efficiently. It's OK to struggle through the beginning and feel like it is hard and takes a lot of time. Assure the group that, like spelling backwards, it gets easier the more they use it. The proof is in the pudding. Having worked with hundreds of teachers who have been introduced to CTS, a majority of the teachers with whom we have kept in contact report that they use it seamlessly and with little effort as a regular part of their practice and professional development.

NEXT STEPS

Now that you have examined what leaders need to know and consider to effectively lead professional development involving CTS, it is time to examine a variety of CTS learning designs, starting with the introductory sections in the next chapter, and to choose ones, or draw from these to design your own, to fit your professional development needs and contexts.

4

Tools, Resources, and Designs for Leading Introductory Sessions on Curriculum Topic Study

This chapter provides three comprehensive modules for introductory Curriculum Topic Study (CTS) sessions that serve to familiarize the new CTS user with the resource books, the CTS guides, and the CTS process. At the end of the chapter, we provide guidelines for developing your own modules based on those in the chapter with science topics specific to your own work. There are three introductory modules we recommend for introducing CTS to new users, including the following:

Module A1. Introduction to CTS Using K–12 Snapshots. This module uses fifteen "snapshots" that are questions or issues a teacher might raise, such as "I will be teaching a new science unit this year; what are the specific ideas students should learn at my grade level?" The specific question within each snapshot can be addressed using one section of a CTS guide, and usually by reading just one CTS resource book. When you introduce CTS through snapshots, your participants do not do a full CTS (i.e., they do not read all six sections of a CTS). Instead the experience provides the new CTS user with an opportunity to engage in learning through some of the sections of a CTS guide and a variety of different curricular topics, using a scaffold (a set of step-by-step instructions) to guide their investigation. This module is especially useful when you have a mixed group that may teach different topics and face different teaching and learning issues and your goal is to introduce the CTS process and give the participants some practice using it to investigate a variety of questions about teaching and learning. Module A1 uses a K–12 snapshots document that includes topics from life science, physical science, Earth and space science, inquiry, implications of science and technology, and unifying themes, making it appropriate for a diverse audience of educators. The CD-ROM at the back of this *Leader's Guide* contains additional snapshots that cut across content and skill areas or are content area specific (e.g., astronomy) that can be used for content specific and grade-level groups such as K–5, 6–8, and 9–12. There are also examples that support specific applications such as curriculum implementation. For example, there is a snapshots version to support leaders who are introducing CTS as a way to support implementation of the Lawrence Hall of Science Great Explorations in Math and Science (GEMS) guides. The GEMS example can be used by leaders to create similar snapshot guides for Full Option Science System (FOSS), Science and Technology for Children (STC), or other curriculum programs, including textbook chapters.

Module A2. Introduction to CTS Resource Scenarios: Models. A constructivist option for introducing and teaching the CTS process is to use resource scenarios. When you use resource scenarios, participants focus on a single topic and explore each of the CTS guide sections for that topic without first having to learn the structure and mechanics of using a CTS guide. In essence, they construct their own knowledge of CTS and its resources by experiencing the CTS process first and then connect their experience to the tools and resources that are central to CTS. The resource scenarios raise particular questions about teaching or learning related to the science topic and guide the user toward the designated reading for that topic from the CTS resources. An added feature of this option is that participants first consider their own prior knowledge of the topic and later develop an appreciation of how CTS can enhance their knowledge and practice by comparing their prior knowledge before doing a CTS with the knowledge they gained from CTS. After participants complete a resource scenario, the facilitator shows them the CTS guide and makes the link between what they just did and how they can study any topic or answer their own questions that may be similar to ones in the scenarios they investigated by using the CTS guides. After experiencing the process firsthand, participants see how different sections of a CTS guide are used for different purposes and that there are different books and sections of the related readings that are linked to the purposes described on a CTS guide. Module A2 uses a K–12 resource scenario on the topic models. The CD-ROM contains additional K–12 resource scenarios the facilitator may substitute in the session to address the grade levels, goals, and topics of interest of the participants.

Module A3. Full-Day Introduction to CTS: Resource Scenarios and Snapshots. The two introductory sessions described above can be combined into one full-day introductory session, starting with the resource scenarios and ending with snapshots as a way to gain more practice in using the CTS guides and familiarity with the resources. Additional snapshots and resource scenarios included on the CD-ROM or the CTS Web site can be substituted for either of the examples included in the module.

MAKE YOUR OWN CTS INTRODUCTORY SESSION

Modules A1, A2, and A3 provide sample topics and questions for the snapshots or resource scenarios. However, sometimes you may be working with an audience on a specific topic (e.g., energy) or a category of topics (e.g., ecology). Sometimes your audience will be grade-level specific. For example, you may be working with a group of middle school teachers who are coming together during a summer institute to learn about teaching horizontal motion in an inquiry-based classroom. You might decide to create your own set of snapshots that include CTS findings related only to horizontal motion and skills of inquiry used to learn about horizontal motion. Or you might have a mixed audience that is interested in learning about motion, and you could decide to create a K–12 resource scenario on motion. In these cases, you can adapt the modules in this chapter by using your own snapshots or resources scenarios. The directions for developing your own snapshots and resource scenarios are found at the end of this chapter.

ESSENTIAL FACILITATOR PREPARATION

For each of the above introductory sessions, it is important for the facilitator to become familiar with the CTS findings from the module used to introduce CTS. Facilitator notes that include the CTS findings are provided for the K–12 snapshots and resource scenarios in Modules A1, A2, and A3. Your familiarity with these findings will help you guide and focus the discussions and draw attention to key points that participants sometimes overlook or fail to mention in the discussion. If you choose the supplementary snapshots or resource scenarios from the CD-ROM, you are encouraged to develop your own facilitator notes before using them by reading all of the CTS resources related to your snapshots or resource scenarios and noting the important ideas that you hope teachers will gain and raise in the discussion, so you are ready to prompt for them or raise them yourself if they are overlooked.

The next sections in this chapter provide the facilitation guides for the three CTS introductory modules (A1, A2, and A3). The handouts, facilitator resources, and PowerPoint presentations for each module are located in the Chapter 4 folder on the CD-ROM at the back of this *Leader's Guide*. Table 4.1 provides an at-a-glance summary of the introductory sessions.

Table 4.1	CTS Introductory Sessions at a Glance			
Module	*Time Needed*	*Grade Level*	*Content*	*When to Use*
Module A1: Snapshots	3.5 hours	K–12	Examples from physical science, life science, and Earth and space science, inquiry, science and technology, and unifying themes	Group is new to CTS; learn the CTS basics—what study guides are and how to use them and what resource books are used

(Continued)

Table 4.1 (Continued)

Module	Time Needed	Grade Level	Content	When to Use
Module A2: Resource Scenarios	3 hours	K–12	Models	Group is new to CTS and participants wish to learn how to conduct a full CTS; learn the CTS sections and the purpose and resource books for each section
Module A3: Combined Snapshots and Resource Scenarios	6.5 hours	K–12	All of the above	Same as above
Make Your Own CTS Introductory Session	Half-day session if based on Modules A1 or A2; full-day session if based on Module A3	Your choice	Your choice	Same as above

Module A1

Introduction to CTS Using K–12 Snapshots, Facilitation Guide

BACKGROUND INFORMATION

Description of the Module

This module is a half-day introductory session to science CTS. It is best used when you have a group that is new to CTS and needs to learn the CTS basics, including what a study guide is, what the resource books are, and how they are used, and the group needs to have some guided practice in using CTS to answer specific questions about teaching and learning. The K–12 snapshots are focused questions that can be answered fairly quickly by reading selected sections of a CTS guide without the need to do a full CTS. The snapshots give CTS newcomers a "taste" of the CTS books and process and prepare them to conduct a full CTS at another time.

Audience

This session is designed for preservice teachers, classroom teachers, and other educators who work with or across Grades K–12 science education and are interested in learning what CTS is and how it can be used in their work. It is used primarily to teach CTS to first-timers.

Purpose and Goals

This introductory module provides a way to introduce and practice using CTS within a variety of K–12 life science, physical science, and Earth and space science examples so that all participants can make a link to a topic and issue relevant to their teaching context. The goals of this module are to

- develop awareness of CTS as a tool that connects standards and research on learning to classroom practice,
- provide guided practice in using CTS, and
- consider a variety of ways to use CTS in curricular, instructional, assessment, and teacher development contexts.

Key Components

Key components of this introductory session include the following:

Time to Get Acquainted With CTS and Resource Books

Starting a full topic study can be overwhelming if teachers have not been introduced to the basics of using the CTS parent book and the resource books used with the CTS guides. To support teachers in learning how to use CTS, this introductory session includes time for exploring the CTS parent book, the structure of a CTS guide, and for becoming familiar with each CTS resource book and its purpose.

Scaffold

The multiple steps of CTS and the need to use different books for different purposes can be a lot to manage the first time through. To address this difficulty in managing multiple resources and sections of a CTS guide, this session features the use of a scaffold with step-by-step directions to guide novices through the process. After a few uses of the scaffold, teachers find they refer to it less and less as they learn the steps. Once they become familiar in the use of the CTS process, it becomes internalized and they no longer need the scaffold.

Reading and Group Processing Strategies

CTS always involves reading and interpreting text from the CTS resources. Strategies that help participants to process what they are reading with a partner or group are embedded in this introductory module, and other strategies for engaging participants in reading and processing are provided in Chapter 3 of this *Leader's Guide*. (Refer to these if you wish to substitute any of the ones suggested in this module.)

Application

For participants to apply what they learn and develop a commitment to using CTS, they need time to look at and discuss the different ways CTS can be applied to their own curriculum, instruction, assessment, and in professional development and preservice or graduate education programs. This module provides time for participants to identify applications that are relevant to their own K–12 science education work or preservice or graduate education context.

SESSION DESIGN

Time Required

Approximately 2.5 to 3.5 hours depending on the introduction to the CTS resources option chosen (includes a 10-minute break).

Agenda

Welcome and Overview (10 minutes)

- Welcome and introductions (5 minutes)
- Goals and overview of CTS (5 minutes)

Engagement—Preparing for CTS (50–65 minutes)

- Getting to know the CTS parent book and study guides (20 minutes)
- Getting to know the CTS resource books (15–30 minutes, depending on the option selected)
- Introduction to CTS scaffold (15 minutes)

Elicitation (10–15 minutes)

- Snapshots: Eliciting prior knowledge (10–15 minutes)

Exploration and Development (55 minutes)

- Snapshots: Exploring and practicing CTS (35 minutes followed by 10-minute break)
- Key learning and CTS process debrief (10 minutes)

Applications (15 minutes)

- Context and applications for CTS (15 minutes)

Reflection (15 minutes)

- Quick write (3 minutes)
- Paired verbal fluency conversation (12 minutes)

Wrap-Up and Evaluation (10 minutes)

MATERIALS AND PREPARATION

Materials Needed by Facilitator

CTS Parent Book

Science Curriculum Topic Study: Bridging the Gap Between Standards and Practice (Keeley, 2005)

Resource Books

One copy of each of the following resource books:

- *Science for All Americans* (American Association for the Advancement of Science [AAAS], 1989)
- *Science Matters* (Hazen & Trefil, 1991 or 2009)
- *Benchmarks for Science Literacy* (AAAS, 1993)
- *National Science Education Standards* (National Research Council [NRC], 1996)
- *Making Sense of Secondary Science* (Driver, Squires, Rushworth, & Wood-Robinson, 1994)
- *Atlas of Science Literacy* (Vol. 1) (AAAS, 2001)
- *Atlas of Science Literacy* (Vol. 2) (AAAS, 2007)

> **Facilitator Note**
>
> It is important for CTS facilitators to have both volumes of the *Atlas of Science Literacy*, but for this module, only Volume 1 is used. For a complete list of the science topics that are addressed in each of the *Atlas* volumes, see Handout 2.2: *Atlas of Science Literacy* (Vols. 1–2) Crosswalk, in the Chapter 2 folder on the CD-ROM at the back of this book.

CTS Module A1 PowerPoint Presentation

The Module A1 PowerPoint presentation is included in the Chapter 4 PowerPoint on the CD-ROM. Review it and tailor it to your needs and audience as needed. Insert your date and location on Slide 1, add additional graphics as desired, and add your own contact information on the last slide. Select which option you will use to introduce the CTS resources (see p. 94) and insert the PowerPoint slide(s) for your option.

Supplies and Equipment

- Computer and LCD projector to show PowerPoint presentation
- Flip chart easel, pad, and markers
- Blank paper for note taking
- Sticky notes
- Highlighter pens (optional)

Wall Charts

Print out signs of letters A through O from Facilitator Resource A1.9 in the Chapter 4 folder Module A.1 handouts. Post these signs on a long wall or around the perimeter of the room where participants can see them (see Figure 4.1). Spread them out so that small groups of participants can gather in front of the signs.

Prepare and post an additional wall chart labeled *CTS Reminders* with these reminders:

When Doing a CTS:

1. Record the *exact language* in Sections III and V.

2. Read only the text that is related to your specific inquiry.

3. Take notes and include the name of the book and page numbers.

Prepare and post another wall chart labeled *CTS Content Questions*.

Facilitator Resources

Facilitator Resource A1.1: K–12 Snapshots Facilitator Summary Notes

Facilitator Resource A1.9 Snapshots Signs

Figure 4.1	Chart 1

Materials Needed for Participants

Distribute the CTS parent book and resource books as follows:

CTS Parent Book

Ideally each participant will receive a book; but if that is not possible, have at least one copy on each table.

CTS Resource Books

If sets of books are available, you will need one copy of each of the following for every four to five participants:

- *Science for All Americans* (AAAS, 1989)

Facilitator Note

If you do not have copies of the CTS parent book for each participant, you will need to make copies of pages 24–27 and page 33 for each participant. If you do not have enough books for at least one per table, you will also need to prepare a table copy of the pages of CTS used during the "Application of CTS Activity" (see pp. 53–90 of the CTS parent book).

- *Benchmarks for Science Literacy* (AAAS, 1993)
- *National Science Education Standards* (NRC, 1996)
- *Making Sense of Secondary Science* (Driver et al., 1994)
- *Atlas of Science Literacy* (Vol. 1) (AAAS, 2001)
- *Atlas of Science Literacy* (Vol. 2) (AAAS, 2007; optional—not used for this module)
- *Science Matters* (Hazen & Trefil, 1991 or 2009; optional)

Handouts

All handouts can be found in the Chapter 4 folder on the CD-ROM.

- Handout A1.1: Facilitator Resource K–12 Snapshots (Optional—This handout is for the *facilitator only unless* the group is made up of people who might use this activity themselves; if so, provide them with a copy of Facilitator Resource A1.1 after they have completed the snapshots.)
- Handout A1.2: Agenda at a Glance
- Handout A1.3: Introduction to Science CTS Snapshots—Scaffold
- Handout A1.4: List of Science Curriculum Topic Study Guides (if participants have their own CTS parent books, you will not need this handout)
- Handout A1.5: Snapshots
- Handout A1.6: Anatomy of a Study Guide
- Handout A1.7: Answer Key
- Handout A1.8: Snapshots Recording Sheet
- Handout A.1.9: Letters for the Wall (only one set needed)
- Handout A.1.10: Introducing Science CTS to PLCs (Snapshots for people working in Professional Learning Communities)
- Optional handouts (if you do not have enough copies of the CTS parent book, make handout copies of pages 24–27 and page 33 of that book)

DIRECTIONS FOR FACILITATING THE MODULE

Welcome and Overview (10 minutes)

Welcome and Introductions (5 minutes)

Have Module A1 PowerPoint Slide 1 showing as people arrive. Welcome participants and explain that this session will introduce them to CTS. (Show PowerPoint Slide 2.) Ask participants to introduce themselves at their tables and share something they hope to learn today. Invite participants to refer to Handout A1.1, a copy of the PowerPoint presentation. Briefly review the agenda using Handout A1.2: Agenda at a Glance.

Goals and Overview of CTS (5 minutes)

Ask participants to review the session goals on Slide 3. Using Module A1 Slides 4 through 7, explain to participants that CTS was developed by a National Science Foundation (NSF)–funded project. The project

Facilitator Note

Print out all handouts and facilitator resources to refer to as you review the directions for facilitating each module.

Facilitator Note

Several of these resources or sections of them are available online; the CTS Web site provides the URLs and a direct link to the online resources including the *National Science Education Standards, Benchmarks for Science Literacy, Science for All Americans,* and sample *Atlas of Science Literacy* maps. If enough resource books to provide a full set for every four to five people are not available, you can create fifteen snapshot folders, each labeled with a large letter (A through O) on the front, into which multiple copies of each individual snapshot reading are placed (see Figure 4.2). These folders replace the books or supplement a limited supply of books. Four copies of each individual snapshot reading placed in each labeled folder work well for a group of up to twenty-five participants. You can add or decrease the number of copies, depending on the size of your group. If CTS parent books are not available for each participant, include copies of the appropriate CTS guide in each folder.

For example, four packets of the readings for Snapshot A are placed inside a manila folder with a large "A" on the outside, and four copies of the CTS guide, copied on different colored paper, are also placed inside the folder. These folders (A through O) are then placed in an area of the room where participants can take a reading and study guide out of the folder and replace it when they are done so that others can use it on their second round of reading. (If you have a very large group or a large room, place snapshot folders in several parts of the room, such as at the front and back, to cut down on the movement in the room and the number of people sharing.) Facilitators can reuse these folders and handouts with other groups.

Figure 4.2 Snapshot Folders

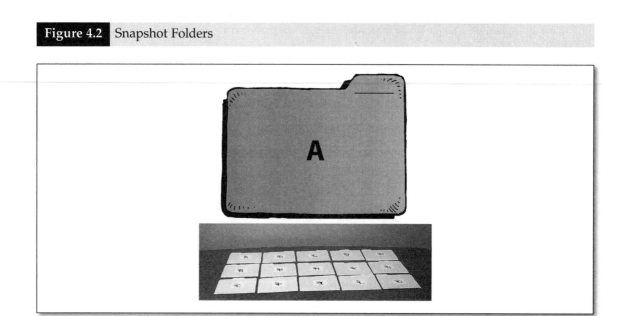

developed a set of tools and processes, using professional resources, to bridge the gap between national standards and research on learning and classroom practice and state standards (Slide 4). Ask for a show of hands of people who may have had some experience with CTS or heard about it prior to this session. Review what CTS is and what it is not (Slides 5 and 6) and describe how it has been called the "missing link" (Slide 7). (Review Chapter 2 of this *Leader's Guide* for more information on the project background.)

 Engagement (50–65 minutes)

Getting to Know the CTS Parent Book and Study Guides (20 minutes)

1. Ask participants to work with a partner or in table groups depending on how many books are available. Have them locate copies of the CTS parent book on their tables. Using Slide 8, start an exercise called *First Glance* and tell the pairs or small table groups to open the book randomly and take a "first glance." Have each pair or small table group open to at least three spots in the book, take a first glance, and discuss what they find.

2. Ask for a few volunteers to report on their first glance on any page up to page 113. Ask the following questions: "What did you find? Why did that interest you?" Then ask if anyone looked at one of the Curriculum Topic Study Guides that start on page 113 in the CTS parent book. Ask everyone to turn to these pages and look at a study guide. Ask participants to report on what they found on the study guides. Elaborate on what they say, explaining that there are 147 study guides and that they are the core resource for using CTS.

3. Explain that Handout A1.4: List of Science Curriculum Topic Study Guides (alternatively, page ix of the CTS parent book if everyone has their own copy) lists the 147 different CTS guides and that the study guides are described in Chapter 2 of the CTS parent book starting on page 19. Using Slides 9 and 10, point out the structure of a study guide (including the Web site link at the bottom) and the six sections to the study guide that are used to answer different questions or issues related to science understanding, teaching, and learning. Refer them to Handout A1.6: Anatomy of a Study Guide as you point out each section of the guide. For example, the readings in Section I address what adult literacy in the topic includes. Section II suggests instructional implications or suggestions of ways to effectively teach the topic. Section III identifies the concepts and specific ideas students learn at different grade bands. Section IV summarizes the cognitive research on student learning. Section V shows how the topic connects with and builds on or leads to other knowledge (as well as includes

updated research on student learning in *Atlas of Science Literacy* [Vol. 2]). Section VI suggests reviewing state and local standards and curriculum documents for clarification. Tell participants they will come back to the study guides and work with them in a few minutes.

4. Use Slide 11, the Swiss Army Knife, to point out the many functions of the versatile CTS process. Describe how there are times when you might use only one tool on your knife for a specific purpose. Other times you might use a combination of tools. What makes CTS so useful, just like a Swiss Army knife, is that the variety of tools ensures that you can find the right tool for the right purpose.

Getting to Know the Resource Books
(15–30 minutes, depending on which option is selected)

There are three options for introducing the CTS resource books. These options are described at the end of this chapter beginning on page 94, and the PowerPoint slide for each of the three options is included in the Chapter 4 PowerPoint folder on the CD-ROM at the back of this book. The process you select will depend on the amount of time you have in your agenda, number of resource books available, and participants' familiarity with the resource books. Options 1 and 2 are highly recommended. Use Option 3 only if you do not have any access to the resource books or you have very limited time.

1. Show Slide 12. Explain that these are the resource books used in CTS. Mention that several of these resources or parts of these resources are available online and that the CTS Web site provides the URLs and a direct link to the online resources. You may wish to post the URLs on chart paper for your participants to record for their use later. Explain how having access to these books, which have significantly informed science education, is like having a science education expert available 24/7.

2. Show Slide 13 to point out that there is a parallel process and resource books for mathematics CTS and that several of the same books are used for both.

Your Selected Option

Follow the directions for the option you selected for introducing the CTS resources on pages 94–102. Be sure the PowerPoint slide(s) for the introduction you selected are inserted in advance to replace Slide 14, which is included just as a placeholder. Note that if you use Option 1 or 2, there is only one slide to insert. If you opt for the more in-depth Option 3, there are eleven slides you will need to insert. Introduce the resources using your selected option.

Introduction to CTS Scaffold (15 minutes)

Show Slide 15. Explain that the group will use a scaffold (a step-by-step guide) to guide their initial use of CTS to answer teaching and learning questions related to different curricular topics. Show Slide 16 and refer to Handout A1.3: Introduction to Science CTS Snapshots—Scaffold. Quickly go through Slides 17 through 32 (these provide examples of what they will see on the scaffold and give them practice choosing the right section on which to focus).

Point out the *CTS Reminders* chart you posted earlier with reminders including the following:

> **Facilitator Note**
>
> This should be a rapid-paced practice session of how to find the right study guide. If everyone gets the process after the first few examples, you can skip over the other ones, although the last one is very useful in pointing out what is meant by reading only the "related sections."

When Doing a CTS:

1. Use *exact language* in Sections III and V. (Remind your group they should not rewrite or paraphrase goals; the standards developers chose the words used in the standards very carefully!)

2. Read only the text that is related to your specific inquiry. (See Slides 31–32 for an example of what is meant by related text.)

3. Write everything down that is related and important; cite the name of the book you used and page numbers so you can refer others back to the text later.

 ## Elicitation (10–15 minutes)

Eliciting Prior Knowledge (10–15 minutes)

Show Slide 33 and refer participants to Handout A1.5: Snapshots. Explain that they will now dip their toes into the water to sample CTS. They won't be doing full topic studies, but rather short, partial studies called snapshots in order to practice using the study guides and the resource books.

Show Slide 34. Explain that before they do their first CTS snapshot, it is important to take a few minutes to activate their prior knowledge so that after they do a CTS snapshot, they can compare what they knew before to what they gained from the CTS. Ask participants to review Handout A1.5: Snapshots and have them choose a snapshot that interests them. Give them a few minutes to write a brief response to the first question on Handout A1.8, based on their prior knowledge and experience. When everyone has finished, ask them to get up and stand by the letter on the wall that represents the snapshot they chose. In small groups or pairs, have participants share how they responded to the snapshot based on their own knowledge of teaching and learning. If some participants are the only ones who chose the snapshot and are unmatched, either have a facilitator converse with them or match them with others who are the only ones standing at their signs and ask them to exchange ideas. Once everyone has shared their pre-CTS ideas, have them return to their tables.

 ## Exploration and Development (55 minutes)

Exploring and Practicing CTS (45 minutes)

Show Slide 35. Tell the group that they will now have time to do the CTS on the snapshot they have chosen and just responded to with their initial ideas. Ask if there are any questions about using the steps on Handout A1.3: Introduction to Science CTS Snapshots—Scaffold or about interpreting the study guides. Address any questions that come up. Remind them to record the CTS guide, section(s), and resource book(s) they used to do their CTS on Handout A1.8, Question 2, and summarize their findings on Question 3 after completing the reading. Ask them to specifically include any new knowledge or insights they gained after doing the CTS.

Invite participants to work individually or in pairs (with someone who chose the same snapshot) and practice using the scaffold going through Steps 1 through 7. Remind them not to skip steps the first time through. Encourage them to take notes as they read, recording information that is relevant to the question posed in the snapshot. Once they finish their snapshot, encourage them to try another one if they have time, recording their pre-CTS ideas, doing the CTS, and then recording new knowledge or insights gained from CTS. Remind them to start by jotting down their initial ideas before doing the CTS reading. Show Slide 36. (Keep this slide projected as they work on the snapshots.) Point out Handout A1.7: Answer Key, which they can use to make sure they are on the right category, topic, and section before they start their readings. Remind them to scan the readings and read only the text directly related to the questions on their snapshots.

As you walk around and check on groups, answer any questions about the resources or the process as they complete their snapshots. If people seem confused or stuck, point them back to the scaffold on Handout A1.3: Introduction to Science CTS Snapshots—Scaffold and to Handout A1.6: Anatomy of a Study Guide. Check that they are finding the right sections and readings referenced in the study guides. If CTS questions come up that you can't address at that time, ask the participant to write the content question on a sticky note and post on the wall chart labeled *CTS Content Questions*.

Key Learning and CTS Process Debrief (10 minutes)

After participants have had time to complete at least one or two of the snapshots, ask the group to stand up again, with Handout A1.8 in hand and meet at the snapshot wall sign with the same group or partner with whom they discussed their pre-CTS ideas. Give them time to discuss what new knowledge or insights they gained from doing CTS. After groups have had time to discuss their CTS findings, ask for a few report outs on how CTS addressed the snapshot questions and how CTS can be a valuable tool, even for experienced teachers, in answering questions about teaching and learning. Probe for the key CTS findings to come out. (Refer to Facilitator Resource A1.1: K–12 Snapshot Facilitator Summary Notes for the CTS findings that should be the main points of discussion at this point.) Ask participants to comment on how the process worked for them and what they found valuable in terms of addressing their snapshot questions.

Applications of CTS (15 minutes)

Context and Applications (15 minutes)

Ask participants to step back from the activity they have just done to think about different reasons why a science educator might use CTS. Using the steps on Slide 37, have participants review the CTS application examples starting on page 33 of the CTS parent book and ask for other examples. If everyone has access to one, have them turn to page 53 in Chapter 4 of the CTS parent book to scan examples of suggestions and support materials for using CTS in

> **Facilitator Note**
>
> If any questions have been posted on the *CTS Content Questions* chart, this is a good time to read those to the group. Address any you can (given time constraints) and refer participants to other resources or people to talk with after the session.

various content, curricular, instructional, assessment, or professional development applications. Point out the collection of vignettes in Chapter 6 that illuminate how CTS is used in various contexts.

Reflection (15 minutes)

Show Slide 38. Ask participants to do a quick write (three minutes of quiet time to write a reflection) that summarizes their thoughts on how CTS could be useful to them.

Use the group reflection exercise, paired verbal fluency, to provide time for participants to share their reflection with one other person in the room. See directions for paired verbal fluency in Chapter 3, page 55 of this *Leader's Guide*.

> **Facilitator Note**
>
> If there is only one CTS parent book per table, ask one person at each table to show these sections to the people at the table, or you can briefly describe to the group what is found in these chapters.

Wrap-Up and Evaluation (10 minutes)

Wrap-Up (5 minutes)

Ask participants what questions they have about using CTS and answer any questions. Refer them to page 2 of the CTS parent book for more information on the outcomes of using CTS, and remind participants to visit the CTS Web site. Remind them that this was "just a taste" of CTS and that for their next experience, they should try a full topic study. (See modules for full topic studies in Chapter 5 of this *Leader's Guide* to plan a full topic study session.)

Evaluation (5 minutes)

Thank everyone for their participation and ask them to provide you with written feedback on the session. You can use one of the evaluation strategies such as Got It/Need It found in Chapter 3 of the CTS parent book or develop your own evaluation questionnaire.

Module A2

Introduction To CTS Resource Scenarios— Models, Facilitation Guide

BACKGROUND INFORMATION

Description of the Module

Module A2 uses mini-cases or what we call "resource scenarios" that raise specific questions teachers might have about a science topic. The scenarios are used to explore how doing a CTS enriches understanding of a science topic. For this session, the scenarios focus on the unifying topic of models. (Both the *National Science Education Standards* and the *Benchmarks for Science Literacy* include unifying concepts or common themes. These are topics that cut across the disciplines of science, mathematics, and technology.) In small groups, participants reflect on what they already know about the topic of models in response to the scenario they are assigned. They read their assigned section from a resource book and record new knowledge gained from the readings. They compare the knowledge they started with to the knowledge they gained to recognize the "added value" of using CTS to enhance and expand upon the individual and collective knowledge of the group. Following completion of that task, they are introduced to a CTS guide and then connect what they did with the resource scenario to the structure of a CTS guide.

Audience

This session is designed for preservice and classroom teachers and other educators who work with or across Grade K–12 science education and are interested in learning what CTS is and how it can be used in their work. The content focus of the session is models, a major unifying theme in science, making the experience applicable for people who work across the different subject areas of science.

Purpose

This CTS module provides a way to teach CTS to first-timers and novice users by focusing on a single topic and making a case for how the resources can lead to powerful learning. Use this session when you have a group of people who are new to CTS and you want them to see what can be gained by exploring the topic of models. The module allows participants to experience the process and learn the basics, including being introduced to the resource books. The approach used in this module leads participants to recognize the value added in using the vetted readings from national source documents and the CTS guides to build or enhance existing knowledge of content, teaching, and learning.

This session is especially useful if you have some participants who have strong content knowledge, but question what else they need to learn about the standards and science teaching and learning. During the CTS field tests, participants with strong content knowledge, such as scientists and teachers entering teaching from science-related careers, reported the great value they found in using CTS to identify what students should know at the different grade spans, and the research on student learning and misconceptions to inform their teaching.

Goals

The goals of this module are to

- develop awareness of the CTS process and the collective resources used for connecting standards and research on learning to classroom practice;
- provide guided practice in conducting a full CTS; and
- consider a variety of ways to use CTS in curricular, instructional, assessment, and professional development contexts.

Key Components

Key components of this introductory session include the following:

Time to Get Acquainted With the CTS Book

To support teachers in learning how to use CTS, this introductory session includes time for exploring the CTS parent book and what is in a CTS guide.

Embedded Practice Using the Resource Books and Selected Readings

Unlike Module A1, where participants explore and learn about the resource books and guides before doing CTS, this module uses a constructivist approach to embed the learning in the study process. After the participants have worked with the scenarios, the facilitator will link what they did to CTS by introducing the components of a CTS guide and connecting the purposes of each section of a study guide to a CTS resource.

Reading and Group Processing Strategies

CTS always involves reading and interpreting text from the CTS resources. Strategies that help participants to process what they are reading with a partner or group are embedded in this introductory module, and others can be found in Chapter 3 of this *Leader's Guide,* beginning on page 51.

Application

For participants to apply what they learn and develop a commitment to using CTS, they need time to look at and discuss the different ways CTS can be applied to curriculum, instruction, assessment, and professional development. This module provides time for participants to identify applications that can be used in their own work.

SESSION DESIGN

Time Required

Approximately 2.75 to 3 hours (includes a 10-minute break).

Agenda

Welcome and Overview (10 minutes)

- Welcome and introductions (5 minutes)
- Overview of CTS (5 minutes)

Engagement—Preparing for CTS (5 minutes)

- Warm-up talk (5 minutes)

Elicitation—What Do We Already Know? (20 minutes)

- Form study groups and record individual responses to scenarios (10 minutes)
- Groups discuss and record initial ideas (10 minutes)

Exploration and Development of Models Ideas (30–40 minutes)

- Reading and studying about models (10–15 minutes)
- Developing models ideas (10–15 minutes)
- Reporting on models ideas (10 minutes)

Break (10 minutes)

Synthesizing Group Findings (15–20 minutes)

- Recapping findings (5 minutes)
- Context questions (10–15 minutes)

Constructing an Understanding of the CTS Guides (20 minutes)

- Connecting back to the study guides (5 minutes)
- Linking the resource books to the study guide sections (10 minutes)
- Summarizing the resources (5 minutes)

Application and Reflection (25 minutes)

- Thinking about what we learned and how we can use it (10–15 minutes)
- Applying CTS in different contexts (10 minutes)
- Partner reflection (5 minutes)

Wrap-Up and Evaluation (5–10 minutes)

MATERIALS AND PREPARATION

Materials Needed by Facilitator

CTS Book

Science Curriculum Topic Study: Bridging the Gap Between Standards and Practice (Keeley, 2005)

Resource Books

One copy of each of the following resource books:

- *Science for All Americans* (AAAS, 1989)
- *Benchmarks for Science Literacy* (AAAS, 1993)

- *National Science Education Standards* (NRC, 1996)
- *Making Sense of Secondary Science* (Driver et al., 1994)
- *Science Matters* (Hazen & Trefil, 1991 or 2009; optional)
- *Atlas of Science Literacy* (Vol. 1) (AAAS, 2001; optional)
- *Atlas of Science Literacy* (Vol. 2) (AAAS, 2007)

CTS Introductory Module A2 PowerPoint Presentation

The Module A2 PowerPoint presentation is included in the Chapter 4 PowerPoint folder on the CD-ROM at the back of this book. Review it and tailor it to your needs and audience as needed. Insert your date and location on Slide 1, add additional graphics as desired, and add your own contact information on the last slide.

> **Facilitator Note**
>
> It is important for CTS facilitators to have their own copies of both volumes of the *Atlas of Science Literacy*; but, for this module only, Volume 2 is used and the map needed for this session is available online.

Supplies and Equipment

- Computer and LCD projector to show PowerPoint presentation
- Flip chart easel, pad, and markers
- Blank paper for note taking
- Sticky notes (small and large)
- Highlighter pens (optional)

Wall Charts

Post in pairs (side by side) fourteen sheets of chart paper around the room to create seven stations. Number each station (one through seven) to correspond to the number of the resource scenarios. Label the first chart in each pair of charts *Before CTS*. Label the second chart in each pair *After CTS*. Post charts around the room where groups can stand by and record on them.

Prepare and post another chart labeled *CTS Content Questions*, where participants can post sticky notes any time there is a content question that can't be answered in their group.

> **Facilitator Note**
>
> As part of your preparation, carefully review the summary notes so you know the scenarios and the background on each before the session and can lead an effective report out and discussion.

Facilitator Resources

- Facilitator Resource A2.1: Summary Notes for the Resource Scenarios, found in the Chapter 4 folder under the A2 Handouts on the CD-ROM at the back of this book.
- Facilitator Resource Signs found in the Chapter 4 folder under the A2 Handouts on the CD-ROM: One set of the $8\frac{1}{2} \times 11$ signs for each of the CTS sections used with the resource scenarios: Models (IA, IB, IIA, IIB, IIIA, IIIB, IVA, IVB, V) along with the name of the resource printed on it.

> **Facilitator Note**
>
> Note that Sections IB and IIB are not used for this session, but the signs for these are included on the CD-ROM for you to use if you make your own resource scenarios or modify the session to include these sections.

Materials Needed by Participants

CTS Parent Book

Science Curriculum Topic Study: Bridging the Gap Between Standards and Practice (Keeley, 2005). Ideally each participant will receive a book; but if that is not possible, provide at least one copy for every table group of about five to six people.

Facilitator Note

If you do not have copies of the CTS parent book for each participant, you will need to prepare copies of the following:

- Pages 24–27: Descriptions of Resources Used
- Page 33: Examples of CTS Applications
- Page 194: Earth, Moon, and Sun Study Guide
- Page 269: Models Study Guide

CTS Resource Books

For every *seven* people, provide the following resources:

- One copy *Science for All Americans*
- Three copies *Benchmarks for Science Literacy*
- One copy *National Science Education Standards*
- One copy *Making Sense of Secondary Science*
- One copy *Atlas of Science Literacy* (Vol. 2)

Copies of the CTS guide reading sections from each of the above can be provided if there are not enough or no resource books available. The *Atlas of Science Literacy* (Vol. 2) map on models is available to print out from the Project 2061 Web site (www.project2061.org).

Handouts

Facilitator Note

Print out the handouts and facilitator resources to refer to as you review the directions for the module.

- Facilitator Resource A2.1: Models Summary Notes (this is for the Facilitator, but you can provide copies to participants if the group is made up of people who will use this session design with others later on)
- Handout A2.2: Agenda at a Glance
- Handout A2.3: Resource Scenarios—Models
- Handout A2.4: Anatomy of a Study Guide
- Copy of PowerPoint presentation (optional)

DIRECTIONS FOR FACILITATING THE MODULE

Welcome, Introductions, and Overview (5–10 minutes)

Welcome and Introductions (5 minutes)

Welcome participants and explain that this module will introduce them to science CTS. If your participants do not know one another, do quick introductions by either having people introduce themselves at their tables or choose your own introduction activity. (Add additional time as needed.)

Overview of CTS (5 minutes)

Review goals on Slide 2 and refer participants to Handout A2.2: Agenda. Explain that this is an introductory session intended to introduce the CTS process, its purpose, and how to use the process. Suggest that once everyone learns the process, they may have a chance to go deeper by using CTS themselves or by attending follow-up sessions with CTS. Using Slides 3 through 6, explain to participants that CTS was a National Science Foundation (NSF)–funded project that developed a set of tools and processes, using professional resources, to bridge the gap between national standards and research on learning and classroom practice and state standards (Slide 3). Ask for a show of hands of people who have had some experience with CTS or heard about it prior to this session. Review what CTS is and what it is not (Slides 4 and 5) and describe how it has been called the "missing link" for implementing standards and research (Slide 6). (Review Chapter 2 of this book for key points to raise during this part of the presentation.)

Engagement—Preparing for CTS (25 minutes)

Warm-Up Talk (5 minutes)

Show Slide 7. Ask participants to have a quick discussion with an "elbow partner" (i.e., a person sitting next to them) using the prompt on Slide 7: "What role do models play in K–12 science teaching and learning?" Allow about three minutes for this warm-up talk and then announce that the group is going to spend some time learning about models in science. Rather than relying solely on the expertise of this group, we are going to access the prior knowledge we have and then build upon it by using readings from the CTS resource books.

Show Slide 8 and tell participants they will use an activity called resource scenarios to learn the CTS process and deepen their knowledge of a major unifying topic in science. Refer them to Handout A2.3: Resource Scenarios—Models. Explain that each resource scenario is an example of a K–12 teaching or learning question related to models about which educators might need to learn more to support their teaching. The scenarios will illustrate the value of using the CTS professional resources to improve and expand upon their existing knowledge and prior experience. The activity will also help them practice the difference between speaking from their own experience, beliefs, and knowledge base and speaking from and referencing a nationally validated, common body of knowledge—what we call "CTS talk." To begin the process, explain that you will start by forming study groups.

Elicitation—What Do We Already Know? (20 minutes)

Form Study Groups and Record Individual Responses to Scenarios (10 minutes)

Use one of the strategies below to form study groups and record initial ideas before looking at any resource books.

- Ask participants to count off by sevens and assign each group to one of the scenarios on Handout A2.3: Resource Scenarios—Models. Show Slide 9 and give participants a few minutes to read their scenarios and jot down any ideas they have about their scenarios before doing the CTS. Remind them not to look in any of the resource books yet. They should record their responses based on their own prior knowledge or experience.

OR

- Have each person refer to Handout A2.3: Resource Scenarios—Models and look over the scenarios. Ask them to mark two scenarios in which they are most interested and jot down their existing ideas for addressing their scenarios in the *Before CTS Resource Reading* box.

Show Slide 9 and ask everyone to follow the directions to write their ideas. Next ask participants to stand by the wall chart for their scenario. If you used the second grouping option above, ask people to stand by their first choice or move to their second choice if there are too many people at their first choice. If some scenarios have few or no participants, ask for volunteers to work on them so that all are covered. Remind them to take their Handout A2.3: Resource Scenarios—Models with them.

Groups Discuss and Record Initial Ideas (10 minutes)

The groups should be standing by their scenario charts and have Handout A2.3 with them. Ask each group to read their scenario again and discuss any ideas they wrote down to address the scenario. Remind them that everyone should have a chance to briefly share ideas and this is not the time to tell their "individual stories." Remind them to focus on the *Before CTS* scenario task. Suggest that if the science topic is unfamiliar to them, they can note that they have no firm ideas at this time or list some things they think they know on the *Before CTS* chart. Ask the group to record one to three key ideas on the chart marked *Before CTS* and then return to their seats.

 ## Exploration and Development of Models Ideas (30–40 minutes)

Reading and Studying About Models (10–15 minutes)

> **Facilitator Note**
>
> Refer the participants to the CTS parent book (or the copies of pages 24–26 you provided). Tell them there is more information on the resource books they are using starting on page 24.

Show Slide 10. Help participants locate the resource book needed for their scenario or a copy of the reading from the selected resource book. Ask participants to read their section quietly, making notes from their reading of information that enhances, adds to, or changes what they discussed and wrote on their *Before CTS* chart. Tell everyone that in a few minutes they will be asked to share ideas from the reading and cite the resource and page number used, so they should take notes or highlight text with underlining or sticky notes.

 ### Developing Models Ideas (10–15 minutes)

When everyone in the group is done reading, show Slide 11 and ask them to meet back at their wall charts and discuss their readings as they relate to their scenarios. Ask groups to examine the ideas they put up earlier and discuss what additional knowledge they would now add from reading their CTS section. Ask them to record at least three new or enhanced ideas they gained from reading the CTS resource. Remind everyone that this is the time that they should switch the discussion from talking about their own ideas or beliefs to citing the information in the resource readings (what we call "CTS talk"). Ask that they make sure that anything that goes up on the second chart can be traced back to the readings and not a personal opinion, belief, or experience. Have each group prepare to report on one new insight they gained from the CTS reading. This could include changing something that they had on their *Before CTS* chart that they no longer think should be there, adding new knowledge they gained, or enhancing or building upon ideas they had previously identified.

Reporting on Models Ideas (10 minutes)

When groups are finished (or if some are taking longer, you may need to ask them to stop where they are), ask each group to stand by their chart. Ask a couple of the members of each group to take no more than one minute to briefly share one or two new insights they gained from CTS (something they did not previously know). Thank the groups and have them return to their seats.

Break (10 minutes)

 ## Synthesizing Group Findings (15–20 minutes)

Recapping Findings (5 minutes)

Provide a short recap of what you heard and observed from the group's learning. For example, you might emphasize a point such as a group learned that there are three different types of models, when they had usually thought of models only as physical models. Refer to Facilitator Resource A2.1: Summary Notes for the Resource Scenarios for each scenario and bring up any key points that may not have been mentioned or recorded by the groups. If you have a cofacilitator, one of you should record these synthesis points on a flip chart as you bring them up or ask a participant to chart the main points. This is a good time to remind everyone that if they have content questions they want to raise, they can write them on sticky notes and put them on the wall chart labeled *CTS Content Questions*.

Context Question (10–15 minutes)

Next draw the group's attention to the question on the bottom of Handout A2.3. Show Slide 12. Ask participants to read and jot down their thoughts on the question and then turn to a partner and discuss their ideas.

Constructing an Understanding of the CTS Guides (20 minutes)

Connecting Back to the Study Guides (5 minutes)

Show Slide 13. Refer participants to the CTS guide, "Models" (on page 269 of the CTS parent book), and the CTS guide, "Earth, Moon, and Sun System" (on page 194), or to the handouts of these pages if everyone does not have a CTS book. Tell them these are the CTS guides that were used to identify the resources and readings they just used for the scenarios. Give them a minute to examine the study guides and ask them what they notice.

Linking the Resource Books to the Study Guide Sections (10 minutes)

Refer them to Handout A2.4: Anatomy of a Study Guide. Use Facilitator Resource A2.2 (these are signs made in advance for each of the scenarios with the book and the CTS section used printed on it). Walk to the first scenario chart. Hold up your copy of the resource book used for that scenario *(Science for All Americans)* and tape the sign with the Section IA above the *Before CTS* and *After CTS* charts (e.g., tape the sign "Section IA—*Science for All Americans*" above the charts for Scenario 1). Direct participants to Handout A2.4: Anatomy of a Study Guide to read about the purpose of Section I as you post the first sign. Do this for each of the scenarios, taping the sign for each section on top of the scenario charts and referring participants to review the purpose on Handout A2.4. When you get to Scenario 6, point out that this resource came from the CTS guide on page 194 of the CTS parent book and is an Earth-space related question about models. This shows how two different science CTS guides can complement each other. After the last scenario, remind them of the purpose of Section VI in that it is always important to take the time to connect findings from a CTS back to their own state or local context.

Summarizing the Resources (5 minutes)

Show Slide 14 to remind them of all the resources they used to complete a CTS. Point out the additional resources they did not use for this topic that may be used in other topic studies (e.g., *Science Matters*). Point out that pages 24–27 of the CTS parent book describe each of these resources.

Explain that the CTS developers think of CTS as our Swiss Army knife (show Slide 15). It has the specific tools we need at different times. Show Slide 16 and say there are parallel resources for mathematics teaching and learning they can explore if they also work in mathematics or with math teachers.

Let everyone know there are supplementary resources always being added to the CTS Web site to expand on the information in the standards and research. Show Slide 17. Point out that Volume 2 of the *Atlas of Science Literacy* was released after the CTS parent book was published, but there is now a "crosswalk" of that resource to all the science and mathematics topics on the CTS Web site (as well as on the CD-ROM at the back of this book). Show an example of a supplementary resource related to models as an example of what they can find on the CTS Web site (Slide 18).

 ## Application and Reflection (25 minutes)

Thinking About What We Learned and How We Can Use It (10–15 minutes)

Ask participants to step back from the exploration they have just done to think about the value of the CTS tool. Refer to questions on Slide 19 and ask them to consider how CTS enhanced their knowledge of models and what people can learn from using CTS. Allow time for a short discussion of ideas (five minutes) at tables. Point out that one value of CTS is that it can make us more aware of content we don't understand as well as we would like or content we may not have considered important to know.

Applying CTS in Different Contexts (10 minutes)

Ask participants to think about different reasons why a science educator might use CTS. Using Slide 20, ask participants to review the CTS application examples on page 33 of the CTS parent book and invite them to share other possible examples. Then ask them to turn to Chapter 4, beginning on page 53, to scan examples of suggestions and support materials for using CTS in various content, curricular, instructional, assessment, or professional development applications (if they do not have access to the CTS parent book, you can describe this briefly). Point out the collection of vignettes in Chapter 6 of the CTS parent book that illuminate how CTS is used in various contexts.

Show Slide 21. Ask participants to do a quick write (three to five minutes of quiet time to write a reflection) that summarizes their thoughts on how CTS could be useful to them using these guiding questions on the slide.

> **Facilitator Note**
>
> If any questions have been posted on the *CTS Content Questions* chart, this is a good time to read those to the group. Address any you can (given time constraints) or refer participants to other resources or people to talk with after the session.

 ## Partner Reflections (5 minutes)

Show Slide 22. Ask everyone to find a partner (either make eye contact with someone across the room or find someone they haven't talked with yet) and share the ideas from their quick writes. Allow five minutes and remind people to "share the talk time."

Wrap-Up and Evaluation (5–10 minutes)

Show Slide 23. Ask participants what questions they now have about using CTS and answer as many as time permits. Remind them they can go to www.curriculumtopicstudy.org for more information and that new tools and information are posted quarterly on the CTS Web site.

Thank everyone for their participation and ask them to complete an evaluation of the session.

Module A3

Full-Day Introduction to CTS Resource Scenarios and Snapshots, Facilitation Guide

BACKGROUND INFORMATION

Description of the Module

This session is a combination of Modules A1 and A2. It provides a longer, in-depth introduction to science CTS. Participants begin by exploring resource scenarios that engage participants in answering specific questions teachers might raise about a science topic. They use the scenarios to experience how CTS enriches a teacher's understanding of a science topic. For this session, the scenarios focus on the major unifying topic of models. (Both the *National Science Education Standards* and the *Benchmarks for Science Literacy* include unifying concepts or common themes. These are topics that cut across the disciplines of science, mathematics, and technology.) In small groups, participants reflect on what they already know about models. They read an assigned section from a CTS resource book and record knowledge gained from the readings. They compare the knowledge they had to the knowledge they gained to recognize the added value of using CTS to enhance and expand upon the individual and collective knowledge of the group. Following this, they use fifteen snapshots that are questions or issues a teacher might raise and a CTS-guided scaffold to explore them, gain additional practice in using different sections of each topic study, and then identify next steps for using CTS in their work.

Audience

This session is designed for preservice and classroom teachers and other education professionals who work with or across Grade K–12 science education and are interested in learning what CTS is and how it can be used in their work. In Part 1 of the module, a unifying theme is used as one of the focus areas so that the experience is applicable across the different subject areas of science. In Part 2 of the module, participants select from among topics applicable to Grades K–12 so they can explore areas of greatest interest to them. To tailor this session for specific grade levels or subject area audiences, see the additional resource scenarios and snapshot examples in the Chapter 4 folder on the CD-ROM at the back of this book or on the CTS Web site at www.curriculumtopicstudy.org. You can also develop your own snapshots and resource scenarios following the directions at the end of this chapter on page 102. You can substitute one of those for the one(s) used in this module.

Purpose

Module A3 provides an in-depth, full-day experience to teach CTS to first-timers and novice users by first focusing on a single topic and making a case for how doing a full CTS on one topic can lead to powerful learning.

Part 1 of the module leads participants to recognize the value added in using the professional readings and CTS guides to build or enhance their existing knowledge of content, teaching, and learning. Part 2 of the module provides extended practice in using the CTS guides with different topics, grade spans, and issues of teaching and learning.

Goals

The goals of this module are to

- develop awareness of CTS and the collective resources it uses for connecting standards and research on learning to classroom practice;
- provide guided practice in conducting a full topic study and several partial topic studies; and
- consider a variety of ways to apply CTS in curricular, instructional, assessment, and professional development contexts.

Key Components

The key components of Module A3 include the following:

Time to Get Acquainted With the CTS Book and Study Guides

Starting a topic study can be quite difficult if teachers have not been introduced to the basics of using the CTS parent book and CTS guides. To support teachers in learning how to use CTS, this introductory session includes time for exploring the CTS parent book and the structure and function of a CTS guide.

Embedded Practice Using the Resource Books and Selected Readings

An opportunity for getting to know the CTS resource books is embedded in the first part of this module. Participants are introduced to the CTS resource books and how they connect to the different parts of a CTS guide after they engage in focused readings and discussions about a topic. This allows participants to move right into the second part of the module with an understanding of the resource books and of the CTS section for which each book is used.

Scaffold

The multiple steps of CTS and the need to use different books for different purposes can be confusing at first, so we provide a scaffold for use in the second part of the module. The scaffold has step-by-step directions to guide novices through the process. After a few uses of the scaffold, teachers find they refer to it less and less as they learn the steps. Once they become familiar in the use of the CTS process, it becomes internalized and they no longer need the scaffold.

Reading and Group Processing Strategies

CTS always involves reading and interpretation of text from the CTS resources. Strategies that help participants to process what they are reading with a partner or group are embedded in this module. Other reading and group processing strategies are described in Chapter 3 of this *Leader's Guide*.

Application

For participants to apply what they learn and develop a commitment to using CTS, they need time to look at and discuss the different ways CTS can be applied to curriculum, instruction, assessment, and professional development. This module provides time for participants to identify applications that can be used in their own work.

SESSION DESIGN

Time Required

Approximately 6 to 6.5 hours (includes a 45-minute lunch break and two 10-minute breaks).

Agenda

PART 1: RESOURCE SCENARIOS

Welcome and Overview (10 minutes)

- Welcome and introductions (5 minutes)
- Overview of CTS (5 minutes)

Engagement—Preparing for CTS (25 minutes)

- Getting to know the CTS parent book and study guides (20 minutes)
- Warm-up talk (5 minutes)

Elicitation—What Do We Already Know? (30 minutes)

- Form study groups and record individual responses to scenarios (10 minutes)
- Discuss and record initial ideas (20 minutes)

Exploration and Development of Ideas About Models (30–40 minutes)

- Reading and studying models (10–15 minutes)
- Developing ideas about models (10–15 minutes)
- Reporting on ideas about models (10 minutes)

Break (10 minutes)

Synthesizing Group Findings (15–20 minutes)

- Recapping findings (5 minutes)
- Context question (10–15 minutes)

Constructing an Understanding of the CTS Guides (20 minutes)

- Connecting back to the study guides (5 minutes)
- Linking the resource books to the study guide sections (10 minutes)
- Summarizing the resources (5 minutes)

Reflection (15 minutes)

- What we learned and how we can use it (10–15 minutes)

Lunch Break (45 minutes)

PART 2: SNAPSHOTS

Engagement (15 minutes)

- Preparing for CTS snapshots (15 minutes)

Elicitation (10–15 minutes)

- Snapshots: Eliciting prior knowledge (10–15 minutes)

Exploration and Development (70 minutes)

- Snapshots: Exploring and practicing CTS (1 hour with built-in 10-minute break)
- Key learnings and CTS process debrief (10 minutes)

Applications (15 minutes)

- Applications for using CTS (5 minutes)
- Content, curricular, instructional, and assessment contexts (5 minutes)
- Vignettes (5 minutes)

Reflection (10 minutes)

- Quick write (3 minutes)
- Give me ten (7 minutes)

Wrap-Up and Evaluation (10 minutes)

MATERIALS AND PREPARATION

Materials Needed by Facilitator

CTS Parent Book

Science Curriculum Topic Study: Bridging the Gap Between Standards and Practice (Keeley, 2005)

Resource Books

One copy of each of the following resource books:

- *Science for All Americans* (AAAS, 1989)
- *Benchmarks for Science Literacy* (AAAS, 1993)
- *National Science Education Standards* (NRC, 1996)
- *Making Sense of Secondary Science* (Driver et al., 1994)
- *Atlas of Science Literacy* (Vol. 1) (AAAS, 2001)
- *Atlas of Science Literacy* (Vol. 2) (AAAS, 2007)
- *Science Matters* (Hazen & Trefil, 1991 or 2009; optional)

> **Facilitator Note**
>
> It is important for facilitators to have access to both volumes of the *Atlas of Science Literacy*, but Volume 2 is optional for the session since the map used is available online.

CTS Module A3 PowerPoint Presentation

The module A3 PowerPoint is included in the Chapter 4 PowerPoint folder on the CD-ROM. Review and tailor the PowerPoint presentation to your needs and audience. Insert your date and location on Slide 1, add additional graphics as desired, and add your own contact information on the last slide.

Supplies and Equipment

- Computer and LCD projector to show PowerPoint presentation
- Flip chart easel, pad, and markers
- Blank paper for note taking
- Sticky notes (small and large)
- Highlighter pens (optional)

Wall Charts

Prepare the following charts and post in the meeting room prior to beginning Part 1 of the session:

- *Before CTS* and *After CTS* charts: fourteen sheets of chart paper posted in pairs (side by side) around the room to create seven stations. Number each station (one through seven) to correspond to the number of the resource scenarios used in the first part of this module. Label the first chart in each pair *Before CTS*. Label the second chart in each pair *After CTS*. Post charts around the room where groups can stand by and record on them.
- *CTS Content Questions* chart: Post a chart labeled *CTS Content Questions* where participants can place sticky notes any time they have content or CTS questions that can't be answered in their group.

Prepare the following charts and post in the meeting room prior to beginning Part 2 of the session:

- *CTS Reminders Chart:*

  When Doing a CTS:

  1. Record the *exact language* in Sections III and V.

  2. Read only the text that is related to your specific inquiry.

  3. Take notes and include the name of the book and page numbers.

- Print out signs of letters A through O from Facilitator Resource A1.9 found in the Chapter 4 A.1 handouts. Post these signs on a long wall or around the perimeter of the room where participants can see them (see Figure 4.1 on p. 66). Spread them out so that small groups of participants can gather in front of the signs.

Facilitator Resources

- Facilitator Resources A3.1: Summary Notes for the Snapshots and A3.1a: Summary Notes for the Resource Scenarios, found in the Chapter 4 folder under the A3 Handouts on the CD-ROM at the back of this book.
- Facilitator Resource Signs (found in the Chapter 4 folder on the CD-ROM): One set of the $8\frac{1}{2} \times 11$ signs for each of the CTS sections used with the resource scenario "Models" (IA, IIA, IIIA, IIIB, IVA, IVB, V) along with the name of the resource printed on it.
- Facilitator Resource A1.9 Snapshots Signs

> **Facilitator Note**
>
> As part of your preparation, carefully review these summary notes so you know the scenarios and snapshots and the background on each before the session and can lead an effective report out and discussion.

Materials Needed by Participants

CTS Parent Book

Science Curriculum Topic Study: Bridging the Gap Between Standards and Practice (Keeley, 2005). Ideally each participant will receive a book; if that is not possible, provide at least one copy for every table.

Facilitator Note

If you do not have copies of the CTS parent book for each participant, you will need to prepare copies of the following: pages 24–27: "Descriptions of the Common Resources Used in CTS"; page 33: Figure 3.3, "Examples of CTS Applications"; page 194: study guide, "Earth, Moon, and Sun"; and page 269: study guide, "Models."

CTS Resource Books

For every *seven* people, provide the following resources for Part 1—Resource Scenarios of this module:

- One *Science for All Americans*
- Three *Benchmarks for Science Literacy*
- One *National Science Education Standards*
- One *Making Sense of Secondary Science*
- One *Atlas of Science Literacy* (Vol. 2)

Alternatively, copies of the sections from each of the above can be provided if there are not enough or no resource books available. The *Atlas of Science Literacy* (Vol. 2) map on models is available to print out from the Project 2061 Web site at www.project2061.org.

For Part 2 of this module, have at least one copy of all of the books used with the CTS guides available for each table group. You need only *Atlas of Science Literacy* (Vol. 1) for the snapshots activity in this part of the module.

If enough resource books are not available, you can create fifteen snapshot folders, each labeled with a large letter (A through O) on the front, into which multiple copies of each individual snapshot reading are placed. These folders replace the books or supplement a limited supply of books. Four copies of each individual snapshot reading placed in each labeled folder work well for a group of up to twenty-five teachers. You can add or decrease, depending on the size of your group. If CTS parent books are not available for each participant, include copies of the appropriate CTS guide in each folder as well. For example, four packets of the readings for Snapshot A are placed inside a manila folder with a large "A" on the outside. Four copies of the CTS guide, copied on different colored paper, are also placed inside the folder. These folders (A through O) are then placed in an area of the room where participants can come up and take a reading and study guide out of the folder, and replace it when done so that others can use it on their second round. See Figure 4.2 on page 68.

Handouts

Facilitator Note

Print out handouts to refer to as you review the directions for the module.

- Handout A3.1: Introduction to Science CTS K–12 Snapshots Facilitator Notes (*Note:* These are for the facilitator but should be given as handouts also when your group includes leaders who will do this session with others later on.)
- Handout A3.1a: Resource Scenarios Models: Facilitator Notes
- Handout A3.2: Agenda at a Glance
- Handout A3.3: Resource Scenarios—Models
- Handout A3.4: Anatomy of a Study Guide
- Handout A3.5: Introduction to CTS Snapshots—Scaffold
- Handout A3.6: List of Science Curriculum Topic Study Guides (if participants have their own copy of the CTS parent book, you will not need to provide copies of this handout)
- Handout A3.7: Snapshots
- Handout A3.8: Snapshots Answer Key
- Handout A3.9 Snapshots Recording Sheet
- Copy of PowerPoint presentation (optional)

DIRECTIONS FOR FACILITATING THE MODULE

PART 1–RESOURCE SCENARIOS

Welcome and Overview (10 minutes)

Welcome and Introductions (5 minutes)

Show Slide 1. Welcome participants and explain that this session will introduce them to science CTS. If your participants do not know one another, do quick introductions.

Overview of CTS (5 minutes)

Review the goals on Slide 2 and refer participants to Handout A3.2: Agenda. Explain that this is an introductory session and that participants may have a chance to go deeper into CTS topics themselves once they learn to use the process and resources. Using Slides 3 through 6, explain to participants that CTS was a National Science Foundation (NSF)–funded project that developed a set of tools that help bridge the gap between national standards and research on learning and classroom practice and state standards (Slide 3). Ask for a show of hands of people who have had some experience with CTS or heard about it prior to this session. Review what CTS is and what it is not (Slides 4 and 5) and describe how it has been called the "missing link" for implementing standards and research (Slide 6) because it helps educators examine the standards and research on specific areas of interest and apply what they learn to their own work. For background information on this part of the presentation, see Chapter 2 of this *Leader's Guide.*

 ### Engagement—Preparing for CTS (25 minutes)

Getting to Know the CTS Parent Book and Study Guides (20 minutes)

Ask participants to work with a partner. Distribute their CTS parent books if they are receiving one at the session or have them locate a copy of the CTS parent book on their tables and do the exercise as a table group. Using Slide 7, start an exercise called *First Glance,* and tell the pairs or small table groups to open the book randomly and take a "first glance." Have each pair or small table group open to at least three spots in the book, take a first glance, and discuss what they find.

Ask for a few volunteers to report on their first glance on any page up to page 113. Ask the following questions: "What did you find?" "Why did that interest you?" Then ask if anyone looked at one of the Curriculum Topic Study Guides that start on page 113 in the CTS parent book. Ask everyone to turn to these pages and look at a study guide. Ask participants to report on what they found in the study guides. Elaborate on what they say, explaining that there are 147 study guides and that they are the core resource for using CTS. Show Slide 8. Explain that the study guides are described in Chapter 2 of the CTS parent book. Using Slides 9 and 10, briefly point out the structure of a study guide (including the Web site link at the bottom) and the six sections to the study guide that are used to answer different questions or issues related to science understanding,

> **Facilitator Note**
>
> You can point out that many schools and education organizations own these books but they do not use them—the CTS process can remedy that because it makes the resources more relevant and easier to use.

teaching, and learning. Explain that they will revisit the structure of the study guides in more detail after they have some experience with a topic study. Use Slide 11, the "Swiss Army Knife," to point out the many functions of the CTS process. Describe how there are times when you might use only one tool on your knife for a specific purpose. Other times you might use a combination of tools. What makes CTS so useful, just like a Swiss Army knife, is that the variety of tools ensures that you can find the right tool for the right purpose. Explain that for each CTS section, you use different professional resources to get the information you need. Show Slide 12. It has images of all the front covers of the CTS resource books to let participants know what books they will be using. Ask people to raise their hands if they own a copy or have used one of these books. Explain to your group that the work today will be to experience a full topic study on the topic models and then to explore several different CTS sections on different topics.

Warm-Up Talk (5 minutes)

Show Slide 13. Ask participants to have a quick discussion with an "elbow partner" (i.e., a person sitting next to them) about the topic models using the prompt on Slide 13 (Prompt: "What role do models play in K–12 science teaching and learning?") Allow about three minutes for this "warm-up talk" and then announce that they are going to spend some time learning about models in science. Rather than relying solely on the expertise of this group, they are going to access the prior knowledge that they have and build upon it by using readings from the CTS resources. Show Slide 14 and say the activity they are going to do is called resource scenarios. Refer them to Handout A3.3: Resource Scenarios. Each scenario is an example of a K–12 teaching or learning question related to models about which an educator might need to know more. The scenarios will raise awareness of the value of using professional resources to improve and expand upon an individual's or group's existing knowledge. It will also help them experience the difference between speaking from their own experience, beliefs, and knowledge base and speaking from and referencing a common body of knowledge—what we call "CTS talk." To begin the process, explain that you will start by forming study groups.

 ## Elicitation—What Do We Already Know? (20 minutes)

Form Study Groups and Record Individual Responses to Scenarios (10 minutes)

Form groups using one of the strategies below.

- Ask participants to count off by sevens and assign each group to one of the seven scenarios on Handout A3.3: Resource Scenarios—Models. Give participants a few minutes to read their scenarios and jot down any ideas they have about them before doing the CTS. Remind them not to look in any of the resource books yet. They should record their response based on their existing knowledge or prior experiences.

OR

- Have each person refer to Handout A3.3 and look over the scenarios. Ask them to mark two scenarios they are most interested in exploring and write down their ideas about those scenarios (before doing CTS readings).

Show Slide 15, which summarizes the directions for recording initial ideas and meeting in study groups to share initial ideas. Once participants have recorded their initial ideas, ask them to stand by the wall charts for their scenarios. If you chose the second grouping option above, ask people to stand by their first choice or move to their second choice if there are too many people at their first one. If some scenarios have few or no participants, ask for volunteers to work on them so that all are covered.

Groups Discuss and Record Initial Ideas (10 minutes)

The groups should be standing by their scenario charts with Handout A3.3. Ask them to focus on the *Before CTS* task by having each group read their scenario again and share and post the ideas they wrote down to address the scenario on the *Before CTS* chart. Ask them to be sure that everyone has a chance to share at least one idea; this is not the time to go into lengthy discussions. Tell the groups that if their science topic is unfamiliar to them and they have no or few ideas, it is OK for them to write "we have no firm ideas at this time" or "some things we think we know are . . ." Ask the group to record one to three key ideas on the chart marked *Before CTS* and then return to their seats.

Exploration and Development of Models Ideas (30–40 minutes)

Reading and Studying About Models (10–15 minutes)

Show Slide 16. Help participants locate their resource or a copy of the reading from the selected resource. Ask participants to read their section quietly, making notes from their reading that enhances, adds to, or changes what they wrote on their *Before CTS* chart. Tell everyone that in a few minutes they will be asked to share ideas from the reading and cite the resource used and page numbers, so they should take notes, highlight text with underlining, or use sticky notes.

Developing Models Ideas (10–15 minutes)

When everyone is done reading, show Slide 17 and say they will be returning to their wall charts for a discussion in a minute. Before the groups return to their charts, tell them when they go back to the charts, they will switch their discussion from talking about their own opinions or beliefs to citing the information in the resource readings (book and page number). Remind them to make sure they can trace anything that goes up on the chart back to the CTS readings and nothing posted on the *After CTS* chart is from a personal opinion, belief, or experience, but instead is what we call "CTS talk." Then give the directions for the task:

1. When participants return to their charts, they are to examine the ideas they put up earlier and discuss what they learned from the CTS readings.

2. They are to record at least three new or enhanced ideas gained on the *After CTS* chart.

3. When they are done, they are to choose one new insight to report out to the larger group, such as something that was on the *Before CTS* chart that they no longer think should be there, new knowledge gained, or how they enhanced or built upon ideas previously identified.

Check to see if anyone has questions about "CTS talk" or these directions and then ask them to meet back at their wall charts and discuss their readings as they relate to their scenarios.

Reporting on Models Ideas (10 minutes)

When groups are finished (if some are taking longer, you may need to ask them to stop where they are), ask each group to remain standing by their chart. Ask each group to take no more than one minute to share one new insight they gained from their readings. Clarify any content that is confusing, using your own knowledge and by referring to Facilitator Resource A3.1: Summary Notes for the Resource Scenarios. After hearing from the seven groups, thank everyone and announce that there will be a short break.

> **Facilitator Note**
>
> This is a good time to remind everyone that if they have CTS or content questions they want to raise, they can write them on sticky notes and put them on the wall chart labeled *CTS Content Questions* during the break.

Break (10 minutes)

Synthesizing Group Findings (15–20 minutes)

Recapping Findings (5 minutes)

Provide a short recap of what you observed from the ideas reported by the scenario study groups. For example, you might emphasize a point such as a group learned that there are three different types of models, when they had usually thought of models only as physical models. Refer to Facilitator Resource A3.1: Summary Notes for the Resource Scenarios for each scenario and bring up any key points that may not have been mentioned or recorded by the groups. Ask how reading the resources added to their understanding of the unifying theme of models and ask for a few responses from the large group.

Context Question (10–15 minutes)

Next draw the group's attention to the context questions on the bottom of Handout A3.3: Resource Scenarios—Models and on Slide 18. Using an activity called *Think-Pair-Share*, ask participants to *think* about the questions themselves and jot down their thoughts. After a few minutes of quiet reflection, invite participants to *pair* up with a partner and spend a few minutes each *sharing* their ideas.

Constructing an Understanding of the CTS Guides (20 minutes)

Connecting Back to the Study Guides (5 minutes)

Show Slide 19. Refer participants to the CTS guide, "Models" (on page 269 of the CTS parent book), and the CTS guide, "Earth, Moon, and Sun System" (on page 194), or to the handouts of these pages if everyone does not have a CTS parent book. Tell them these are the CTS guides that were used to identify the resources and readings they just used for the scenarios activity. Give them a minute to examine the study guides.

Linking the Resource Books to the Study Guide Sections (10 minutes)

Refer participants to Handout A3.4: Anatomy of a Study Guide. Starting with Section 1A, ask participants to review the purpose of that CTS section. Use Facilitator Resource A3.2 (which are signs for each of the scenarios with the resources used printed on them). Walk to the Scenario 1 wall chart. Hold up the resource book used for that scenario and tape the sign "*Section IA—Science for All Americans*" above the Scenario 1 wall chart. Continue this for every scenario until all the signs are posted and the descriptions of each section on Handout A3.4 have been reviewed.

When you get to Scenario 6, point out that this resource came from a different CTS guide and have them look at the study guide on page 194. Say that this scenario is an Earth-space related question about models. This shows how two different science CTS guides can complement each other. After the last scenario, remind them that the purpose of CTS Section VI is to connect findings from the national resources explored in CTS Sections I through V back to state and local contexts.

> **Facilitator Note**
>
> You might ask a few participants to hang the signs after you hold them up and connect them to the right scenario.

Summarizing the Resources (5 minutes)

Show Slide 20 to remind your group of the resources they used to complete the CTS (this is the same as Slide 12 shown earlier). Point out the additional resources they did not use for this topic that may be used in other topic studies (e.g., *Science Matters*). Refer them to pages 24–27 of the CTS parent book for a description of each of these resources if they want to read more about them after the session.

Show the Swiss Army knife again (Slide 21). Point out that just like the army knife, CTS has the specific tools we need at different times. Show Slide 22 and say there are parallel resources for mathematics teaching and learning they can explore if they also work in mathematics.

Show Slide 23. Let everyone know there are supplementary resources always being added to the CTS Web site to expand on the information in the standards and research. Point out that Volume 2 of the *Atlas of Science Literacy* and the second edition of *Science Matters* were released after the CTS parent book was published, but there is a "crosswalk" of these resources to all the topics on the Web site (as well as a crosswalk of Volume 2 of the *Atlas* and CTS in the Chapter 2 folder on the CD-ROM at the back of this book). Show an example of a supplementary resource you can find on the Web site (Slide 24).

Reflection (10 minutes)

Thinking About What We Learned and How We Can Use It (10 minutes)

Ask participants to step back from the exploration they have just done to think about the value of CTS. Refer to questions on Slide 25: What did you learn and what can other people learn from using CTS? Allow time for a short discussion of ideas (5 minutes) at tables. Ask a few volunteers to share an example of what can be learned from CTS. Point out that one value of CTS is that it can make us more aware of content we don't understand as well as we would like.

Lunch Break (45 minutes)

PART 2–SNAPSHOTS

Engagement (15 minutes)

Introduction to CTS Scaffold (15 minutes)

- Welcome everyone back and show Slide 26. Explain that they will now do Part 2 of the session in which they will use a scaffold (a step-by-step guide) to learn to use CTS to answer teaching and learning questions related to different curricular topics. Point out what is meant by a scaffold on Slide 26. Show Slide 27 and show the seven steps on the CTS scaffold. Ask participants to pull out the following handouts: Handout A3.4: Anatomy of a Study Guide (they already used this one in Part 1 of the session); Handout A3.5: Introduction to CTS Snapshots—Scaffold; and Handout A3.6: 147 Science Curriculum Topic Study Guides. (*Note:* Use this handout if participants do not have a copy of the CTS parent book; the list of CTS guides is found on pages ix–x of the CTS parent book.) They will use these handouts to learn to use the scaffold.

- Tell everyone they are going to practice using the scaffold with several examples. Show Slide 28 and read the question. Ask participants to look at pages ix–x in their CTS parent book or at Handout A3.6: 147 Science Curriculum Topic Study Guides and scan the eleven categories. In what category should they look to find a topic to address the question? (Take responses; correct response is "Biological Structure and Function.") Then ask what topic guide they would consult. (Take responses; correct response is "Cells.") Ask participants where they would find that topic guide in the CTS parent book. (Take responses; correct response is page 135.) Show Slide 29 and continue to engage the group in answering questions. Ask at what CTS section they would look to answer the question on Slide 29. (Take responses; correct

> **Facilitator Note**
>
> This should be a rapid-paced practice session of how to find the right study guide. If everyone gets the process after the first few examples, you should skip over the other ones, although the last two are very useful in pointing out what is meant by reading only the "related sections."

response is Section II, which is focused on the outcome of raising instructional implications.) Quickly go through Slides 30 through 43 to provide examples of what they will see on the scaffold and practice choosing the right section to focus on.

- Show Slide 44 and point out that CTS takes practice and that is what the group will be doing next. Point out the chart you posted earlier with reminders including the following:

When Doing a CTS:

1. Use *exact language* in Sections III and V. (Don't paraphrase goals; the language used in the standards documents was chosen very deliberately.)

2. Read only the text that is related to your specific inquiry.

3. Write everything down that is related and important; cite the name of the book you used and page numbers.

Elicitation (5 minutes)

Eliciting Prior Knowledge (5 minutes)

Show Slide 45 and refer participants to Handout A3.7: Snapshots. Tell them they will now do some short, partial studies called snapshots in order to practice using the guides and the books. Explain that before they do their first CTS snapshot, it is important to take a few minutes to activate their prior knowledge so that after they do a CTS snapshot, they can compare what they knew before to what they gained from the CTS. Point participants to the snapshots on Handout A3.7 and have them choose one that interests them and answer the question on the snapshot. Give them a few minutes to write a brief response to the first question on Handout A3.0, based on their prior knowledge and experience. When they are done, have them set their responses aside.

When everyone has finished, ask them to stand and move to the letter on the wall that represents the snapshot they chose. In small groups or pairs, have participants share how they responded to the snapshot based on their own knowledge of teaching and learning. If some participants are the only ones who chose the snapshot and are unmatched, either have a facilitator converse with them or match them with others who are the only ones standing at their signs and ask them to exchange ideas. Once everyone has shared their pre-CTS ideas, have them return to their tables.

 Exploration and Development (55 minutes)

Exploring and Practicing CTS (45 minutes)

- Show Slide 46. Tell the group that they will now have time to do the CTS on the snapshot they chose. Ask if there are any questions about using the steps on the scaffold or interpreting the study guides. If you do not have books available for everyone and are using the snapshot folders system described on page 86 for all the readings, point out where the folders are and how to use them. Remind participants they will be sharing these resources and to return them when they are done so others can use them.

> **Facilitator Note**
>
> If you have assembled snapshot folders with the CTS guides and the readings in them, the participants will not have to find the CTS guide in the book or locate the correct resource book, but we still encourage them to use the scaffold to gain practice in how they would do this if they were using the materials on their own.

- Invite participants to work individually or in pairs (with someone who chose the same snapshot) and use the scaffold going through Steps 1 through 7 to find the study guide and the resources they need. Remind them not to skip steps this first time through and to record the CTS guide, section(s), and resource book(s) they used to do their CTS on Handout A1.8, Question 2 and summarize their findings on Question 3 after completing the reading. Ask them to specifically include any new knowledge or insights they gained after doing the CTS.

- Encourage participants to take notes as they read, recording information that is relevant to the question posed in the snapshot. Once they finish their snapshot, encourage them to try another one if they have time, recording their pre-CTS ideas, doing the CTS, and then recording new knowledge or insights gained from CTS. Point out that there is an answer key they can check to make sure they are on the right category, topic, and section before they start their readings. (See Handout A3.8: Snapshots Answer Key.) Remind them to scan the reading and read only the text directly related to the question on their snapshot.

- Walk around and check on groups as they work and answer any questions they might have about the process or the resources as they complete their snapshots. If people seem confused or stuck, point them back to the scaffold and check that they are finding the right sections and readings referenced in the study guides. If CTS questions come up that you can't address at that time, ask the participant to write the question on a sticky note and post on the wall chart labeled *CTS Content Questions*.

Key Learning and CTS Process Debrief (15 minutes)

After participants have had time to complete at least one or two of the snapshots, ask the group to stand up again, with Handout A3.9 in hand, and meet at the snapshot wall signs with the same group or partner they discussed their pre-CTS ideas with. Show Slide 47. Give them time to discuss what new knowledge or insights they gained from doing CTS. After groups have had time to discuss their CTS findings, ask for a few report outs on how CTS addressed the snapshot questions and how CTS can be a valuable tool, even for experienced teachers, in answering questions about teaching and learning. Ask volunteers to share what they learned from their snapshot. Probe to elaborate on what the standards and research suggested about the snapshot. Depending on the size of your group and the amount of time, you may get to only three or four snapshots. Clarify information and add in important information using Facilitator Resource A1.1: Introduction to CTS K–12 Snapshots Facilitator Notes.

> **Facilitator Note**
>
> If any questions have been posted on the *CTS Content Questions* chart, this is a good time to read those to the group. Address any you can (given time constraints) or refer participants to other resources or people to talk with after the session.

Applications of CTS (15 minutes)

Context and Applications (15 minutes)

Ask participants to step back from the snapshot activity they have just finished to think about different reasons why a science educator might use CTS. In table groups, if you only have one CTS parent book per table, or in trios if you have enough books, use the steps on Slide 48 to have participants review the CTS application examples starting on page 33 of the CTS parent book and ask for other examples. Then if they have their own books or a copy on the table, have everyone turn to Chapter 4, page 53, to scan examples of suggestions and support materials for using CTS in various content, curricular, instructional, assessment, or professional development applications. If they do not have access to the books, you can describe the applications briefly. Point out the collection of vignettes in Chapter 6 that illuminate how CTS is used in various contexts.

Reflection (10 minutes)

- Show Slide 49. Ask participants to do a quick write (three minutes of quiet time to write a reflection) that summarizes their thoughts on how CTS could be useful to them.
- Show Slide 50. Have participants stand and find a partner and have a conversation about how CTS could be useful to them.

Wrap-Up and Evaluation (10 minutes)

Wrap-Up (5 minutes)

Ask participants to return to their seats. Answer any questions they have about using CTS and refer them to page 2 of the CTS parent book for more information on the outcomes of using CTS. Remind participants they can visit the CTS Web site at www.curriculumtopicstudy.org for more information and that new tools and information are posted quarterly. Suggest that as a next step they may want to do a full topic study. (See modules for leading full topic studies in Chapter 5 of this *Leader's Guide.*)

> **Facilitator Note**
>
> See suggestions of reflection strategies in Chapter 3 of this *Leader's Guide,* starting on page 54.

Evaluation (5 minutes)

Thank everyone for their participation and ask them to complete an evaluation.

OPTIONS FOR INTRODUCING THE CTS RESOURCE BOOKS

The introductory Module A1 in this chapter includes an introduction to the CTS resource books. Some of these books may be familiar to participants; others may be new to them. It is important to include enough time for participants to become acquainted with each of the resources before they engage in the CTS process. This can be done during Module A1, or you may want to provide an introduction as a separate session, for example, with a preservice class or as an introduction with the full topic studies in Chapter 5 if there are participants who do not know the books.

Three options are provided to introduce the books. The option you select will depend on the number of resource books you have available, the familiarity your participants have with the books, and the amount of time you have for your introductory workshop. Each option includes a set of slides to insert into the Module A1 PowerPoint slides or with one of the full topic studies in Chapter 5 if most of your participants are new to CTS. Check the slides and the script to see where to insert the slides. Note that if you choose Option 3, inserting the slides will change the numbering of the subsequent slides in the facilitator scripts.

Introduction to CTS Resources Option 1

Overview

This option provides participants with a highly engaging, interactive way to explore the CTS resource books firsthand before using them to do CTS.

When to Use This Option

Use this option when you have multiple sets of the resources for participants and the time for them to explore the books firsthand.

Resources Needed

Ideally all six resources will be available for every six to twelve participants. *Science Matters* can be optional. If *Making Sense of Secondary Science* is not available, be sure *Benchmarks* is available as it contains summaries of the research used in Section IV.

Approximate Time

Time needed is 25 to 30 minutes.

Preparation

Provide a set of available materials at each table group. The number of books determines the number of participants at each table, ideally a one-to-one match between a book and a participant. For example, if all six books are available, form table groups of six participants. If there are fewer resources available, participants can form pairs for this activity. If participants do not have their own copies of the CTS parent book, make copies of the resource descriptions on pages 24–27 of that book. Insert the PowerPoint slide for this option in place of the Introduction to CTS Resources slide placeholder in the Module A1 facilitator's script or insert it into a Chapter 5 module if you are using it there.

Process

1. Ask groups to divide the resources on their tables among people at the table and follow directions on the PowerPoint slide (found on the CD-ROM under Option 1). At their tables, they should have stacks of the various resource books they will use for CTS.

2. Each person or pair will take one book and look through it, read the description of it from pages 24–27 of the CTS parent book, and become an expert on this book to later introduce it to the other people at the table. Allow about five minutes to review the book, examining its features and content described on pages 24–27. Tell everyone they will have just two to three minutes to introduce the books to each other, pointing out the main features and how they are used in CTS.

3. When they are done sharing the resources in their small groups, answer any additional questions the large group might have—you can refer to the descriptions of the resources provided in Option 2 below to provide additional information on the resources. Announce that participants will now use the resources they just explored to engage in the CTS process. Resume with the next slide in the module you are using.

Introduction to CTS Resources Option 2

Overview

This option provides a quick way to introduce the resources. It is primarily a mini-lecture and does not provide an opportunity for much interaction.

When to Use This Option

Use this option when you do not have copies of the books available for table groups to use or there is limited time in your agenda. It can also be used if you know that many of the participants already own or have seen the books and it can serve as a review for the group.

Resources Needed

The facilitator should have a set of the resources to hold up. After introducing each of the resources, the facilitator should put them on display at a "library table" where participants can go up during the break or after the workshop to look through them.

Approximate Time

Time needed is 10 to 15 minutes, depending on the detail you want to provide.

Preparation

Have each of the books ready to hold up and describe. If participants do not have their own copies of the CTS parent book, make copies of the resource descriptions on pages 24–27 of that book. Have a library area set up where the books can be on display for participants to look through during the break or after the session. Insert the PowerPoint slide for this option in place of the Introduction to CTS Resources slide placeholder in the facilitator's script.

Process

1. Refer participants to the descriptions of the resources on pages 24–27 of the CTS parent book as you lead them through a tour of the books. Hold up *Science for All Americans* (AAAS, 1989). Ask for a show of hands for the following questions: "How many people have used this book?" "How many have heard of it, but haven't used it?" "How many are learning about it for the first time?" Explain that *Science for All Americans* was the first tool developed by the American Association for the Advancement of Science's Project 2061 to establish what all high school graduates should know in order to be prepared to be a science literate citizen. It was developed with the input and consensus of hundreds of distinguished science, technology, engineering, and mathematics (STEM) professionals and educators. It provides an overall picture of what all adults (not just those who go on to higher education and STEM careers) should know after their K–12 experiences in order to be considered science literate. It includes science, mathematics, technology, and social science and the interconnections among them. *Science for All Americans* is used with CTS Section IA, which identifies adult science literacy.

2. Next show the adult trade book, *Science Matters* (Hazen & Trefil, 1991, 2009). Ask for a show of hands for the following questions: "How many people have used this book?" "How many have heard of it, but haven't used it?" "How many are learning about it for the first time?" Explain that this book is coauthored by two of our nation's highly respected scientists who are committed to helping the public fill in the gaps left by their K–12 and post-graduate formal education so they have the information they need to participate in the public discourse and understanding related to science. It translates scientific concepts in ways the general public can understand when they encounter science in the media or public arena. Besides addressing fundamental science literacy ideas that are also part of *Science for All Americans,* it goes beyond the concepts we would expect every high school graduate to know and be able to use. For example, someone may encounter the concept of a particle accelerator in a newspaper article. *Science Matters* helps the reader, who may not have a background in high-energy physics, understand enough about particle accelerators to make sense of the newspaper article. Likewise, *Science Matters* can help teachers comprehend science concepts they encounter that they do not understand. Emphasize to participants that although science knowledge has expanded considerably since the first edition of this publication in 1991, the fundamental principles of science it explains still pertain today. In the revised 2009 edition, the authors state the following:

> We could not have foreseen many of the remarkable developments of the past two decades—nanotechnology, archaea, LEDs, cloning, dark energy, ancient microbial fossils and deep microbial life, evidence for oceans of water on Mars and lakes of methane on Titan, ribozymes, carbon nanotubes, extrasolar planets and so much more. But all of these unanticipated findings fit into the existing framework of science. The core concepts of science have not changed and we are unable to point to any fundamentally new scientific principle that has emerged during the 1990's or 2000's. Accordingly, while every chapter has been updated, we have added only a single new chapter on the explosion of advances in biotechnology. We conclude that the experience of the past two decades underscores the value of the great ideas approach to achieving scientific literacy. (Hazen & Trefil, 2009, pp. xxi–xxii)

(Note that a new edition of the book was released in 2009. A crosswalk between the new edition and the CTS guides is available on the CTS Web site. The new edition of CTS will include the 2009 *Science Matters* readings.) *Science Matters* is used with Section IB, identifying adult science literacy, in the CTS guides.

3. Next hold up *Benchmarks for Science Literacy* (AAAS, 1993). Ask for a show of hands for the following questions: "How many people have used this book?" "How many have heard of it, but haven't used it?" "How many are learning about it for the first time?" Explain that *Benchmarks* (like its companion, *Science for All Americans*) isn't just about science. It addresses science, mathematics, social science, and technology and lays out explicit learning goals for students to achieve in Grades K–12 in order to achieve the adult science literacy described in *Science for All Americans*. The sections in *Benchmarks* parallel sections in *Science for All Americans*. Point out that the bulleted text lists specific benchmark learning goals at K–2, 3–5, 6–8, and 9–12 grade bands. These bullets are used with Section IIIA in the CTS guides. In addition, several of the benchmarks have been slightly revised, a few moved to other grade levels, and several new benchmarks have been added. These changes in the 1993 benchmarks can be seen on the online version of *Benchmarks for Science Literacy* available on the Project 2061 Web site at www.project2061.org. Besides the bulleted lists of suggested learning goals, there are also grade span essays as well as an overview essay. These essays describe the implications for teaching the ideas and provide suggestions for curricular and instructional contexts. The essays are used with Section IIA, "Consider Instructional Implications," of a CTS guide. *Benchmarks for Science Literacy* also offers a summary of research about students' misconceptions, in the back of the book in Chapter 15. These readings are used with Section IVA, "Examine Research on Student Learning," of a CTS guide.

4. Show the *National Science Education Standards* (NRC, 1996). Ask for a show of hands for the following questions: "How many people have used this book?" "How many have heard of it, but haven't used it?" "How many are learning about it for the first time?" Say that while many of us are much more familiar with our own state and local standards, this document is important because it constitutes a national consensus of what is essential to know and be able to do in science. Most state standards are based on the *National Science Education Standards* or the *Benchmarks*. Point out that most of the content in the *National Science Education Standards* overlaps with the *Benchmarks*; however, it takes a different approach to inquiry and technological design with its emphasis on doing inquiry or design in addition to the understandings of each process. Like the *Benchmarks*, it also includes bulleted lists of learning goals used for Section IIIB as well as essays used in Section IIB.

5. Next show *Making Sense of Secondary Science: Research Into Children's Ideas* (Driver et al., 1994). Ask for a show of hands for the following questions: "How many people have used this book?" "How many have heard of it, but haven't used it?" "How many are learning about it for the first time?" This book also provides summaries of children's ideas, including misconceptions. Even though the title mentions secondary science, it includes children's ideas from elementary school through high school. Although the research included in this book as well as *Benchmarks* dates from before 1995, most of it is still applicable today. Mention that the Web site for CTS has a searchable database where you can find more recent research articles. This book is used for CTS Section IVB, to examine research on student learning.

6. Hold up Volumes 1 and 2 of *Atlas of Science Literacy* (AAAS, 2001–2007; if both are available). Ask for a show of hands for the following questions: "How many people have used this book?" "How many have heard of it, but haven't used it?" "How many are learning about it for the first time?" Explain that the *Atlas* organizes the explicit learning goals found in *Benchmarks* (as well as new or revised ones that are in the *Atlas*, but not in *Benchmarks*, statements from *Science for All Americans,* and learning goals from the *National Science Education Standards* that do not appear in *Benchmarks*) into visual maps that show growth in understanding over a K–12 span, including prerequisite ideas that form the basis for subsequent ideas that build in sophistication over time. "Storylines" at the bottom of the maps show how the ideas related to that storyline unfold over time as you move up through each grade level. The maps show connections within and across topics. The *Atlas* is used with CTS Section V to examine coherency and articulation. The narrative that precedes each map can also be used with Sections II, III, and IV and includes updated summaries of the research on learning. Explain how Volume 2 of the *Atlas* was not available at the time the first edition of the CTS parent book was written. The second edition of the CTS parent book will contain links to the Volume 2 maps. In the meantime, there is a crosswalk to the Volume 2 maps that can be downloaded from the CTS Web site, and this crosswalk is also on the CD-ROM in the back of this book.

7. Finally, hold up a copy of the state standards for the group you are working with, or if it is a multistate group, hold up a copy of any state standards document as an example. Explain how all the resources just mentioned are used to clarify state standards or the learning goals in local curriculum documents. These state or district-specific standards are used in Section VI of the CTS guides.

8. Invite any additional questions and point out that the books will be on display in the room if anyone would like to browse through them. Announce that participants will now use the resources to engage in the CTS process. Resume with the next slide in the module you are using.

Introduction to CTS Resources Option 3

Overview

This option provides an interactive, guided tour through each of the books, giving participants an opportunity to identify and examine the content of the books and how they are used in CTS. The examples provide a comprehensive tour across content areas of life, physical, Earth, and space science, and inquiry. The slides and examples can be modified by the facilitator to focus on one content area if the group is subject specific (e.g., all Earth science teachers).

When to Use This Option

Use this option when you have enough of each book for one to three people in a group and more time on your agenda. This option can be extended into an even longer session if desired. It is particularly helpful to participants, such as preservice, new teachers, and experienced teachers new to teaching science who are not familiar with the major science education standards and research documents. For preservice faculty, this introduction can be used as a separate class or session, prior to introducing the CTS process in the next session or class.

Resources Needed

The facilitator should have a set of the resources to hold up. There should also be one book per individual, pair, or triad of participants to use during the interactive search for information.

Approximate Time

Time needed is 30 to 45 minutes.

Preparation

Have each of the books ready to hold up and describe. If participants do not have their own copies of the CTS parent book, make copies of the resource descriptions on pages 24–27 of that book. Insert the PowerPoint slide for this option in place of the Introduction to CTS Resources slide placeholder in the facilitator's script. Be aware that the additional slides for this option will change the numbering of the subsequent slides. You may want to print out the main slides first, before inserting these, so that you can refer to the same slide numbers in the facilitator's script.

Process

1. Show Slide 1 for Option 3 and refer participants to the descriptions of the resources on pages 24–27 of the CTS parent book as you lead them through a tour of the books. Hold up *Science for All Americans* (AAAS, 1989). Ask for a show of hands for the following questions: "How many people have used this book?" "How many have heard of it, but haven't used it?" "How many are learning about it for the first time?" Have participants quickly read through the description of *Science for All Americans* on page 24 of the CTS parent book. Ask someone to summarize the reading by volunteering an answer to the question, "What is the purpose of *Science for All Americans*?" Feel free to add any additional information drawing from the description above in Option 2. Now show Slide 2 for Option 3. If the purpose of *Science for All Americans* is to define what science literate adults should know about science, suggest that participants find an example of a science literacy idea all adults should know after their K–12 education. Have them individually, in pairs, or in triads, turn to the section on *Forces in Nature* on page 55 and find an idea related to gravity that every adult should know, regardless of whether they work in a science field or not. Give participants two minutes to share and discuss their finding with a partner. Ask for a volunteer from the group to share one idea.

2. Hold up *Science Matters*. Ask for a show of hands for the following questions: "How many people have used this book?" "How many have heard of it, but haven't used it?" "How many are learning about it for the first time?" Have participants quickly read through the description of *Science Matters* on page 25 of the CTS parent book. Ask someone to summarize the reading by volunteering an answer to the question, "What is the purpose of *Science Matters*?" Feel free to add any additional information from the description in Option 2. Now show Slide 3 for Option 3. If the purpose of *Science Matters* is to help nonscientist adults understand basic ideas about science presented in the media or other public venues, suggest

that participants find an example of a science concept that is explained in *Science Matters*. Have them individually, in pairs, or in triads browse through *Science Matters* to find an explanation of a scientific concept that a person without a background or degree in science could use to make sense of an article or a television broadcast that included that concept. Give participants two minutes to share and discuss their finding with a partner. Ask for a volunteer from the group to share one thing they found.

3. Next hold up *Benchmarks for Science Literacy*. Ask for a show of hands for the following questions: "How many people have used this book?" "How many have heard of it, but haven't used it?" "How many are learning about it for the first time?" Have participants quickly read through the description of *Benchmarks for Science Literacy* on page 25 of the CTS parent book. Ask someone to summarize the reading by volunteering an answer to the question, "What is the purpose of *Benchmarks for Science Literacy*?" Feel free to add any additional information from the description in Option 2. Now show Slide 4 for Option 3. If the purpose of *Benchmarks for Science Literacy* is to describe the steps along the way from kindergarten to Grade 12 in achieving the science literacy described in *Science for All Americans*, suggest that participants look at those steps that will lead to the goal of "science literacy for all Americans." Repeat that *Science for All Americans* describes on page 55 how

> everything in the universe exerts gravitational force on everything else, although the effects are readily noticeable only when at least one very large mass is involved (such as a star or planet). Gravity is the force behind the fall of rain, the power of rivers, the pulse of tides; it pulls the matter of planets and stars toward their centers to form spheres, holds planets in orbit, and gathers cosmic dust to form stars. (AAAS, 1989, p. 55)

Ask participants to turn to the parallel section on *Forces of Nature* on pages 94–97 in *Benchmarks*. Find an example of a learning goal from the bulleted list that describes one of the steps along the way to achieving an adult science literacy understanding of gravity. Give participants two minutes to share and discuss a finding with a partner. Ask for a volunteer from the group to share one learning goal.

4. Explain how *Benchmarks* also includes a K–12 overview essay on page 93 and grade span essays on pages 94–96 that help teachers consider instructional implications and sometimes suggest ways to teach the science ideas. Show Slide 5 for Option 3. Ask participants to find an example from either the overview essay or one of the grade level essays that addresses instructional implications. Give participants two minutes to share and discuss a finding with a partner. Ask for a volunteer from the group to share one thing they found.

5. Explain that *Benchmarks* also includes summaries from the Research Base in Chapter 15 that alert teachers to learning difficulties and common misconceptions. Show Slide 6 for Option 3. Have participants turn to page 340 of *Benchmarks* and ask them to find an example of a learning difficulty or misconception related to gravity. Give participants two minutes to share and discuss a finding with a partner. Ask for a volunteer from the group to share one thing they found.

6. Next hold up a copy of the *National Science Education Standards* (NRC, 1996). Ask for a show of hands for the following questions: "How many people have used this book?" "How many have heard of it, but haven't used it?" "How many are learning about it for the first time?" Have participants quickly read through the description of the *National Science Education Standards* on pages 25–26 of the CTS parent book. Ask someone to summarize the reading by volunteering an answer to the question, "What is the purpose of the *National Science Education Standards*?" Feel free to add any additional information from the description in Option 2. Now show Slide 7 for Option 3. Say that most of the content in *National Science Education Standards* overlaps with the *Benchmarks*. Like *Benchmarks*, the *National Science Education Standards* provides lists of learning goals. These learning goals are described as fundamental abilities, concepts, and principles. However, one of the key distinctions between the *Benchmarks* and *National Science Education Standards* is how the *National Science Education Standards* approach the abilities to do inquiry. Ask participants to examine the list of Abilities to Do Scientific Inquiry for their grade level. Choose one ability and share how the *National Science Education Standards* describe that ability. Give participants two minutes to share and discuss a finding with a partner. Ask for a volunteer from the group to share and describe one ability.

7. Describe how the *National Science Education Standards*, like the *Benchmarks*, also includes essays that can be used to inform instruction. Show Slide 8 for Option 3. Ask participants to quickly scan the essay on "Science as Inquiry" at their grade level and share just one example of how the essay can inform instruction. Give participants two minutes to share and discuss a finding with a partner. Ask for a volunteer from the group to share one thing they found.

8. Describe that unlike the *Benchmarks*, one of the features contained in the *National Science Education Standards* that are used in CTS Section II is the vignette. Show Slide 9 for Option 3. The vignettes show what the teaching and learning related to a science topic is actually like in a classroom. Give participants a few minutes to quickly scan through Chapter 6 of the Content Standards (pages 121–203) and find an example of a vignette. Give participants two minutes to share and discuss a finding with a partner. Ask for a volunteer from the group to share one idea.

9. Next hold up a copy of *Making Sense of Secondary Science*. Ask for a show of hands for the following questions: "How many people have used this book?" "How many have heard of it, but haven't used it?" "How many are learning about it for the first time?" Have participants quickly read through the description of *Making Sense of Secondary Science* on page 26 of the CTS parent book. Ask someone to summarize the reading by volunteering an answer to the question, "What is the purpose of *Making Sense of Secondary Science*?" Feel free to add any additional information from the description in Option 2. Now show Slide 10 for Option 3. Ask participants to turn to Chapter 24 of that book, "The Earth in Space." Ask them to find one example of a misconception related to the Earth-moon-sun or Earth-sun system. Give participants two minutes to share and discuss a finding with a partner. Ask for a volunteer from the group to share one finding.

10. Hold up the last resource, *Atlas of Science Literacy*, Volume 1, 2, or both. Ask for a show of hands for the following questions: "How many people have used this book?" "How many have heard of it, but haven't used it?" "How many are

learning about it for the first time?" Have participants quickly read through the description of the *Atlas of Science Literacy* on page 26 of the CTS parent book. Let participants know that the second volume of the *Atlas* came out in 2007, after the CTS parent book was published. Ask someone to summarize the reading by volunteering an answer to the question, "What is the purpose of the *Atlas of Science Literacy*?" Feel free to add any additional information from the description in Option 2. Now show Slide 11 for Option 3. Ask participants to browse through the *Atlas* and find a map related to a topic they teach. Note the conceptual strand names ("storylines") at the bottom and follow the map from bottom to top to see how the learning goals coherently progress from kindergarten to Grade 12 and the connections made within and across a topic. Give participants a few minutes to share and discuss a map with a partner. Ask for a volunteer from the group to share one interesting thing they found on an *Atlas* map.

11. Hold up a copy of a state's standards (preferably the state you are in). Show Slide 12 for Option 3. Have participants read through the description on page 27 of the CTS parent book. Ask someone to summarize the reading by volunteering an answer to the question, "What is the purpose of including state standards or district curriculum guides in CTS?" Feel free to add any additional information from the description in Option 2.

12. Answer any additional questions participants might have about the resources. Announce that participants will now use the resources they just explored to engage in the CTS process. Resume with the next slide in the module you are using.

DEVELOPING YOUR OWN SNAPSHOTS AND RESOURCE SCENARIOS

As discussed in the introduction to this chapter, you may wish to develop your own snapshots and resource scenarios tailored directly to the needs and interests of your participants. The following instructions can be used to create your own session-specific snapshots or resource scenarios. Templates for designing your own are included in the Chapter 4 folder, Design Your Own Templates, on the CD-ROM at the back of this book.

Designing Your Own Snapshots

Select the concepts or skills and grade level that address the needs of your audience.

1. Identify the CTS guide(s) that matches the concepts or skills.

2. Do your own brief CTS, using only the grade level readings for the audience with which you will be working. Take notes on CTS findings for each section that could be useful to your audience.

3. Print out and review the templates for designing your own snapshots found in the Chapter 4 folder, Design Your Own Templates, on the CD-ROM. Decide whether you want to use a snapshot format for questions similar to the one provided in Module A1, Handout A.1.5 (Snapshot Template 1—insert your own examples where indicated), or make up your own question starters (Snapshot Template 2).

4. Decide how many snapshots you would like to provide for your audience to investigate. A minimum of eight and no more than fifteen is suggested. Delete boxes on the Snapshot Templates (1 or 2) that you aren't using. Make sure you have a balance of questions across CTS Sections I through V.

5. If you are using Snapshot Template 2, scramble the snapshots so questions that use different sections of a CTS guide (and different topics if using more than one) are placed on the CTS snapshot template in different squares so they don't sequentially follow a CTS guide. The idea is to provide practice in identifying which CTS guide to use, which CTS section (I–V) to read, and what resource reading will answer the snapshot question. (*Note:* If you develop a set of snapshots on one topic, your participants will use only one CTS guide, so they can skip the step that involves identifying which CTS guide to use.)

6. Use the questions on pages 37 through 39 of the CTS parent book to guide your development of snapshot questions for Template 2, paying close attention to the purposes of CTS Sections I through V.

7. Create an answer sheet similar to Handout A1.7 with the Category, Topic Study Guide, and CTS Section used so participants can check whether they selected the appropriate section(s) as they work through the scaffold.

Designing Your Own Resource Scenarios

Select a CTS guide that matches the needs of your audience.

1. Do the CTS readings for the topic yourself, taking notes from the readings that you might use in the scenarios.

2. Print out and review the templates for designing your own scenarios found in the Chapter 4 folder on the CD-ROM. Decide whether you want to use a resource scenario format for questions similar to the one provided in Module A2 (see Resource Scenario Template 1) or make up your own question starters (see Resource Scenario Template 2).

3. Make sure you have questions that target the first five sections of the CTS (I–V), different grade levels as appropriate, and use the different resource books (or include only the books you have available). Insert your own examples where indicated.

4. Use the guiding questions on pages 37–39 of the CTS parent book to help guide your development of the questions used in your resource scenario.

5. Try out your questions by going back to the CTS reading to be sure the CTS findings will address your questions adequately. If not, modify your questions.

NEXT STEPS AFTER CTS INTRODUCTIONS

After participants have been introduced to CTS, they are ready to use the tools and resources to conduct a full topic study. Examples of designs for leading a full topic study are described in the next chapter.

5

Leading Full Topic Studies

INTRODUCTION TO FULL TOPIC STUDIES

In the previous chapter, we described ways to introduce the CTS process and the resources used with CTS. This chapter provides seven comprehensive modules for leading full topic study sessions that afford the opportunity to delve deeply into important science topics and directions for developing your own topic studies on science content important to your work. The modules are designed to be used with participants who already have some familiarity with CTS; however, they can be adapted to use with audiences who have not experienced one of the Chapter 4 introductory sessions. Each module is intended to show the variety of ways a full CTS can be designed and facilitated (see summary of modules in Table 5.1). We have provided a sample script you can follow or adapt accordingly for your audience or presentation style. In addition, the CD-ROM at the back of this book includes a folder with the PowerPoint slides to use with each of the modules as well as the accompanying handouts and facilitator resources.

Each of the module scripts, B1 through B7, uses the CTS Learning Cycle of Inquiry, Study, and Reflection described on pages 40–44 of Chapter 3 in this *Leader's Guide* as well as pages 36–37 in the CTS parent book. Icons are used in the script and the agenda to indicate the stage of the CTS Learning Cycle (see Table 5.2).

It is important to note that a grade span has been designated for each module. This grade span indicates where the topic is most likely to appear in the K–12 curriculum. As such, the study process includes the grade levels for the designated grade spans. However, all the modules can be used with a K–12 audience to help start a conversation among K–12 teachers about vertical articulation. In addition, state standards may differ from national standards in relation to where a topic appears. For example, in the national standards, the idea of conservation of matter culminates in Grade 8. In some state standards, conservation-of-matter ideas are included in the high school standards. Ultimately it is up to the facilitator to determine the appropriate grade level based on the K–12 progression of the state standards and the audience attending the session.

Table 5.1 Summary of CTS Topic Modules

Module	Content Domain	Grade Levels*	Time (hours)*	Application Used
B1: Experimental Design	Inquiry	3–12	3.5	Instructional scaffold
B2: Evidence and Explanation	Inquiry	K–12	3	Instructional framework
B3: Earth, Moon, and Sun System	Earth science and astronomy	K–8	3	Assessment probe with examples of student thinking
B4: Conservation of Matter	Physical science	K–8	2.75	Assessment probe with student work
B5: Atoms and Molecules	Physical science	K–12	2.75	Identifying phenomena
B6: Photosynthesis and Respiration	Biology	K–12	2.75	Pre- and post-content questionnaire
B7: Life Cycles	Biology	K–4	2.75	Comparison of curriculum units

* NOTE: Grade levels designate the grades where the knowledge and skills are emphasized in the standards. Grade levels can be modified according to the need and context of the audience. Times are approximate. Audiences with no or little experience with CTS will need additional time.

Table 5.2 CTS Learning Cycle Icons

Icon	CTS Learning Cycle Stage
	Engagement
	Elicitation
	Exploration
	Development

(Continued)

Table 5.2 (Continued)

Icon	CTS Learning Cycle Stage
	Synthesis
	Application
	Reflection and Self-Assessment

In addition to the seven modules, this chapter provides suggestions for developing your own full topic study as well as ways to combine two topics for an integrated study. These are not included as modules since they are not topic specific. The following are brief summaries of the seven modules in this chapter that are described in Table 5.1 and the sections that provide information for designing your own full and combined topic studies.

Full CTS Modules (Half-Day Sessions)

Module B1: Experimental Design

In this session, participants will use the CTS guide, "Experimental Design," to examine the skills and understandings of designing scientific experiments. The module helps participants understand that experiments are a particular type of scientific investigation, examine an important skill of inquiry that can be difficult for students—identifying and controlling variables—and recognize that "fair tests" are elementary science precursors to formal experimentation. Participants will connect their CTS findings to an instructional scaffold used in upper elementary grades through high school (with modifications) that guides students through the steps of designing their own experiments that involve identifying and controlling variables.

Module B2: Evidence and Explanation

In this session, participants will use the CTS guide, "Evidence and Explanation," to understand what constitutes evidence in science and how the use of explanations, supported by evidence, are a central feature of scientific inquiry. The module helps participants examine students' conceptions of explanations (or theories) and the evidence for them. Participants will apply their CTS findings to an instructional framework that can be used to support middle and high school students' construction of scientific explanations.

Module B3: Earth, Moon, and Sun System

In this session, participants will use the CTS guide, "Earth, Moon, and Sun System," to examine Earth-sun and Earth-moon-sun system concepts, including relative size and position of the Earth, sun, and moon; seasons; rotations and orbits; phases of the moon; tides; and eclipses. An example of student work is used to show common misconceptions students have about what causes the phases of the moon.

Module B4: Conservation of Matter

In this session, participants engage in a full topic study of an important principle in science—conservation of matter. The module helps participants see how a coherent sequence of conservation-related ideas develops in students from Grades K to 8. Participants also examine students' commonly held ideas related to conservation of matter.

Module B5: Atoms and Molecules

In this session, participants will use the CTS guide, "Particulate Nature of Matter," to examine an unfolding K–12 conceptual curriculum strand beginning with objects and materials and working up to an understanding of substances, atoms and molecules, and parts of atoms and molecules. The module also includes an application of identifying phenomena that can be explained using a particle model.

Module B6: Photosynthesis and Respiration

In this session, participants engage in a full topic study of photosynthesis and respiration. This study will develop an appreciation for the connections between plants and animals, their nutritional needs, the role of these processes in ecosystems, and the inextricable link between structure, function, and needs of organisms. The session will help teachers at all grade levels make better informed choices about the key ideas, from simple to complex, taught at different grade levels. It also helps identify the instructional decisions that may help challenge students' misconceptions and move them toward a science literate understanding of these processes. The module includes a pre- and post-assessment questionnaire to elicit teachers' preconceptions related to these processes.

Module B7: Life Cycles

In this CTS module, participants engage in a K–4 topic study using a modified version of the CTS guide, "Reproduction, Growth, and Development (Life Cycles)." The module helps participants gain an understanding of the important ideas for teaching about life cycles. Participants use their findings to make and discuss comparisons between two different elementary science life cycle units and connect their findings to their own instructional materials.

Additional Suggestions for Designing and Leading Full Topic Studies

This section of the chapter provides guidance and suggestions for how to use CTS templates and guidelines to develop your own full topic studies as well as how two topics can be combined into one full, combined study.

The next sections in this chapter provide the facilitation guides for the seven comprehensive CTS full topic study modules (B1–B7). The handouts, facilitator resources, and

PowerPoint presentations for each module are located in the Chapter 5 folder on the CD-ROM at the back of this book. The following table (Table 5.3) provides an at-a-glance summary of the full topic study modules.

Table 5.3 CTS Full Topic Study Sessions at a Glance

Module	Time Needed	Grade Level	Content	When to Use
Module B1: Experimental Design	3–4 hours	3–12 (may also be used with K–2 teachers with a focus on how the inquiry concepts can be developed in primary grades)	Experimental design and identifying and controlling variables	Group wishes to enhance instruction and student understanding in the area of experimental design; teachers need instructional tools on experimental design.
Module B2: Evidence and Explanation	3 hours	K–12	What constitutes scientific evidence and how to construct explanations supported by evidence	Group wishes to enhance their own and their students' understanding of the importance of using evidence and constructing evidence-based explanations.
Module B3: Earth, Moon, and Sun System	2.5–3 hours	K–8	Earth, moon, and sun systems concepts, including relative size and position of the Earth, sun, and moon; seasons, rotations and orbits; phases of the moon; tides; and eclipses	Teachers wish to deepen understanding of the scientific ideas around the Earth, moon, and sun systems and enhance teaching of these ideas in Grades K–8.
Module B4: Conservation of Matter	2.5–3 hours	K–8	How the understanding of the scientific idea of conservation of matter develops over Grades K–8	Teachers wish to clarify the K–8 content for conservation of matter and examine the research on learning to understand difficulties students have with this topic.
Module B5: Atoms and Molecules	2.5–3 hours	K–12	The particulate nature of matter, moving from understanding of parts and wholes of objects to idea of atoms and molecules and parts of atoms	Teachers wish to expand their knowledge of the particulate model of matter and consider the instructional implications for helping students understand the arrangement and behavior of particles.

Module	Time Needed	Grade Level	Content	When to Use
Module B6: Photosynthesis and Respiration	2.5–3 hours	K–12	The connections between plant and animals, the role of photosynthesis and respiration in ecosystems, the links among structure, function, and needs of organisms	Teachers wish to understand how understanding of these processes develops over the K–12 grades and the conceptual difficulties students have related to photosynthesis and respiration.
Module B7: Life Cycles	2.5–3 hours	K–4	Explore content of life cycles, including the concept that plants and animals have life cycles and life cycles are different for different organisms	Elementary teachers who wish to focus instruction on important ideas for teaching about life cycles and learn what to look for in instructional materials that support development of the important science ideas.

Module B1

"Experimental Design" Facilitation Guide

BACKGROUND INFORMATION

Description of the Module

In this CTS module, participants use the CTS guide, "Experimental Design," to distinguish scientific experiments from other types of scientific investigations and consider the implications for teaching and learning. The module helps participants examine an important skill of inquiry that research indicates can be difficult for students: identifying and controlling variables. The "Experimental Design" topic study clarifies the understandings and skills that are developed at different grade spans and levels of complexity. Participants have an opportunity to connect their CTS findings to the use of an instructional scaffold that guides students through the process of designing their own experiments (Goldsworthy, 1997). The scaffold serves as an application in which participants can see how CTS results are reflected in the use of an instructional resource.

Audience

This session is primarily for Grade 3–12 teachers. The instructional scaffold used in the session is designed to be used with upper elementary students (after they have been exposed to the idea of a fair test) through high school students. K–2 teachers can benefit from learning how the inquiry concepts, skills, and experiences they provide their students form the foundation for designing controlled experiments.

Purpose and Goals

The overall purpose of this module is to learn about the important skills and knowledge students need to be able to design experiments and understand the concept of experimentation. The goals for this module are to

- clarify content and teaching implications related to an understanding of experimentation and the abilities necessary to design scientific experiments,
- examine research on learning to anticipate and understand difficulties students have in designing their own experiments, and
- link CTS findings to an instructional scaffold that can be used to guide students in designing experiments.

SESSION DESIGN

Time Required

Allow 3 to 4 hours (include a 15-minute break). This is an approximate time and will be adjusted based on the experience and needs of your group.

Agenda at a Glance

Welcome and Engagement (5 minutes)

- Welcome and goals of the session (1 minute)
- Engagement table talk (4 minutes)

Elicitation (15 minutes)

- Frayer model (5 minutes)
- Report out (10 minutes)

Exploration (40–50 minutes)

- Brainstorm: Components of scientific experiments (10 minutes)
- Preparing for the CTS (10 minutes)
- Quiet read (20–30 minutes plus allow a 10- to 15-minute break)

Development and Synthesis (40–55 minutes)

- Clarifying discussion (20–30 minutes)
- Synthesis of key points (20–25 minutes)

Applications (40–55 minutes)

- Revisiting the Frayer model (10 minutes)
- Examining the instructional scaffold (20–30 minutes)
- Applying findings to state standards (10–15 minutes)

Reflection and Evaluation (10 minutes)

- New insights (5 minutes)
- Evaluation (5 minutes)

MATERIALS AND PREPARATION

Materials Needed by Facilitator

CTS Parent Book

Science Curriculum Topic Study: Bridging the Gap Between Standards and Practice (Keeley, 2005)

Resource Books

One copy of each of the following resource books:

- *Science for All Americans* (American Association for the Advancement of Science [AAAS], 1989)
- *Benchmarks for Science Literacy* (AAAS, 1993)
- *National Science Education Standards* (National Research Council [NRC], 1996)
- *Atlas of Science Literacy* (Vol. 1) (AAAS, 2001)

CTS Module B1 PowerPoint Presentation

The Module B1 PowerPoint presentation is located in the Chapter 5 folder on the CD-ROM at the back of this book. Review it and tailor it to your needs and audience. Insert your date and location on Slide 1, add additional graphics as desired, and add your own contact information on the last slide.

Supplies and Equipment

- Computer and LCD projector to show PowerPoint presentation.
- Flip chart easel, pads, and markers.
- Paper for participants to take notes.
- At least two different colored pads of small (2×1.5 inch) sticky notes per table. These are used to place over the text boxes on Handout B1.5.

Wall Charts

- Prepare and post five sheets of chart paper in the meeting room prior to the session labeled as follows: *CTS Reminders*, *CTS Content Questions* (for description of these two charts, see Module A1 in Chapter 4, p. 66), *Key Words*, *Science Experiments*, and *Frayer Model* (a chart of Handout B1.2).

Facilitator Preparation

- Conduct your own "Experimental Design" CTS prior to leading the session and make a list of key points from the study or create a summary sheet to guide your facilitation.
- Identify connections to state standards (as necessary for your group).
- Choose the option for distributing and sharing readings (see Chapter 3 of this *Leader's Guide*, pp. 48–50) that best matches your audience.
- Decide whether to use the seed germination example with the scaffold or substitute your own example relevant to the curriculum used by participating teachers.
- Prepare an evaluation form to collect feedback from participants.

Materials Needed for Participants

(*Note:* All handouts can be found in the Chapter 5 folder on the CD-ROM.)
- Handout B1.1: Agenda.
- Handout B1.2: Scientific Experiment Frayer Model.
- Handout B1.3: Experimental Design Reading Assignments.
- Handout B1.4: To Hypothesize or Not to Hypothesize.
- Handout B1.5: Student Experiment Scaffold.
- Handout B1.6: Linking CTS Findings to the Scaffold.
- Copies of the PowerPoint presentation for Module B1 (on CD-ROM). Insert your own graphics and additional information you wish to add. (Optional)
- If participants do not have their own CTS parent books, make copies for each participant of the CTS guide, "Experimental Design," on page 235 of the CTS parent book.
- If enough resource books are not available, make copies of the selected readings from the CTS guide, "Experimental Design," on page 235 of the CTS parent book.

DIRECTIONS FOR FACILITATING THE MODULE

Welcome, Introductions, and Engagement (5 minutes)

Engagement Table Talk (5 minutes)

Show Slide 1. Welcome the participants and explain the purpose of the session. Provide an opportunity for quick introductions. Show Slide 2. Review the goals for the session. Ask participants to briefly share why experimental design is an important topic for science teachers to study and what they hope to gain.

Elicitation (15 minutes)

Frayer Model (15 minutes)

Show Slide 3 and refer to Handout B1.2: Scientific Experiment Frayer Model. Tell participants that in order to examine the skills and understandings of designing scientific experiments, we need to develop a common understanding of the term *scientific experiment;* often people have different interpretations of this concept. Explain that we will use an instructional strategy called the Frayer Model to document our prior knowledge about the term scientific experiment (Frayer, Frederick, & Klausmeier, 1969). Explain that a Frayer Model is a type of graphic organizer that helps students develop conceptual relationships and examples associated with a vocabulary word. In science, it provides students an opportunity to explain and elaborate with examples their understanding of a scientific concept or technical term (Allen, 2007). The scientific term or word (scientific experiment) is placed in a circle in the center of the worksheet. Participants provide their own operational definition of a scientific experiment in the upper left box and characteristics or features of a scientific experiment in the upper right box. Participants then list examples of scientific experiments, contrasted with examples that are not considered scientific experiments. Ask everyone to look at Handout B1.2: Scientific Experiment Frayer Model. Point out that a Frayer Model has four sections: operational definition, characteristics, examples, and nonexamples. Give the group five minutes to fill in the four sections on the concept scientific experiments. They may want to divide the sections and have a pair take one section each to work on and then exchange ideas at their table on each section. After about five minutes, ask for each group to give an answer to one section. Record their answers in the appropriate section on the *Frayer Model* chart you prepared and posted. Keep the chart posted and explain that the group will go back to their clarification of the term scientific experiment after they gain more information from the CTS readings.

Exploration (40–50 minutes)

Brainstorm (10 minutes)

Show Slide 4, leading the group through a brainstorm. Explain that before they begin the CTS, they need to be clear about the components of a scientific experiment: which components involve designing only, which involve carrying out the experiment, and which could be applied to both designing and carrying out experiments. Explain that the reason they are doing this is to focus on just the design aspects during the study, not the full experiment that would include carrying it out. For example, it is sometimes important in an assessment or classroom activity for students to be able to describe the design of an experiment without necessarily carrying it out. Begin by having participants call out all the things students do when they engage in scientific experiments. List them on the chart you prepared labeled *Science Experiments.* Guide participants to include components that represent the entire experimentation process (e.g., asking a question, identifying variables, collecting data, communicating results, etc.). Point out that some of these activities relate to *designing* experiments, some to *conducting* experiments, and some may apply to both. By quick consensus, explain that they will place a letter *D* next to those components involved in design, a letter *C* next to those involved in conducting an experiment, and letters *D and C* next to those involved in both. After participants assign the letters *D, C, or D and C* as a group to the components listed on the chart, tell the group that for this topic study, they will be focusing just on those aspects related to designing an experiment—those with the letter *D* (and some *D and Cs*—use your judgment to determine which of these to include) next to them. This will help participants focus their readings on the design elements of experimentation only.

Preparing for the CTS (10 minutes)

Explain that in order to examine the skills and understandings of designing scientific experiments and to consider the difficulties students have with experimental design, we will conduct a curriculum topic study. Have participants turn to the CTS guide, "Experimental Design," on page 235 in the CTS parent book or refer to the copy of page 235 you prepared and point out that the CTS guide provides the resources and page numbers for the readings on the topic experimental design. Point out that the guide, "Experimental Design," was used to create Handout B1.3: Experimental Design Reading Assignment in order to guide participants who are new to doing a full CTS. Show Slide 5 and go over each of the reading assignments on Handout B1.3. Make sure participants know that they will not be reading everything listed on the guide; for example, they will not read the vignettes in Section IIB of their study guide during this session (they may read them at their leisure after the session). Describe how the readings will be distributed and shared, based on the option for organizing and discussing readings you selected from Chapter 3 of this *Leader's Guide* (pp. 48–50). Make sure all participants are clear about their reading assignments.

> **Facilitator Note**
>
> If the group is large and you chose to do a jigsaw, it is suggested that you break up the Section II and Section III readings by grade spans.

> **Facilitator Note**
>
> You may wish to take a 15-minute break after the readings or build break time into the time allocated for reading.

Quiet Read (20–30 minutes plus a 15-minute break)

Ask participants to read their sections quietly. Adjust the time according to the option selected (e.g., if participants wish to read all the sections for their grade level plus Sections I, IV, and V, you will need to build in more time). Encourage participants to take notes that are relevant to the purpose of the section they are reading that apply only to the experimental design process. Point out the brainstorm chart to remind them they are not focusing on skills and understandings involved in conducting the experiment (e.g., they will skip over the sections that refer to organizing or analyzing the data, as that involves carrying out the experiment).

 ## Development and Synthesis (40–55 minutes)

Clarifying Discussion (20–30 minutes)

Depending on the option selected for distributing and sharing the readings, participants will work in groups to discuss their readings. Remind participants to refer to Handout B1.3, which describes what they should focus on during their reading.

 ### Synthesis of Key Points (20–25 minutes)

Show Slide 6. After groups have discussed their readings, ask each group to develop a list of at least five of the most important points that came from their CTS readings on experimental design. The key points should focus on major ideas that came from the study as a whole. Ask each table group to write their key points on chart paper and post them on the wall. Then invite everyone to walk around the room in small groups to do a "gallery walk." Ask them to examine the key points on each chart and look for similarities and differences of the key points from chart to chart. After the gallery walk, ask the large group what they noticed, what the similarities and differences were in the charts, and what key words were explicitly included in the standards in Section III that are part of the scientific language of experimentation and at what grade level. Post these words on the wall chart labeled *Key Words*, along with the grade level, and explain that they will come back to them later.

> **Facilitator Note**
>
> "Key words" might include *experiment*, *hypothesis*, *variable*, *control*, and so forth.

 ## Application (40–55 minutes)

Revisiting the Frayer Model (10 minutes)

Show Slide 7. Have participants look back at what they wrote on Handout B1.2: Scientific Experiment Frayer Model. In their small groups, ask them to make revisions based on any new knowledge gained from doing the CTS. Share a few ideas and, as a group, decide on a working definition of *scientific experiment* and share some of the characteristics, examples, and nonexamples. Make sure participants understand that an experiment is a type of scientific investigation that involves systematically testing a hypothesis or prediction but that not all investigations are experiments (e.g., observational field studies).

Examining the Instructional Scaffold (20–30 minutes)

Show Slide 8. Tell the group that they will use what they learned from their CTS readings and discussions to examine an instructional scaffold that can be used to guide novices through the process of designing an experiment. Explain that this scaffold was adapted from the inquiry wall posters developed by Ann Goldsworthy (1997) in the United Kingdom (it was further adapted by the Colorado Department of Education for their Goals 2000 Inquiry Toolkit). Explain that a scaffold is an instructional tool, not a methodology that may imply a rigid set of steps in doing science. Be sure participants know that the scaffold is intended to help students who struggle with the steps involved in setting up an experiment and controlling variables, but that they as educators need to make sure students understand this does not mean there is a single lock-step method all scientists follow when engaged in investigation. Reinforce this concern by pointing out the findings from CTS that alert teachers to the danger of teaching a rigid scientific method. The scaffold is a type of representation for modeling the steps in designing an experiment, but like any model or representation, it does not fully capture the process and has some drawbacks. This scaffold guides students through the basic process of experimentation. It does not address more complex types of experiments that involve multiple variables or control groups such as those used for comparison in clinical trials. It also does not capture how scientists often cycle back and forth when testing their ideas.

Explain that germinating (or sprouting) bean seeds is a common activity used to help students understand how living things are dependent on physical factors in their environment. We will use this as our example for exploring the use of the scaffold and connecting it to our CTS findings.

Refer participants to Handouts B1.5 and B1.6. Explain that the group will work through the scaffold using the bean seed

> **Facilitator Note**
>
> The germinating bean seeds example is one that most teachers are familiar with. Since this part of the CTS does not involve doing the experiment, it is best to select a familiar example. Depending on the background of your group and the context in which they teach, feel free to substitute the germinating bean seeds example with one that ties to the instructional materials, curricular goals, or grade level of your teachers.

germination example. As they go through the sections of the scaffold, encourage participants to make notes on Handout B1.6 of any linkages they see between the steps in the scaffold and the findings from the CTS readings and discussion, including the experimental design words listed on the *Key Words* chart you generated earlier.

Explain how inquiry often begins with wondering about some object, event, or process in the natural world. This wondering is often coupled with accessing prior knowledge or experiences related to the phenomenon at hand. Show Slide 9 to illustrate an example of a student's "wondering" about seeds that initiates the inquiry and leads to the design of an experiment.

Show Slide 10. Explain that this part of the scaffold helps students identify the different things that might affect seed germination. In Part A of Handout B1.5, have them list all the things that could be changed when investigating what seeds need to germinate.

Ask participants to write these factors on colored sticky notes, one note for each factor (use the same color sticky note for all factors—we will call it "Color A"). Place the sticky notes on the boxes in the scaffold on Handout B1.5, as shown by the blue boxes on Slide 11.

Ask participants what kinds of things could be measured or observed during seed germination.

Ask participants to place four things they identified that can be measured or observed in each of the four boxes using a different colored sticky note ("Color B") as shown by the yellow boxes on Slide 12. Debrief this part by explaining that students are brainstorming possible things that can be measured or observed. They have not yet refined their experiment to focus on two experimental variables.

Show Slide 13 and refer to the "Identifying Variables" section of the scaffold. Explain that now students are beginning to isolate a manipulated and responding variable. Ask them to imagine that students decided they wanted to know how the amount of water affects germination. Guide participants into moving the Color A sticky note, amount of water, into the box under "I will change." Ask them to imagine the students decided they would count how many seeds in a given sample sprouted. Guide participants to move the Color B sticky note, number of seeds sprouted, into the box under "I will measure or observe." Now have participants pick up and move the other three Color A sticky notes into the three boxes under "I will not change." Explain how the type of soil, type of seeds, and temperature will all stay the same (as well as other factors that may have been identified). The only thing that will change is the amount of water. Have them repeat this with the three remaining Color B sticky notes, showing that they do not need to observe or measure these things.

Pause and ask participants to reflect on their CTS readings, findings, and discussion and make the link between what they have done so far with the scaffold and how it reflects the CTS findings. Have them record their ideas on Handout B1.6. It is important for them to take the time to reflect on these linkages between the instructional tool and their findings from CTS so that they can better understand the value and appropriate use of the scaffold. Encourage them to cite the research from CTS Section IV that describes the difficulty students have identifying and controlling variables and how the scaffold guides them in developing this skill.

Facilitator Note

Guide participants to include amount of water on their lists. Other examples might include type of seed, type of soil, temperature, amount of time for sprouting, and amount of light.

Facilitator Note

You need not be limited by the four boxes on the handout—use more if desired by extending the layout of the boxes.

Facilitator Note

Guide participants to make sure the number of bean seeds sprouted is included. Other examples are the size of the sprouts, mass of the sprouts, color of sprouts, and the number of days they take to sprout.

Facilitator Note

Again, you need not be limited by only four boxes—use more if desired by extending the layout of the boxes.

Have participants refer to the "Asking a Testable Question" section of their scaffold on page 3 of Handout B1.5. Explain that they will now move the two colored notes from the previous steps into these boxes to frame a testable question. They will use the question frame to write their question that will guide the experiment. Show Slide 14 as an example of framing a testable question. Point out that this is a time when they access their prior knowledge and experience to think about what they already know in relation to their question. Pause and have participants link this step (on Handout B1.6) to their CTS findings. Ask what they found in their CTS reading that supports this important step for students?

Show Slide 15. Ask participants to pull out Handout B1.4: To Hypothesize or Not to Hypothesize, which they read or discussed during the CTS reading time. Ask them to consider the language they examined in the standards (Handout B1.3, Section II reading) and recorded on the *Key Words* chart. Ask the following questions:

- At what grade level would you expect students to make a hypothesis versus a prediction?
- When should they do neither and just ask a question?
- How do students' prior experiences influence what you would expect from them?

Discuss whether participants think their own students would be making a hypothesis or prediction or posing a question. Pause to complete the linkage to CTS on Handout B1.6: Linking CTS Findings to the Scaffold on the row labeled "Developing a Prediction or a Hypothesis."

Show Slide 16 and refer to the "Develop the Procedure" part of the scaffold. Have them move the colored sticky notes from the previous section onto the boxes provided and complete the procedure. Point out that for older students, new terminology is introduced in the scaffold (*dependent* and *independent variable*). Ask participants why the terminology wasn't introduced earlier (refer to *Key Word* chart developed earlier). Ask your group to discuss in small groups why a conceptual understanding of the two types of variables is needed before bringing in the scientific terminology. After a few minutes of discussion, ask for ideas from the groups. Elaborate on what was said as needed and suggest that these two terms (*dependent* and *independent* or *responding* and *manipulated* variable) are often used in textbooks and some activities without developing conceptual understanding of the difference between the two types of variables and that the terminology goes beyond the language used in the standards but is appropriate if teachers feel their students have had the opportunity to conceptually distinguish between the two terms. Memorization does not ensure students can conceptually understand the words.

Explain that they will stop at that point since the next part of the scaffold deals with carrying out the experiment (the rest of the scaffold can be used with other CTS inquiry-related topics). Go back to the initial *Science Experiments* wall chart that shows where they drew the boundaries for the study "Experimental Design."

Give participants a few minutes to look over and complete Handout B1.6 and make any additions. Show Slide 17 and ask for comments on the scaffold. Suggested questions for the discussion:

- How did the scaffold reflect the findings from CTS?
- What cautions should you be aware of when using this scaffold? (It is important for teachers to make sure students understand that every investigation does not have to be an experiment; an experiment is a particular kind of investigation that includes testing ideas that often involve identifying and controlling variables; there is not one particular method all scientists use, nor is it a linear process. For example, astronomers are not able to do experiments like physicists but often rely on observing large samples of objects.)

> **Facilitator Note**
>
> It is particularly important to emphasize that the scaffold addresses a common difficulty indicated in the research—identifying and controlling variables. It is also important to draw out from participants the importance of developing the notion of a fair test with younger students before asking them to design controlled experiments that involve identifying variables.

- At what grade level(s) would you use this scaffold? What modifications would you make for younger students, based on the CTS findings and your own context?

Point out that the understandings of experimental design are just as important as the abilities to design an experiment. For example, students might be able to use the scaffold to identify and control variables, but it is also important that they understand why this is necessary in an experiment. The CTS involves examining both the skills and the understandings.

Additional Facilitator Notes

1. Consider having teams create CTS summaries as a record of their session. Examples of the "Experimental Design" summary notes can be found in the Chapter 3 examples of CTS summaries folder on the CD-ROM at the back of this book.

2. See Chapter 7 for an example of how the CTS guide, "Experimental Design," was used to introduce participants to the CTS study group professional development strategy.

3. This module can be adapted to focus on a single grade span, 3–5, 6–8, or 9–12, keeping in mind the importance of examining the precursor ideas (CTS Sections III and V) and instruction (CTS Section II) that happen in the previous grade span.

4. Encourage participants to search the CTS supplementary database for additional articles, Web sites, and materials that can further extend their understanding of experimental design.

5. The *Understanding Science* Web site at http://undsci.berkeley.edu/ is a comprehensive resource funded by the National Science Foundation that addresses how science is done, including experimentation. There is extensive material on this site that may be useful to facilitators who wish to extend this module.

Applying Findings to State Standards (10–15 minutes)

Show Slide 18. If participants have their state standards available or are using common instructional materials, go to CTS Section VI. Ask participants to turn to the section in their state standards that addresses experimental design. What are students in their state expected to know and do at different grade levels? How does this reflect (or not reflect) the CTS findings? How could the scaffold be used to support student achievement of these learning goals? In the alternative, have participants find an example of an experiment students do in their instructional materials. Ask how the study of experimental design clarifies what students are expected to do in the instructional materials. Discuss how the scaffold may be used to support the instructional materials. Discuss how the overall CTS guide, "Experimental Design," helped them clarify the meaning and intent of their state standards or instructional objectives stated in their curriculum or instructional materials. Ask for any other observations or comments.

 ### Reflection and Evaluation (10 minutes)

New Insights (10 minutes)

Show Slide 19. Ask participants to share a few new insights they gained from the CTS and examination of the scaffold. Address any questions, thank participants for coming, and ask them to complete a session evaluation form.

Module B2

"Evidence and Explanation" Facilitation Guide

BACKGROUND INFORMATION

Description of the Module

In this session participants use the CTS guide, "Evidence and Explanation," to understand what constitutes evidence in science and how the use of explanations, supported by evidence, is a central feature of scientific inquiry.

Audience

This session can be used with a K–12 audience. The instructional framework used in the application part of this module is intended for use by middle and high school teachers. However, K–4 teachers will find it helpful in identifying the skills and knowledge that form the elementary level foundation for understanding evidence and explanation.

Purpose and Goals

The overall purpose of this module is to use CTS to examine the central role evidence and explanations play in supporting and communicating ideas and developing an understanding of scientific principles. The goals for this module are to

- clarify content and instructional implications related to identifying and using evidence and constructing evidence-based explanations,
- examine research on learning to anticipate and understand difficulties students have in identifying evidence, analyzing arguments, and constructing explanations, and
- link CTS findings to a framework that can be used to help students construct better explanations.

SESSION DESIGN

Time Required

Allow about 3 hours (include a 15-minute break). This is an approximate time and will be adjusted based on the experience and needs of your audience.

Agenda at a Glance

Welcome and Engagement (5 minutes)

- Welcome and goals of the session (1 minute)
- Engagement table talk (4 minutes)

Elicitation (10 minutes)

- Accessing prior knowledge and experience (10 minutes)

Exploration (45 minutes)

- Preparing for the CTS (10 minutes)
- Quiet read (20–30 minutes plus allow a 10- to 15-minute break)

Development and Synthesis (45–60 minutes)

- Clarifying discussion (20–30 minutes)
- Synthesis of key findings (25–30 minutes)

Application (45 minutes)

- What happens to the mass? (10 minutes)
- Explanation framework (15 minutes)
- Linking back to CTS (10 minutes)
- Applying findings to state standards (10 minutes)

Reflection and Evaluation (15 minutes)

- I used to think…(10 minutes)
- Evaluation (5 minutes)

MATERIALS AND PREPARATION

Materials Needed by Facilitator

CTS Parent Book

Science Curriculum Topic Study: Bridging the Gap Between Standards and Practice (Keeley, 2005)

Resource Books

One copy of each of the following resource books:

- *Science for All Americans* (American Association for the Advancement of Science [AAAS], 1989)
- *Benchmarks for Science Literacy* (AAAS, 1993)
- *National Science Education Standards* (National Research Council [NRC], 1996)
- *Atlas of Science Literacy* (Vols. 1–2) (AAAS, 2001–2007)

CTS Module B2 PowerPoint Presentation

The Module B2 PowerPoint presentation is located in the Chapter 5 folder on the CD-ROM at the back of this book. Review it and tailor it to your needs and audience. Insert your date and location on Slide 1, add additional graphics as desired, and add your own contact information on the last slide.

Supplies and Equipment

- Computer and LCD projector to show PowerPoint presentation
- Flip chart easel, pads, and markers
- Paper for participants to take notes

Wall Charts

- Prepare and post three sheets of chart paper in the meeting room prior to the session labeled as follows: *CTS Reminders, CTS Content Questions* (for description of these two charts, see Module A1 in Chapter 4, pp. 63–71), and *CTS Key Words.*

Facilitator Preparation

- PowerPoint slides for Module B2 (on CD-ROM). Insert your own graphics and additional information as desired. (Optional: Print out copies of the PowerPoint slides for participants.)
- Conduct your own "Evidence and Explanation" CTS prior to leading the session.
- Identify connections to state standards if necessary.
- Choose the option for distributing and sharing readings (see Chapter 3, pp. 48–50) that best matches your audience.
- Decide whether to use the "What Happens to the Mass?" scenario or substitute your own example related to the content teachers are learning in your professional development setting.
- Prepare an evaluation form to collect feedback from participants.

Materials Needed for Participants

(*Note:* All handouts can be found in the Chapter 5 folder on the CD-ROM.)

- Handout B2.1: Agenda at a Glance
- Handout B2.2: Evidence and Explanation Reading Assignments
- Handout B2.3: What Happens to the Mass?
- Handout B2.4: Explanation Framework
- PowerPoint slides for Module B2 (optional)
- If participants do not have their own CTS parent books, make copies of the CTS guide, "Evidence and Explanation," on page 234 for each participant.
- If enough resource books are not available, make copies of the selected readings from the CTS guide, "Evidence and Explanation," on page 234 in the CTS parent book that are used in Handout B2.2.

DIRECTIONS FOR FACILITATING THE MODULE

Welcome, Introductions, and Engagement (5 minutes)

Engagement Table Talk (5 minutes)

Show Slides 1, 2, and 3. Welcome the participants and explain the purpose of this session. Review the goals for the session and review the agenda. If participants in the group do not know one another, provide an opportunity for quick introductions. Ask participants to briefly share why evidence and explanation are important topics for science teachers to study and what they hope to gain from engaging in the CTS.

Elicitation (10 minutes)

Accessing Prior Knowledge and Experience (10 minutes)

Show Slide 4. Ask participants to brainstorm a list of instances in their everyday lives, not related to science, where students are asked to explain something. Record their responses on chart paper. Now ask them to brainstorm instances where students may be asked to explain something scientifically, that is, provide a scientific explanation. Chart examples. Ask the group for differences between the two types of explanations. What distinguishes a *scientific explanation* from the other types of explanations? How does the everyday meaning of *explain* differ from the meaning used in science? Tell the group that CTS may help them gain new ideas and considerations for teaching about evidence and explanation that will help students (and teachers) distinguish a scientific explanation from other types of explanations, particularly the difference between opinion and evidence-based explanations or the difference between description and explanation.

 ## Exploration (45 minutes)

Preparing for the CTS (10 minutes)

Have participants look at the CTS guide, "Evidence and Explanation," on page 234, and Handout B2.2, which designate the CTS readings and provide focus for the study. Explain that in order to examine the skills and understandings related to evidence and explanation and consider the difficulties students have with these skills and concepts, we will conduct a CTS on this topic. Show Slide 5 and go over each of the reading assignments. Describe how the readings will be distributed and shared, based on the option you selected in Chapter 3 (pp. 48–50). If the group is large and you choose to do a jigsaw for dividing the readings among participants, it is suggested that you break up the Section II and Section III readings by grade spans. Make sure everyone is clear about which section(s) they are reading and remind them that the page numbers are listed on their CTS guide.

> **Facilitator Note**
>
> Adjust the reading time according to the option selected (e.g., if participants wish to read all the sections for their grade level plus Sections I, IV, and V, you will need to build in more time). You may provide a 15-minute break after the readings or build a break into the reading time.

Quiet Read (20–30 minutes plus 15-minute break)

Ask participants to read their sections quietly using Handout B2.2 as a reminder of their assignment and focus. Encourage them to take notes that are relevant to the purpose of the section they are reading that apply only to the skills and concepts related to evidence and explanation.

 ## Development and Synthesis (45–60 minutes)

Clarifying Discussion (20–30 minutes)

After the readings are completed, ask participants to work in small groups of about five to six people to discuss their readings and clarify understandings related to the topic. Remind participants to refer to Handout B2.2, which describes what they should focus on during their reading.

 ### Synthesis of Key Findings (25–30 minutes)

> **Facilitator Note**
>
> Key words might include *theory, claim, logical argument*, and so forth.

Show Slide 6. After groups have discussed their readings, ask participants to form "role-alike" groups of three to four people who have similar roles, such as teachers who teach at the same grade level or within a few grades (e.g., seventh- and eighth-grade teachers) or people with nonclassroom roles (e.g., professional developer, informal educator). Have each role-alike group develop a list of the most important key findings for their role that came from their CTS on evidence and explanation. The key findings can come from each or any of the five sections of the CTS guide. Ask each role-alike group to write their key findings on chart paper and post them on the wall (organize sections by K–2, 3–5, 6–8, 9–10, and nonclassroom teachers). Have everyone get up and in small groups do a "gallery walk" to examine the key findings on the posted charts, looking for similarities and differences that emerged from their CTS readings. After the gallery walk, invite the whole group to share any observations. Ask the group to report out the key words and grade level explicitly included in the standards in Section III that are part of the scientific language of evidence and explanation. Post these words on the *CTS Key Words* chart, along with the grade level at which these words are introduced and explain that they will go back to them later.

 ## Application (45 minutes)

Analyzing Claims in the Media (10 minutes)

Show Slide 7. Based on their readings, how would they analyze this claim and respond to it? What flaws do they detect in this argument? Give participants time to link their findings on detecting flaws in arguments to the statements on Slide 7.

> **Facilitator Note**
>
> The context used in this example is appropriate for Grade 6 and up.

Explain that claims in the media need to be supported by an explanation. Note that now they will look at what is involved in constructing explanations. Refer participants to Handout B2.3. Describe the scenario and ask participants to write an explanation based on how they think a science literate middle school student might respond. Provide a few minutes for participants to write their responses and share their examples at their tables.

Explanation Framework (15 minutes)

Ask participants to share how and when they think students usually learn how to identify evidence and use it to construct explanations. Ask the following questions:

- Is the skill of constructing a good explanation ever explicitly taught?
- If so, when and how?
- What three main components of constructing a scientific explanation should be explicitly taught to students?

Show Slide 9 and ask participants to share their responses with a partner. Now show Slide 10 and explain that these are the three components of a scientific explanation that are used in a framework for constructing explanations. Show Slide 11 and explain that furthermore, Project 2061 developed a new benchmark, which is in the *Atlas of Science Literacy* (Vol. 2) "Communication Skills" map that explicitly addresses the idea of a claim, evidence, and reasoning to construct an explanation. Refer participants to Handout B2.4: Explanation Framework. Explain that a group from the University of Michigan originally developed this work (Sutherland et al., 2006). Point out the citation at the bottom of the handout and post the following Web site to let people know they can find more information on this framework in a chapter by Katherine McNeill and Joseph Krajcik through the NSTA Learning Center at http://learningcenter.nsta.org/product_detail.aspx?id=10.2505/9781933531267.11. Ask participants to go back to Handout B2.4: Explanation Framework and invite them to revise their response using this framework. Ask participants to share at their tables how the framework, combined with their CTS knowledge, helped them think differently and revise the explanation they wrote on Question 3 on Handout B2.3. Share examples of sufficient and appropriate evidence (e.g., more than one type of phase change, mass data before and after the change, more than one type of substance, recognition of a closed system). Show Slide 12 as an example of an explanation constructed using the framework as a guide. Ask for evidence that the student made a claim, selected sufficient and appropriate evidence, and linked the evidence to the claim through reasoning.

Additional Facilitator Notes

1. Consider having teams create CTS summaries as a record of their session. Examples of the "Evidence and Explanation" summary notes can be found in the Chapter 3 folder on the CD-ROM.

2. This module can be adapted to focus on a single grade span, K–2, 3–5, 6–8, or 9–12, keeping in mind the importance of examining the precursor ideas (CTS Sections III and V) and instruction (CTS Section II) that happen in the previous grade span. If used with teachers below fifth grade, use the Explanation Framework for teacher learning. Explain that the framework is not intended for students below middle school.

3. Encourage participants to search the CTS supplementary database for additional articles, Web sites, and materials that can further extend their understanding of evidence and explanation.

4. This topic study can be followed by a book study of *Ready, Set, Science!* (Michaels, Shouse, & Schweingruber, 2008), which can be used to further delve into CTS Sections II and IV based on recent research intended for the classroom. *Ready, Set, Science!* describes the types of instructional approaches, based on current research, that help K–8 students learn and understand science. It describes four strands of proficiency that build upon the *Benchmarks for Science Literacy* and the *National Science Education Standards.* Two of these four strands are Strand 1: Understanding Scientific Explanations and Strand 2: Generating Scientific Evidence.

Linking Back to CTS (10 minutes)

Show Slide 13. Give participants time to discuss how the findings from CTS are reflected in the framework and how they might use the framework with their students (or teachers).

Applying CTS Findings to State Standards (10–15 minutes)

Show Slide 14. If participants have their state standards available or are using common instructional materials, link to CTS Section VI and discuss how the overall curriculum topic study on evidence and explanation helped them clarify the meaning and intent of their state standards. Ask them to identify learning goals related to evidence and explanation in their state standards and compare them to the specific ideas and skills described in the CTS. How does the CTS help teachers better understand or interpret their state standards? Which ideas and skills from CTS are included in the state standards? Which ideas and skills are missing from the state standards that are important to teach? How can the framework help students attain the knowledge and skills related to evidence and explanation that are required in the state standards? Ask them to look for examples of activities in their instructional materials that require students to use evidence and develop explanations. Discuss how the CTS findings and use of the framework can help teachers teach more effectively with their instructional materials. Ask for any other comments about how CTS can help them implement their state standards or curriculum materials.

 Reflection and Evaluation (15 minutes)

Used to Think (10 minutes)

Show Slide 15. Ask participants to fill in the blanks to complete the statement, based on what they gained from this module. Ask for a few examples to share with the group.

Evaluation (5 minutes)

Answer any questions, thank participants, and ask them to complete a session evaluation.

Module B3

"Earth, Moon, and Sun System" Facilitation Guide

BACKGROUND INFORMATION

Description of the Module

In this CTS module, participants engage in a K–8 full topic study on the Earth, moon, and sun system. Participants examine Earth-moon-sun system concepts, including relative size and position of the Earth, moon, and sun; seasons; rotations and orbits; phases of the moon and tides; eclipses; and gravity. The CTS creates greater awareness of the types of experiences and representations that can help students understand these challenging concepts. The information from this CTS will help teachers make informed choices about what important ideas should be taught at different grade levels and the instructional strategies that may help students change their strongly held alternative conceptions.

Audience

This session is primarily for teachers in Grades K–8 who teach concepts related to the Earth, moon, and sun. Most standards address the Earth-moon-sun ideas up through middle school. Although they have traditionally been deemed appropriate for K–8, the concepts continue to challenge high school students and even teachers. Depending on your audience, this module can be extended to K–12 or used to build teachers' knowledge of these important concepts.

Purpose and Goals

The overall purpose of this module is to help participants gain an understanding of the important Earth, moon, and sun concepts that are recommended for instruction at the different grade levels. The goals for this module are to

- deepen conceptual understanding of the scientific ideas related to the Earth, moon, and sun system in the K–8 curriculum;
- use standards and research to examine teaching and learning implications related to the Earth, moon, and sun system; and
- consider how results from the CTS can be used to inform curriculum and instruction.

SESSION DESIGN

Time Required

Allow 2.5 to 3 hours. This is an approximate time and will be adjusted based on the experiences and needs of your audience.

Agenda at a Glance

Welcome and Engagement (25 minutes)

- Welcome, introductions, and goals (5 minutes)
- Establishing a purpose (10 minutes)
- Examining students' ideas about moon phases (10 minutes)

Elicitation (20–30 min)

- Concept topic categories (5–10 minutes)
- Brainstorm "before CTS" teaching and learning goals (15–20 minutes)

Exploration (30 min)

- Preparing for the CTS (10 minutes)
- CTS individual readings (20 minutes)

Development (40–55 minutes)

- Developing an understanding of the topic through CTS discussion (25–35 minutes)
- Revisiting initial ideas (15–20 minutes)

Synthesis (15–20 minutes)

- Gallery walk (10 minutes)
- Synthesis of key points (5–10 minutes)

Application (15–20 minutes)

- Applying findings (15–20 minutes)

Reflection and Evaluation (15 minutes)

- Quick write and discussion (10 minutes)
- Evaluation (5 minutes)

MATERIALS AND PREPARATION

Materials Needed by Facilitator

CTS Parent Book

Science Curriculum Topic Study: Bridging the Gap Between Standards and Practice (Keeley, 2005)

Resource Books

One copy of each of the following resource books:

- *Science for All Americans* (American Association for the Advancement of Science [AAAS], 1989)
- *Benchmarks for Science Literacy* (AAAS, 1993)

- *National Science Education Standards* (National Research Council [NRC], 1996)
- *Making Sense of Secondary Science* (Driver, Squires, Rushworth, & Wood-Robinson, 1994)
- *Atlas of Science Literacy* (Vol. 1) (AAAS, 2001)

CTS Module B3 PowerPoint Presentation

The Module B3 PowerPoint presentation is located in the Chapter 5 folder on the CD-ROM at the back of this book. Review it and tailor it to your needs and audience as needed. Insert your date and location on Slide 1, add additional graphics as desired, and add your own contact information on the last slide.

Supplies and Equipment

- Computer and LCD projector to show PowerPoint presentation
- Flip chart easel, pads, and markers
- Paper for participants to take notes

Wall Charts

- Prepare and post the *CTS Reminders* and *CTS Content* Questions charts in meeting room prior to session (for description of these two charts, see Module A1 in Chapter 4, pp. 63–71).
- Post five charts around the room to create five stations. Label with the following:

  1. Earth's Rotation: Day/Night Cycle

  2. Solar and Lunar Eclipses

  3. Tides

  4. Phases of the Moon

  5. Seasons

Facilitator Preparation

- Conduct your own "Earth, Moon, and Sun System" CTS prior to leading the session, including Section VI to identify connections to state standards if necessary.
- Review Facilitator Resource B3.4: CTS Facilitator Summary for the Earth, Moon, and Sun System, included in the Chapter 5 folder on the CD-ROM.
- Choose the option for distributing and sharing readings (see Chapter 3, pp. 48–50) that best matches your audience.
- Prepare an evaluation form to collect feedback from participants.

Materials Needed for Participants

(*Note:* All handouts can be found in the Chapter 5 folder on the CD-ROM.)

- Handout B3.1: Agenda at a Glance
- Handout B3.2: Sample Eighth Grade Responses: Going Through a Phase
- Handout B3.3: Earth, Moon, and Sun CTS Reading Assignments
- PowerPoint slides for Module B3 (on CD-ROM). Insert your own graphics and additional information you wish to add. (Optional)

Facilitator Note

The *Science for All Americans* Chapter 10 reading in Section IA and *Benchmarks* Chapter 10 readings in Sections IIA and IIIA are not used in this module but can be added if this module is used primarily with middle and high school teachers.

- If participants do not have their own CTS parent books, make copies of the CTS guide, "Earth, Moon, and Sun System," on page 194 for each participant.
- If enough resource books are not available, make copies of the selected readings from the CTS guide, "Earth, Moon, and Sun System," on page 194 that are used in Handout B3.3.

DIRECTIONS FOR FACILITATING THE MODULE

 ### Welcome, Introductions, and Engagement (30–35 minutes)

Welcome and Introductions (5 minutes)

Show Slide 1. Welcome the participants, do quick introductions if participants do not know each other, and explain that the session will engage them in using curriculum topic study to deepen their knowledge about the Earth, moon, and sun system and consider how to apply what they learn to improve the teaching and learning of this important and often misunderstood topic.

Establishing a Purpose and Examining Students' Ideas (25–30 minutes)

Show Slide 2. Point out the goals of the session. Mention that although most of the focus of this study will be on K–8, since that is where the emphasis is in the national standards, there are high school curricula and state standards that include the ideas they will examine that may be relevant to high school teachers in the audience. Furthermore, high school students and even adults still struggle with these concepts, even after being taught them. Ask participants to write down two or three personal learning goals they would like to achieve during the session to improve their understanding of the Earth, moon, and sun system. Ask them to save their list of personal goals to refer to at the end of the session.

Show Slide 3. Tell participants in this study they will be examining ideas and teaching suggestions that involve the Earth, moon, and sun system. The Earth-sun system is also included in this study. Show Slide 4 and explain that phases of the moon are just one topic students encounter when learning about the Earth, moon, and sun system. Begin by asking, "How many remember learning about the phases of the moon when they were in elementary school? Middle school? High school?" Briefly discuss the questions on the slide as a warm-up prior to looking at the student responses.

Refer everyone to Handout B3.2: Sample Eighth Grade Responses: Going Through a Phase (from *Uncovering Student Ideas in Science*, Keeley, Eberle, & Farrin, 2005). Show Slide 5. Ask participants to read the probe and select the explanation they think is the best answer (Sophia's response). Show Slide 6 and refer participants to the student responses on Handout B3.2. Have participants turn to a partner at their tables and discuss the students' ideas using the questions on Slide 6. Point out that students can even choose the right answer but have flawed reasoning. For example, Student B selected the best explanation (Sophia's) and seems to have a partial understanding of the positional relationship. However, the student still has a major misconception about shadows causing the phases of the moon, compounded by ideas about the angle of reflection. Have participants briefly share why they think students have different ideas about what causes the phases of the moon. Explain that CTS can help us further examine this learning problem as well as other Earth, moon, and sun learning difficulties.

Elicitation of Prior Knowledge (20–30 minutes)

Concept and Topic Categories (5–10 minutes)

Show Slide 7. Ask participants what they think the major K–8+ concepts or topics are that make up the Earth, moon, and sun system topic in the standards. Start them off with the example, phases of the moon. Ask them to generate a few concepts related to Earth, moon, and sun and call them out while you record them on chart paper. Then reveal the list of five concepts and topics on Slide 8 that will be the focus of this CTS. Explain that for the purpose of this topic study, they are going to focus on five major concepts of the Earth, moon, and sun system: (1) Earth's rotation, including the day/night cycle; (2) solar and lunar eclipses; (3) tides; (4) phases of the moon; and (5) seasons. Explain that our study of the Earth, moon, and sun system will focus on these major five concepts because they include or are related to the key ideas in the standards and have available research on children's ideas.

Brainstorm "Before CTS" Teaching and Learning Goals (15–20 minutes)

Divide participants evenly into groups by the five major Earth, moon, and sun concepts listed. Have participants meet with their groups and generate brainstormed lists following the directions on Slide 9 (without using the CTS resources yet) of what they think the important teaching suggestions and learning goals are in their category based on their own beliefs and prior knowledge about standards and research on learning. Encourage them to think about ideas not only explicit to the concept but also ideas that contribute to an understanding of the concept (e.g., gravity, light reflection, orbits). After they generate their lists, ask them to indicate where in the K–12 sequence they think students should be expected to learn the ideas or experience the teaching suggestions the groups generated. Have them develop a *Before CTS* wall chart that lists the ideas about teaching and learning they generated for their concept category by grade span K–2, 3–5, 6–8, and 9–12. Remind them to watch the time and record as many ideas as possible in the time given, honoring all voices without passing judgment on others' ideas. This is a brainstorm. Later, after completing the CTS, they will reexamine their brainstormed lists. Facilitators may want to remind participants of the norms for brainstorming.

Show Slide 10. Announce that they are now going to explore the Earth, moon, and sun system topic by doing a modified CTS to examine content, curricular, and instructional considerations related to the Earth, moon, and sun system. Refer participants to the CTS guide on page 194 of the CTS parent book (or use a handout copy) and explain that they will use selected (not all) readings from the CTS guide for this work. To guide the readings, refer participants to Handout B3.3: Earth, Moon, and Sun CTS Reading Assignments, which identifies the selected sections from the CTS guide they will focus on during this session. Remind them that if at any time they have content or CTS process questions, they can write them on sticky notes and post them on the wall chart labeled *CTS Questions*.

> **Facilitator Note**
>
> Depending on your goals, you may suggest that your participants do the readings for the other CTS sections listed on the guide on their own after the session. The historical episodes may be of particular interest and relevance to middle and high school teachers.

 ## Exploring the Topic Using CTS Readings (35–50 minutes)

CTS Individual Readings (15–20 minutes depending on option selected)

Have participants stay in their concept category group. Select one of the six options for distributing and sharing readings described on page 48 in Chapter 3 of this *Leader's Guide*. Make sure groups know that all eight of the readings must be covered in their group. (*Note:* The connection to state standards and instructional materials will be discussed by the whole group after examining the selected CTS sections.) Give participants time to read their assigned sections on Handout B3.3. Remind them to look for findings related to the five concept categories representing the Earth, moon, and sun system ideas. Remind participants as they read to look for CTS findings that directly describe the concept or include ideas that contribute to an understanding of the concept (e.g., gravity, light reflection, orbits, etc.).

 Developing an Understanding of the Topic Through CTS Discussion (20–30 minutes)

CTS Discussion

After finishing their assigned readings, have participants discuss each of the readings in their small groups as they relate to their concept category. Remind participants to refer to Handout B3.3, which describes what they should focus on during their reading. Time and format of the discussion will depend on which of the options for distributing and sharing the readings was selected. Spend about 5 to 10 minutes to pull out the main ideas from the study as a whole group discussion. (Refer to Facilitator Resource B3.1, Summary Chart of CTS Findings for the main points that should be addressed.)

> **Facilitator Note**
>
> Point out that when participants decide to discard ideas, it does not mean these ideas are not important or can't be taught after students have achieved the basic ideas, but rather they are not the ideas that are first needed for science literacy.

> **Facilitator Note**
>
> Provide markers a different color from those that participants used to make their brainstormed list so they can easily see changes they made in the chart, or provide colored dots for groups to use to mark their *Before CTS* charts. For example, they can use colored markers to make a large green dot next to the idea that they want to keep, a red dot for something they will discard, and orange for ideas they will keep but move to a different grade span or change the wording. Or use the following key: K for keep as is, X for discard, and M for modify.

Revisiting Initial Ideas (15–20 minutes)

After concluding a debrief discussion of the five concepts, ask the small groups to return to their brainstormed concept charts posted on the wall and reconsider their initial ideas about teaching and learning related to their assigned concept. Show Slide 11. Ask the group to reconsider their list of learning goals and teaching ideas based on their CTS readings and discussions. Ask the groups to decide which things on the list to keep at a grade level, based on the study results, which ones to move to a different grade level, and which ones to discard as not necessary for science literacy on this topic.

Ask groups to decide which terminology to leave in or change, as well as any changes in the wording of the goals to better match the ideas in the standards. Also ask them to add any new ideas they gained from doing CTS. Have each group be prepared to give a short presentation to summarize how CTS changed or enhanced their initial thinking about the Earth, moon, and sun category they were originally assigned.

 Topic Synthesis (15–20 minutes)

Gallery Walk (15–20 minutes)

Have each category group appoint a spokesperson or two to share one or two insights they gained from their CTS that changed or enhanced their initial ideas. Gather the larger group around each chart (or break into smaller groups if you have more than twenty participants). Allow about two minutes of talk at each chart and then move the group to the next one until all have visited all five categories.

Do a short recap of what you observed in terms of the groups' learning. Bring up any key points that may not have been raised by the groups. Ask participants to share any other questions or comments about what they did and what they learned.

Application of CTS Findings and Reflection (15–20 minutes)

Applying Findings to the Participants' Own Work (10–15 minutes)

Ask participants to step back from the CTS they have just completed to think about the value of CTS. Allow time for several people to share their insights. If needed, point out that one value of CTS is that it can make us more aware of science content we don't understand as well as we would like and affords the chance to explore what we need to know to better serve our students. Address any of the CTS questions that may have been posted on the *CTS Questions* chart. Refer to the question on Slide 11. Allow time for a short discussion of ideas at tables. What are some ideas for using this CTS and results in their work?

Reflection (5 minutes)

Revisit the individual professional development goals identified at the beginning of the session. Show Slide 12. Have participants do a quick write and share with a partner the responses to the questions on the slide. Ask for a few final comments from the group.

Wrap-Up and Evaluation (5 minutes)

Evaluation (5 minutes)

Invite any questions and then ask participants to complete an evaluation.

Module B4

"Conservation of Matter" Facilitation Guide

BACKGROUND INFORMATION

Description of the Module

In this session, participants engage in a full topic study of an important principle in science, conservation of matter. The module helps participants see how a coherent sequence of conservation-related ideas develops from kindergarten to Grade 8. Participants also examine students' commonly held ideas related to conservation of matter.

Audience

This session is designed for a Grade K–8 audience. Although the national standards do not extend conservation-of-matter learning goals beyond Grade 8, this session can be used with a K–12 audience so that high school teachers can examine K–8 science literacy ideas that prepare students to apply conservation-of-matter reasoning in biology, chemistry, and Earth science.

Purpose and Goals

The overall purpose of this module is to examine the progression of K–8 ideas about conservation of matter and the teaching and learning implications related to this important scientific principle. The goals for this module are to

- clarify K–8 content and instructional implications related to conservation of matter,
- examine research on learning to anticipate and understand difficulties students have related to conservation of matter,
- identify a coherent sequence of K–8 conservation-related ideas and their connections to other topics, and
- apply CTS findings to teachers' instructional contexts.

SESSION DESIGN

Time Required

Allow about 2.5 to 3 hours. This is an approximate time and will be adjusted based on the experience and needs of your audience.

Agenda at a Glance

Welcome and Engagement (10 minutes)

- Welcome and goals of the session (2–3 minutes)
- Engagement table talk (7 minutes)

Elicitation (10 minutes)

- Accessing prior knowledge and experience (10 minutes)

Exploration (50 minutes)

- Preparing for the CTS (10 minutes)
- Quiet read (20–30 minutes plus allow a 10- to 15-minute break)

Development and Synthesis (40 minutes)

- Clarifying discussion (25 minutes)
- Synthesis of important points (15 minutes)

Application (35–40 minutes)

- Looking at student work: Ice Cubes in a Bag (20 minutes)
- Looking at other contexts (15–20 minutes)

Reflection and Evaluation (15 minutes)

- Quick write (10 minutes)
- Evaluation (5 minutes)

MATERIALS AND PREPARATION

Materials Needed by Facilitator

CTS Parent Book

Science Curriculum Topic Study: Bridging the Gap Between Standards and Practice (Keeley, 2005)

Resource Books

One copy of each of the following resource books:

- *Science for All Americans* (American Association for the Advancement of Science [AAAS], 1989)
- *Benchmarks for Science Literacy* (AAAS, 1993)
- *National Science Education Standards* (National Research Council [NRC], 1996)
- *Making Sense of Secondary Science* (Driver, Squires, Rushworth, & Wood-Robinson, 1994)
- *Atlas of Science Literacy* (Vol. 1) (AAAS, 2001)

CTS Module B4 PowerPoint Presentation

The Module B4 PowerPoint presentation is located in the Chapter 5 folder on the CD-ROM. Review it and tailor it to your needs and audience. Insert your date and location on Slide 1, add additional graphics as desired, and add your own contact information on the last slide.

Supplies and Equipment

- Computer and LCD projector to show PowerPoint presentation.
- Flip chart easel, pads, and markers.
- Paper for participants to take notes.
- Sentence strips (one for every pair or participants). Either buy sentence strips or make them by cutting multiple sheets of chart paper into thirds.

Wall Charts

- Prepare and post the *CTS Reminders* chart and the *CTS Content Questions* chart in meeting room prior to session (for description of these two charts, see Module A1 in Chapter 4, pp. 63–71).

Facilitator Preparation

- Conduct your own "Conservation of Matter" CTS prior to leading the session.
- Identify connections to state standards if necessary.
- Collect student work using the probe Ice Cubes in a Bag (Handout B4.2) or use the optional student work in Handouts B4.4 and B4.5.
- Choose the option for distributing and sharing readings (see Chapter 3, pp. 48–50) that best matches your audience.
- Prepare an evaluation form to collect feedback from participants.

Materials Needed for Participants

(*Note:* The handouts are available in the Chapter 5 folder on the CD-ROM.)

- Handout B4.1: Agenda at a Glance.
- Handout B4.2: Ice Cubes in a Bag Probe.
- Handout B4.3: Conservation of Matter Reading Assignments.
- Handout B4.4: Ice Cubes in a Bag—Grade 4 Responses (or use your own student work).
- Handout B4.5: Ice Cubes in a Bag—Grade 7 Responses (or use your own student work).
- Handout B4.6: Looking at Other Contexts.
- PowerPoint slides for Module B3 (on CD-ROM). Insert your own graphics and any information you wish to add. (Optional)
- If participants do not have their own CTS parent books, make copies of the CTS guide, "Conservation of Matter," on page 163 for each participant.
- If enough resource books are not available, make copies of the selected readings from the CTS guide, "Conservation of Matter" on page 163 that are used in Handout B4.3.

DIRECTIONS FOR FACILITATING THE MODULE

Welcome, Introductions, and Engagement (10 minutes)

Welcome and Engagement Table Talk (10 minutes)

Show Slides 1, 2, and 3. Welcome the participants and explain the purpose of this session. Go over the agenda. If participants in the group do not know one another, provide an opportunity for quick introductions. Review the goals for the session. Ask participants to briefly share why conservation of matter is an important topic for science teachers to study and what they hope to gain from studying this topic.

Elicitation (10 minutes)

Accessing Prior Knowledge (10 minutes)

Refer to Handout B4.2: Ice Cubes in a Bag Probe. Have participants discuss the probe using the questions on Slide 4. Remind them to draw upon their own prior knowledge. Explain that this CTS session will build understanding of how ideas related to conservation of matter coherently develop from kindergarten to Grade 8 and how conservation of matter can be a challenging topic for students of all ages and adults to apply in various contexts they encounter in the natural world. Let them know that later on in the session they will be looking at student responses to the Ice Cubes in a Bag Probe and examining different contexts in which conservation of matter ideas are applied.

 Exploration (50 minutes)

Preparing for the CTS (10 minutes)

Have participants turn to the CTS guide, "Conservation of Matter," on page 163, and Handout B4.3, which provides focus for the readings. Show Slide 5 and review each of the reading assignments and the purpose of each CTS guide section (I–VI). Point out that the learning goals for this topic culminate in eighth grade in the national standards (although some state standards include conservation of mass at the high school level). The conservation of matter (or mass) principle should be revisited and reinforced throughout high school in connection with other learning goals in biological, chemical, and Earth science contexts. Describe how the readings will be distributed and shared, based on the option you selected.

Quiet Read (40 minutes including a 15-minute break)

Participants will now conduct a quiet read, depending on which option for distributing readings was selected. Adjust the time according to the option selected (e.g., if participants wish to read all the sections, you will need to build in more time).

Encourage participants to take notes that are relevant to the purpose of the section they are reading that apply only to the knowledge needed to understand conservation of matter. Handout B4.3: CTS Conservation of Matter Reading Assignments reminds them of their focus. If some participants finish their reading(s) before others, encourage them to read other sections.

> **Facilitator Note**
>
> You may wish to take a 15-minute break after the readings or build it in to the reading time by telling participants they will have 40 minutes to use to complete the readings and take a break.

 Development and Synthesis (40 minutes)

Clarifying Discussion (25 minutes)

Depending on the option selected for distributing and sharing the readings, participants will discuss each of the readings in small groups and clarify findings related to the topic. Remind participants to refer to Handout B4.3, which describes what they should focus on during their reading.

 Synthesis of Important Points (15 minutes)

Show Slide 6. After groups have discussed each of the readings, ask them to break into pairs or triads and list three to five important key points that emerged from the "Conservation of Matter"

CTS that are important to their work. Ask each pair or triad to pick one key point related to their findings from the CTS that they will be sure to consider in their own work and write it on the sentence strips of paper provided. Have each group briefly share their sentence strip and then post the sentence strip on the wall for all to read. Group similar strips. Point out commonalities or unique ideas that emerged that will inform how participants approach this topic with their students or teachers with whom they work.

 ## Application (30–40 minutes)

Looking at Student Work (20 minutes)

Show Slide 7 and refer participants back to Handout B4.2: Ice Cubes in a Bag Probe. Based on their study of conservation of matter, ask them to identify the learning goal(s) from CTS Section III or V related to this probe. Ask them which learning goal(s) seem to be most aligned with the context of this probe. Based on the CTS results, ask them what commonly held ideas they might anticipate from the student work. Refer them to Handouts B4.4 and B4.5 or have them use their own student work if they brought it to this session. Give them time to examine the student responses and note any differences between the elementary and middle school responses. Have participants discuss how the students' responses reflect the research findings. Ask them to suggest how they would use the results of the CTS to address the students' ideas if these were their own students.

Looking at Other Contexts (15–20 minutes)

Show Slide 8 and refer to Handout B4.6, which shows five other examples of probes that address conservation of matter from the books, *Uncovering Student Ideas in Science* (Vols. 1–4) (Keeley et al., 2005–2009). Explain how analyzing these probes will help show how to connect the appropriate context to a CTS learning goal as well as identify phenomena that can be explained using conservation-of-matter ideas. Give the groups time to examine and discuss the probes and questions on the slide. To get them started, use the Ice Cubes in a Bag Probe (Handout B4.2), which was previously examined. Point out how the probe is set in the context of a physical change involving a change in state of a substance (water). The phenomenon is ice melting in a closed container. If participants do not know what phenomena are, explain that phenomena are real-world objects, systems, and events that provide evidence of key ideas. In this case, observing that the mass stays the same after the ice melts provides evidence that mass does not change during a change in state. The probe is related to the K–2 4B learning goal on the *Atlas* map: "Water can be a liquid or solid and can go back and forth from one form to another. If water is turned into ice and then the ice is allowed to melt, the amount of water is the same as it was before freezing" (AAAS, 2001, p. 57). Even though this learning goal targets the basic idea that the mass does not change when ice melts or water freezes, the example is at a higher level of sophistication that matches the 6–8 benchmark, "No matter how substances within a closed system interact...the total mass of the system remains the same. The idea of atoms explains the conservation of matter. If the number of atoms stays the same no matter how they are rearranged, then their total mass stays the same" (AAAS, 1993, p. 79). Although this phenomenon does not apply to a rearrangement, such as what would happen during a chemical change, it is at a higher level of sophistication because it involves the concept of mass, targets a substance rather than an object or material, involves a closed system, and can be explained using the idea of atoms. The CTS results indicate the importance of developing the concept of atoms by Grade 8 in order to explain phenomena. The research also points out that students often mistake the features of a change, such as the water spreading out in the bag, or the appearance of ice as being more compact with a change in mass. The *Atlas* map helps us see how the conceptual strand "changes in state" develops from kindergarten to Grade 8 in relation to the topic of conservation of matter with the simpler Grade K–2 idea that the amount of water is the same when it changes from ice to water, to the Grade 3–5 idea that some features change when other features stay the same, to the Grade 6–8 idea that addresses change within a closed system explained using the idea of atoms. Discuss how this probe could be modified for the K–2 idea by simply asking students if the amount of water stays the same, rather than using a concept like mass that may be unfamiliar. Have them link the context of the example probes to key ideas from CTS Section III and Section IV research on learning.

Reflection and Evaluation (15 minutes)

Show Slide 9. Ask participants to do a quick written response to the question, *How will you use the results of this CTS?* Ask for a few volunteers to share their next steps. Ask participants to reflect back on the personal session goal they identified. End with Slide 10, an early contemplation on conservation of matter, especially as it applied to gases. Answer any questions, thank participants, and ask them to complete an evaluation.

Additional Facilitator Notes

One issue that will often come up with these examples is the use of the word *mass* versus *weight*. It is not the intent of the assessment probes to use these interchangeably. It is important to point out that for the purpose of understanding students' conservation reasoning about matter conservation, and the fact that we are on Earth, the familiar term weight is appropriate in this context as we are not assessing whether students know the difference between weight and mass. Point out how the research indicates students sometimes confuse the word *mass* with *massive*, which is why the probes and the learning goals are careful not to use the word *mass* until students have a conceptual understanding of what mass is. For the purpose of this CTS, it is not worth getting into a long, protracted conversation of the merits of using *weight* versus *mass*. To understand the key ideas in this topic, it is not necessary to distinguish between the two terms. Ultimately it is up to the teacher to decide which terminology to use based on their understanding of their students.

Module B5

"Atoms and Molecules" Facilitation Guide

BACKGROUND INFORMATION

Description of the Module

In this session, participants use the CTS guide, "Particulate Nature of Matter (Atoms and Molecules)," to examine the K–12 growth in understanding related to the particle model of matter, beginning with basic parts of objects to the abstract notion of atoms and molecules to even smaller parts of atoms. Participants examine the importance of helping students develop a particle model of matter that builds in sophistication from one grade span to the next. Participants use the research on commonly held ideas to compare differences between students' preconceived ideas and misunderstandings about particulate matter and a scientific particle model of matter that is used to explain various natural phenomena in science.

Audience

This session can be used with a K–12 audience or modified to focus on a specific grade span. Although atoms and molecules are not formally introduced until late middle school, a basic notion of dividing matter into smaller pieces begins in elementary school with parts of objects. The module also helps elementary teachers clarify the grade level content that helps them set the stage for students to learn about atoms and molecules in later grades.

Purpose and Goals

The overall purpose of this module is to examine the important role the particle model of matter has in K–12 science teaching and learning. The goals for this module are to

- understand the central role the particle model of matter has in predicting the behavior and properties of matter,
- identify instructional contexts and specific ideas related to atoms and molecules that begin in K–2 with parts of objects and culminate at the high school level with parts of an atom,
- examine research on learning to anticipate and understand the conceptual models students use to explain the arrangement and behavior of particles, and
- apply CTS findings to participants' own curriculum.

SESSION DESIGN

Time Required

Allow about 2.5 to 3 hours (include a 15-minute break). This is an approximate time and will be adjusted based on the experience and needs of your audience.

Agenda at a Glance

Welcome and Engagement (10 minutes)

- Welcome and goals of the session (2 minutes)
- Aluminum foil questions (8 minutes)

Elicitation (10 minutes)

- Activating prior knowledge: Matter is made of atoms (10 minutes)

Exploration (40–50 minutes)

- Preparing for the CTS (10 minutes)
- Quiet read (20–30 minutes plus allow a 10-minute break)

Development and Synthesis (60–75 minutes)

- Clarifying discussion (30–45 minutes)
- Synthesis of key findings (30 minutes)

Application (15 minutes)

- Applying CTS to participants' own contexts (15 minutes)

Reflection and Evaluation (15 minutes)

- Think-Pair-Share (10 minutes)
- Evaluation (5 minutes)

MATERIALS AND PREPARATION

Materials Needed by Facilitator

CTS Parent Book

Science Curriculum Topic Study: Bridging the Gap Between Standards and Practice (Keeley, 2005)

Resource Books

One copy of each of the following resource books:

- *Science for All Americans* (American Association for the Advancement of Science [AAAS], 1989)
- *Science Matters* (Hazen & Trefil, 1991, 2009)
- *Benchmarks for Science Literacy* (AAAS, 1993)
- *National Science Education Standards* (National Research Council [NRC], 1996)
- *Making Sense of Secondary Science* (Driver, Squires, Rushworth, & Wood-Robinson, 1994)
- *Atlas of Science Literacy* (Vol. 1) (AAAS, 2001)

CTS Module B5 PowerPoint Presentation

The Module B5 PowerPoint presentation is located in the Chapter 5 folder on the CD-ROM. Review it and tailor it to your needs and audience. Insert your date and location on Slide 1, add additional graphics as desired, and add your own contact information on the last slide.

Supplies and Equipment

- Computer and LCD projector to show PowerPoint presentation
- Flip chart easel, pads, and markers
- Paper for participants to take notes
- Small strips of aluminum foil (about 6 inches by 1 inch)—one strip per pair of participants

Wall Charts

- Prepare and post the *CTS Reminders* chart and the *CTS Content Questions* chart in meeting room prior to session (for description of these two charts, see Module A1 in Chapter 4, pp. 63–71).

Facilitator Preparation

- Conduct your own "Particulate Nature of Matter (Atoms and Molecules)" CTS prior to leading the session.
- Identify connections to state standards if necessary.
- Choose the option for distributing and sharing readings (see Chapter 3, pp. 48–50) that best matches your audience.
- Place strips of aluminum foil on the tables.
- Prepare an evaluation form to collect feedback from participants.

Materials Needed for Participants

(*Note:* Handouts are in the Chapter 5 folder on the CD-ROM.)

- Handout B5.1: Agenda at a Glance.
- Handout B5.2: Matter Is Made of Atoms.
- Handout B5.3: Reading Assignments.
- Handout B5.4: Explaining the Natural World Using Particle Ideas.
- PowerPoint slides for Module B5 (on CD-ROM). Insert your own graphics and any information you wish to add. (Optional)
- If participants do not have their own CTS parent books, make copies of the CTS guide, "Particulate Nature of Matter (Atoms and Molecules)," on page 169 for each participant.
- If enough resource books are not available, make copies of the selected readings from the CTS guide, "Particulate Nature of Matter (Atoms and Molecules)," on page 169 that are used in Handout 3.
- Prepare your own agenda sheet with time built in for breaks or other components unique to your context.
- Make up a session evaluation form using your own format or one of the suggested examples on the CD-ROM.

DIRECTIONS FOR FACILITATING THE MODULE

Welcome, Introductions, and Engagement (10 minutes)

Welcome Participants (5 minutes)

If participants do not know one another, provide an opportunity for quick introductions. Show Slides 2 and 3. Share the goals for the session and have participants briefly discuss what they hope to gain from this session. Review the agenda.

Aluminum Foil Questions (5 minutes)

Working in pairs, ask participants to pick up a strip of aluminum foil from their tables. Show Slide 4. Ask them to describe the foil. Do they see any parts that make up the foil? If they were to put on glasses that could magnify their view billions of times, ask them what they think they would see. Ask them to draw a "particle picture" (i.e., if they could see particles what would they look like?). Ask them to keep dividing the foil in half until they can no longer divide it. Ask them to consider if they were using the finest tools, could they keep dividing it and still have aluminum foil? Would there ever be a point when they couldn't divide the aluminum anymore and still have aluminum? Have participants share their ideas, identifying someone in the group to point out that you could keep dividing the foil even though the pieces are too small to see. A single atom would be the point where the aluminum could no longer be divided and still be aluminum. Ask a few of the pairs to share their particle pictures and explain that the CTS will help us understand different ways students visualize particles in a substance. Explain that this CTS session will help us answer these questions and surface teaching implications related to the particle nature of matter.

Elicitation (10 minutes)

Activating Prior Knowledge: Matter Is Made of Atoms (10 minutes)

Show Slide 5. Explain that this quote came from a famous, beloved Nobel Prize winning physicist and educator, Richard Feynman. Refer to Handout B5.2: Matter Is Made of Atoms. Have participants read and discuss the quote in small groups and answer the questions based on their prior knowledge before using the CTS resources. After participants have had time to surface and discuss their beliefs about the importance of the topic and implications for curriculum and instruction, explain that a study of the topic may change or build upon their initial ideas. Later, as they engage in the CTS, encourage them to look back at these questions to see if any of their initial ideas have changed.

Exploration (50 minutes)

Preparing for the CTS (10 minutes)

Have participants refer to the CTS guide, "Particulate Nature of Matter (Atoms and Molecules)," on page 169, and Handout B5.3: Reading Assignments. Explain that in order to examine the K–12 content in the standards as well as the curricular and instructional considerations that build toward and contribute to an understanding of atoms and molecules, they will conduct a curriculum topic study. The topic study will also help clarify some of the questions they had related to Handout B5.2: Matter Is Made of Atoms. Show Slide 6 and point out that they will not be reading everything listed on the CTS guide on page 169. Explain they will use Handout B5.3: CTS Atoms and Molecules Reading Assignments to focus on specific pages within the CTS guide readings for this session. Explain that they can go back and read the other sections on their own. Describe how the readings will be distributed and shared based on the option you selected. Make sure everyone is clear about their reading assignments and remind them that the page numbers for the reading assignments are listed on their CTS guide on page 169.

> **Facilitator Note**
>
> If you are using the 2009 edition of *Science Matters*, change the page numbers on Handout B5.3 to Chapter 4, pages 67–79 and Chapter 7, pages 116–121. The pages listed on Handout B5.3 are for the 1991 edition of *Science Matters*.

Quiet Read (40 minutes, including time for a break; adjust time according to option chosen)

Participants will read their sections quietly. Adjust the time according to the option selected. Encourage participants to take notes that are relevant to the purpose of the CTS section they are

> **Facilitator Note**
>
> The time indicated for this portion reflects the approximate time to read and make notes on one section.

reading. If they have CTS books, refer them to Figure 3.7 on pages 37–39 for questions they can use to process their text. Remind them that Handout B5.3 can help focus their reading. You may wish to take a 15-minute break after the readings or build a break into the reading time.

 ## Development and Synthesis (60–75 minutes)

Clarifying Discussion (30–45 minutes)

After participants complete the readings, ask them to discuss them in small groups and clarify content and curricular and instructional considerations related to the topic. Encourage participants to cite their text readings as they discuss the topic. Circulate and answer questions as needed. Remind them to focus their part of the discussion on their task described on Handout B5.3.

Synthesis of Key Findings: Explaining Phenomena and Characteristics (30 minutes)

Refer everyone back to the aluminum foil activity. Ask for comments on how the CTS findings provide insight into the commonly held ideas students might have if they were asked this question. Ask for examples of findings from CTS, such as learning goals and instructional considerations that would support students' being able to apply particle ideas to the aluminum foil question. Show Slide 7 and refer to Handout B5.4. Announce that participants will now have an opportunity to further apply the CTS particle idea findings. Have participants form small groups of three to four at similar grade levels (elementary, middle, or high school). Ask them to choose one of the phenomena or characteristics listed on Slide 7 that can be explained using the "particle ideas" they discussed earlier. Tell the groups to use the CTS findings for their grade level to inform their answers to the questions on the handout. Give them time to complete the handout and discuss their ideas about how the CTS findings apply to the example they selected. Ask them to briefly explain the phenomenon or characteristic the way their students, at their grade level, would be expected to explain it, based on the CTS findings recorded on their worksheet. Have each group briefly share one insight they gained from their task.

 ## Application (15 minutes)

Applying CTS to Participants' Own Contexts (15 minutes)

Ask participants to step back from what they have just done to think about the value of this CTS session in their own context. Refer to the questions on Slide 8, "What new ideas or insights did you gain from this CTS that you will take back to your setting? What will you do with these new ideas or insights?" Allow time for a short discussion of ideas at tables. Point out that one value of CTS is that it can make us more aware of K–12 content appropriate at different grade levels as well as science content we don't understand as well as we would like. For example, point out how CTS helps us understand why details about the parts of the atom are not necessary to teach before high school. There are many other important ideas that need time to develop first that form a foundation for chemistry ideas taught in high school. The CTS also helps us be aware of misconceptions that students struggle with that may affect their learning. Caution them that when they look only at state standards without considering all the important contributing ideas and research, they run the risk of fragmenting students' knowledge and creating incoherency. Remind them "standards are not a curriculum." Allow time to share a few ideas with the full group.

Reflection and Evaluation (10 minutes)

Show Slide 9 and remind the group that the CTS Web site contains several supplementary resources for participants who want to further study this topic after the session. The book, *Ready, Set, Science! Putting Research to Work in K–8 Science Classrooms* (Michaels, Shouse, & Schweingruber, 2008) is listed in the CTS supplementary database for this topic study. It includes an example of a learning progression that can be used with this topic. Show Slide 10 and provide a few minutes for everyone to do a written reflection. Ask them to look back on their responses to the questions on Handout B5.2. Would they change or add anything? Ask for some volunteers to share their reflection in the large group or ask everyone to turn to a neighbor and share their thinking. End with Slide 11 and thank participants for their hard work and thinking during this session. Answer any lingering questions. Ask participants to complete a worksheet evaluation form.

Additional Facilitator Notes

You may want to show and explain the resource, *Ready, Set, Science! Putting Research to Work in K–8 Science Classrooms* (Michaels et al., 2008). This book provides further information about developing students' conceptual understanding of the atomic-molecular theory, including recent research that informed the development of a learning progression on this topic. If you have a copy of this book, put it on display during your session and put it out as a resource.

Module B6

"Photosynthesis and Respiration" Facilitation Guide

BACKGROUND INFORMATION

Description of the Module

In this session, participants engage in a full topic study of photosynthesis and respiration. This interconnected topic will help participants develop a deep appreciation for and an understanding of the connections between plants and animals; their nutritional needs; the role of these processes in ecosystems; and the inextricable link between structure, function, and needs of organisms. The module helps teachers identify the early foundational ideas developed in elementary school, built upon in middle school, and eventually culminating at the end of Grade 12 in a sophisticated understanding of the abstract physiological and ecological link between these two processes. Participants will explore the many commonly held ideas associated with this topic. The session will help teachers at all grade levels make better informed choices about the key ideas taught at different grade levels and the instructional decisions that may help challenge students' misconceptions and move them toward a scientific understanding of these major processes.

Audience

This session is designed for a K–12 audience. Although formal ideas related to the processes of photosynthesis and respiration begin in middle school, the module helps K–5 teachers understand the important foundation they build at the elementary level in support of these concepts.

Purpose and Goals

The overall purpose of this module is to examine the important K–12 content necessary for science literacy related to these two processes and the challenges posed in helping students (and teachers) change their misconceptions related to the topic. The goals for this module are to

- deepen adult understanding of the physiological and ecological ideas related to photosynthesis and respiration,
- clarify the K–12 content and instructional implications related to photosynthesis and respiration,
- examine research on learning to anticipate and understand conceptual difficulties related to photosynthesis and respiration, and
- identify the interconnections among K–12 learning goals in the curriculum.

SESSION DESIGN

Time Required

Allow 2.5 to 3 hours. This is an approximate time and will be adjusted based on the experience and needs of your audience.

Agenda at a Glance

Welcome and Engagement (20 minutes)

- Overview (5 minutes)
- Engagement table talk (15 minutes)

Elicitation (15 minutes)

- Accessing prior knowledge (15 minutes)

Exploration (40–55 minutes)

- Preparing for the CTS (10 minutes)
- Quiet read (20–30 minutes, allow a 10- to 15-minute break)

Development and Synthesis (55 minutes)

- Clarifying discussion (25 minutes)
- Subtopic key points (15 minutes)
- Interconnections of content, teaching, and learning (15 minutes)

Application (30 minutes)

- Revisiting the photosynthesis and respiration questions (20 minutes)
- Report out (10 minutes)

Reflection and Evaluation (15 minutes)

- Quote reflection and report out (10 minutes)
- Evaluation (5 minutes)

MATERIALS AND PREPARATION

Materials Needed by Facilitator

CTS Parent Book

Science Curriculum Topic Study: Bridging the Gap Between Standards and Practice (Keeley, 2005)

Resource Books

One copy of each of the following resource books:

- *Science for All Americans* (American Association for the Advancement of Science [AAAS], 1989)
- *Benchmarks for Science Literacy* (AAAS, 1993)
- *National Science Education Standards* (National Research Council [NRC], 1996)
- *Making Sense of Secondary Science* (Driver, Squires, Rushworth, & Wood-Robinson, 1994)
- *Atlas of Science Literacy* (Vol. 1) (AAAS, 2001)
- *Science Matters* (Hazen & Trefil, 1991, 2009) (Optional)

CTS Module B6 PowerPoint Presentation

The Module B6 PowerPoint presentation is located in the Chapter 5 folder on the CD-ROM. Review it and tailor it to your needs and audience. Insert your date and location on Slide 1, add additional graphics as desired, and add your own contact information on the last slide.

Supplies and Equipment

- Computer and LCD projector to show PowerPoint presentation
- Flip chart easel, pads, and markers
- Paper for participants to take notes

Wall Charts

- Prepare and post five charts with the following titles: (1) *Needs of Plants and Animals;* (2) *Biological Structure and Function;* (3) *Plant and Animal Nutrition—Acquiring, Making, and Using Food;* (4) *Nature of Food;* and (5) *Transformation and Movement of Matter and Energy (organism and ecosystem level).*
- Prepare and post the *CTS Reminders* chart, the *CTS Content Questions* chart (for description of these two charts, see Module A1 in Chapter 4, pp. 63–71), and a chart labeled *CTS Words* in meeting room prior to session.

Facilitator Preparation

- PowerPoint slides for Module B6 (on CD-ROM). Insert your own graphics and additional information.
- Conduct your own "Photosynthesis and Respiration" CTS prior to leading the session.
- Identify connections to state standards if necessary.
- Choose the option for distributing and sharing readings (see Chapter 3, pp. 48–50) that best matches your audience.
- Prepare an evaluation form to collect feedback from participants.

Materials Needed for Participants

(*Note:* All handouts are in the Chapter 5 folder on the CD-ROM.)

- Handout B6.1: Agenda at a Glance.
- Handout B6.2: Photosynthesis and Respiration Questions.
- Handout B6.3: Photosynthesis and Respiration Reading Assignments.
- Handout B6.4: CTS Note Taking Guide.
- PowerPoint slides for Module B3 (on CD-ROM). Insert your own graphics and additional information you wish to add. (Optional)
- If participants do not have their own CTS parent books, make copies of the CTS guide, "Photosynthesis and Respiration," on page 143 for each participant.
- If enough resource books are not available, make copies of the selected readings from the CTS guide, "Photosynthesis and Respiration," on page 143 that are used in Handout B6.3. If you are using the 2009 edition of *Science Matters*, change the readings to Chapter 15, pages 263–266 and Chapter 18, pages 329–332.

DIRECTIONS FOR FACILITATING THE MODULE

Welcome, Introductions, and Engagement (20 minutes)

Overview (10 minutes)

Show Slides 1, 2, and 3. Welcome the participants and explain the purpose of this session. Review the goals for the session, give participants a moment to share what they would like to gain from the

CTS, and then go over the agenda. If participants in the group do not know one another, provide an opportunity for quick introductions. Show Slide 4 and explain that photosynthesis and respiration are topics that include many interconnected ideas in the K–12 curriculum and state and national standards. Ask participants for their working definition of each process—photosynthesis and respiration. (*Note:* They do not need to provide a detailed scientific definition—encourage them to describe what each process is in their own words.) Provide a few minutes to clarify each bullet by asking participants for an example (e.g., for the first bullet, participants might say that plants need water, carbon dioxide, and sunlight). Explain how this CTS will provide an opportunity for them to delve into these ideas and use their findings to inform teaching and learning in their own settings.

Engagement Table Talk (10 minutes)

Show Slide 5. Give small groups time to discuss the questions on the slide one at a time. The questions are as follows: "Why do you think this topic is important to K–12 science literacy? Why do you think older students can provide a 'definition' of photosynthesis and respiration, use the technical vocabulary, and yet struggle with the fundamental ideas? What connection does this topic have to learning goals in your K–12 science curriculum?" Ask for a few comments from the group. If you have a K–12 audience, encourage comments from the full range of grade levels.

 ## Elicitation (15 minutes)

Accessing Prior Knowledge (15 minutes)

Show Slide 6. Refer to Handout B6.2: Photosynthesis and Respiration Questions. Have participants work in pairs to fill out the questionnaire and discuss the items. Remind them to draw upon their own prior knowledge and not turn to the CTS resources for information yet. Explain that these ideas will surface from the content readings, standards, and research on student learning as we engage in the CTS. Encourage them to check back on the ones about which they are unsure after completing the CTS.

> **Facilitator Note**
>
> Be sensitive to the fact that some teachers with weak content background may feel uncomfortable during this activity. Make it safe for everyone to participate by ensuring that no one is put on the spot to reveal their own misconceptions if they do not wish to do so. Remind everyone that there is a wide range of knowledge in the room related to the topic and to encourage and support their colleagues who may have less knowledge about the topic than they do.

 ## Exploration (55 minutes)

Preparing for the CTS (10 minutes)

Have participants turn to the CTS guide, "Photosynthesis and Respiration," on page 143, and Handout B6.3: Photosynthesis and Respiration Reading Assignments, which provides a focus for the readings. Show Slide 7 and review each of the reading assignments (A–F) and the purpose of each section (I–VI). Describe how the readings will be distributed and shared, based on the option you selected from Chapter 3. One suggestion is that you start by having participants select B, C, or D, depending on their grade level first, then assign the remaining A, E, and F readings. It is also important for this CTS to try to have K–12 representation in each small group if possible. Remind participants that everyone will review Section VI, the connection to state standards, frameworks, and curriculum materials, as a whole group after completing the study.

Quiet Read (45 minutes including a 15-minute break)

Show Slide 8 to remind participants to focus their reading on these subtopics as they relate to the photosynthesis and respiration CTS. Depending on their assignment, they may not find information related to all of the subtopics. Remind them that the readings contain text that does not relate directly to this topic and that they should read selectively. Refer them to Handout B6.4 as a note-taking guide to use as they read (distribute additional copies if they are reading more than one section). Take a minute to have everyone identify their assigned reading on Handout B6.3 and write down the purpose of their reading (this is described on Handout B6.3 as well as generally described on the CTS guide's left-hand column on page 143 in the CTS parent book). Participants will now conduct a quiet read, depending on which option for distributing readings was selected. Remind them to do a quick scan of the reading first, and then take notes on related sections. Adjust the time according to the option selected (e.g., if participants wish to read their grade level and the A, E, and F readings, you will need to build in more time). You may wish to take a fifteen-minute break after the readings or build it in to the reading time.

Development and Synthesis (55 minutes)

Clarifying Discussion (25–30 minutes)

Depending on the option selected for distributing and sharing the readings, ask participants to work in small groups to discuss each of the readings and clarify findings related to the topic. Encourage them to use Handout B6.4 to guide their discussion, summarizing key relevant parts of their reading. Once everyone in the group has shared their section, ask groups to discuss the connection to their own state standards, frameworks, or curriculum materials. How does the CTS provide insight into the teaching and learning context in their own setting?

Subtopic Key Points (15 minutes)

Refer to Slide 8 again. After groups have discussed each of the readings as a whole, ask the small groups to have each person in the group select one of the five subtopics (make sure all five subtopics are represented in each group). Using sticky notes at their table, ask each person to record three to five key points from readings A–G related to that subtopic. Encourage them to work together and seek further clarification from others who read and shared a section. The intent here is to show the range of information CTS reveals for each of the subtopics as they relate to photosynthesis and respiration. Show Slide 9. A key point can be a content statement, an important grade-level learning goal, a grade-level instructional consideration to be aware of, an important research finding, or important considerations indicated by the *Atlas*. Write one key point per sticky note. When they are done, have each person stand at the chart that represents the subtopic they were assigned along with others who selected the same subtopic and share their sticky notes, grouping them on the chart by CTS sections and purposes I, II, III, IV, or V.

Content, Teaching, and Learning Interconnections (15 minutes)

Have participants do a "gallery walk," examining the CTS summaries grouped by CTS section on each of the five charts. After they have walked by each of the charts and examined the findings, have them return to their small groups and show Slide 10. Use a think-pair-share strategy to discuss the following question from the slide: "How do the five combined subtopics give you a better understanding of the interconnections between the content, teaching, and learning related to photosynthesis and respiration?" (Ask participants to take a few minutes to *think* about their own ideas, and then *pair* up with someone in the room and discuss their ideas, then call on several pairs to *share* their ideas with the whole group.)

Application (20 minutes)

Revisiting the Photosynthesis and Respiration Questions (20 minutes)

Show Slide 11 and refer participants back to Handout B6.2. With a partner, have them go through the questions again, either noting where they would change their answers and why, questions with which they are still struggling, or discussing the insights they gained about teaching and learning related to each question.

Reflection and Evaluation (15 minutes)

Show Slide 12. Explain that this quote came from a famous Nobel prize-winning scientist and teacher, Richard Feynman. Ask them to reflect on the significance of this quote based on what they learned today. How does this "big idea" differ from what students typically learn about photosynthesis and respiration? Ask for a few volunteers to share their thoughts. Answer any lingering questions, thank participants, and ask them to complete an evaluation.

> **Additional Facilitator Notes**
>
> There are several supplementary materials listed on the CTS Web site that can be used to extend this session, including assessment probes that can be used to collect and examine student work and videos of clinical interviews with children that reveal their misconceptions. The Harvard Smithsonian Digital Video Library Web site at http://hsdvl.org provides an *Atlas* map interface that can be used to access videos of clinical interviews related to the key ideas identified in CTS Sections IIIA and V.

Module B7

"Life Cycles (K–4)" Facilitation Guide

BACKGROUND INFORMATION

Description of the Module

In this CTS module, participants engage in a Grade K–4 topic using a modified version of the CTS guide, "Reproduction, Growth, and Development (Life Cycles)" to examine the topic of life cycles at the K–4 level. The module helps participants gain an understanding of the important ideas and implications for teaching about life cycles. Participants use their findings to make and discuss comparisons between two different elementary science life cycle units and connect their findings to their own instructional materials. The session creates greater awareness of the limitations an instructional context can pose if "big ideas" that apply across contexts are not explicitly developed. Additionally, it can help teachers see how CTS can be used to make informed choices that can strengthen curricular units. This session also demonstrates how CTS guides can be modified for a specific audience and purpose.

Audience

This session is for teachers in Grades K–4 who teach concepts related to life cycles of organisms, including humans. In addition, this module is a good example to use when introducing CTS to an elementary audience. Since the topic is a familiar one for K–4 teachers, the script provides more guidance than other examples, and since the strategies used are familiar to elementary teachers, this module may be used for initially introducing CTS to elementary teachers for the purpose of engaging them in using the CTS tools and resources on other topics later on.

Purpose and Goals

The overall purpose of this module is to examine the important ideas related to life cycles. The goals of this module are to

- deepen teachers' content understanding of life cycle ideas in the elementary science curriculum,
- use standards and research to examine teaching and learning implications related to life cycles, and
- consider how results from the CTS can be used to inform curricular goals and the selection, use, and modification of instructional materials.

SESSION DESIGN

Time Required

Allow 2.75 to 3 hours (this will vary depending on the experience and needs of the group and any breaks taken).

Agenda at a Glance

Welcome and Engagement (10 minutes)

- Welcome and introductions (3 minutes)
- Establishing a purpose (7 minutes)

Elicitation of Prior Knowledge and Exploration of Students' Ideas (35–40 minutes)

- What do we know and want to know? (10 minutes)
- Does it have a life cycle? (20–25 minutes)
- What else do we want to know? (5 minutes)

Topic Exploration Using CTS (35–45 minutes)

- CTS individual readings (25–30 minutes plus allow a 10- to 15-minute break)

Topic Development (20 minutes)

- CTS small group discussions (20 minutes)

Topic Synthesis (15 minutes)

- Key take-home points (10 minutes)
- Summary (5 minutes)

Application of CTS Findings (30 minutes)

- Comparing and contrasting two instructional units (20 minutes)
- Implications for local curriculum and instructional materials (10 minutes)

Reflection (10 minutes)

- Revisit the K-W-L chart (5 minutes)
- Group reflection (5 minutes)

Evaluation (10 minutes)

- Vignette review (5 minutes)
- Evaluation (5 minutes)

MATERIALS AND PREPARATION

Materials Needed by Facilitator

CTS Parent Book

Science Curriculum Topic Study: Bridging the Gap Between Standards and Practice (Keeley, 2005)

Resource Books

- Benchmarks for Science Literacy (American Association for the Advancement of Science [AAAS], 1993)

- *National Science Education Standards* (National Research Council [NRC], 1996)
- *Making Sense of Secondary Science* (Driver, Squires, Rushworth, & Wood-Robinson, 1994)
- Optional: *Uncovering Student Ideas in Science, Volume 3: Another 25 Formative Assessment Probes* (Keeley, Eberle & Dorsey, 2008), which contains the "Does It Have a Life Cycle?" probe used in this module and the teacher notes for using the probe

CTS Module B7 PowerPoint Presentation

The Module B7 PowerPoint presentation is located in the Chapter 5 folder on the CD-ROM. Review it and tailor it to your needs and audience. Insert your date and location on Slide 1, add additional graphics as desired, and add your own contact information on the last slide.

Supplies and Equipment

- Computer and LCD projector to show PowerPoint presentation
- Flip chart easel, pads, and markers
- Paper for participants to take notes

Wall Charts

- Prepare and post the *CTS Reminders* chart and the *CTS Content Questions* chart in the meeting room prior to the session (for description of these two charts, see Module A1 in Chapter 4, pp. 63–71). Prepare a third chart labeled *What We Want to Know About Life Cycles* (part of Handout B7.2 and elicitation activity).

Facilitator Preparation

- Print out and review Facilitator Resources B7.1 and B7.2 from the Chapter 7 folder on the CD-ROM.
- Conduct your own "Life Cycles" CTS prior to leading the session. Become familiar with the summary on Facilitator Resource Handout B7.1.
- Identify connections to state standards or local curriculum materials if necessary.
- Choose the option for distributing and sharing readings (see Chapter 3, pp. 48–50) that best matches your audience.
- Prepare an evaluation form to collect feedback from participants.

Materials Needed for Participants

(*Note:* All Module B7 handouts are in the Chapter 5 folder on the CD-ROM. The optional Handout A1.6 can be found in the Chapter 4 folder of the CD-ROM.)

- Handout B7.1: Agenda at a Glance.
- Handout B7.2: K–4 Life Cycles K-W-L.
- Handout B7.3: K–4 Life Cycles Modified CTS Guide.
- Handout B7.4: Does It Have a Life Cycle? Content Explanation.
- Handout B7.5: Does It Have a Life Cycle? Student Responses.
- Handout B7.6: Curriculum Unit 1—Life Cycles.
- Handout B7.7: Curriculum Unit 2—Life Cycles.
- Copy of Vignette 3, pages 96–97 in the CTS parent book (for participants who do not have the book).
- PowerPoint slides for Module B7 (on CD-ROM).

- If enough resource books are not available, provide copies of the selected readings from the resources listed in Handout B7.3.
- Handout A1.6 (in the Chapter 4 folder on the CD-ROM) if participants are relatively new to CTS and need more guidance in using the study guide. (Optional)

DIRECTIONS FOR FACILITATING THE MODULE

Welcome, Introductions, and Engagement (10 minutes)

Welcome and Introductions (5 minutes)

Show Slide 1. Welcome the participants and explain that the session will engage them in using CTS to deepen their knowledge about teaching and learning related to a common elementary life science topic, life cycles. If your participants do not know one another, do quick introductions by either having people introduce themselves at their tables or choosing your own introduction activity and adding additional time as needed.

Engagement: Establishing a Purpose (5 minutes)

Show Slide 2. Refer participants to Handout B7.1: Agenda and describe the flow of the day, emphasizing that the session will move quickly and it is important to stick to the time for each activity in order to meet the goals of the session. Show Slide 3. Ask participants to write down their personal goal(s) for the session. Ask participants to turn to a person near them and exchange one thing they hope to gain from this session.

Elicitation of Prior Knowledge and Exploration of Students' Ideas (25 minutes)

What Do We Know and Want to Know? (10 minutes)

Show Slide 4. Refer participants to Handout B7.2: K–4 Life Cycles K-W-L. Briefly describe the K-W-L strategy as a pre- and post-reflection exercise. (Many elementary teachers are familiar with this as a reading strategy.) Ask participants to fill out the first two columns as they relate to teaching and learning about life cycles. Ask them to select one "I want to know" to write on a sticky note and attach on the wall chart *What We Want to Know About Life Cycles*. Note the different things participants want to know about and point them out later as they come up during the CTS. Explain how they will be using a CTS guide on K–4 life cycles to learn more about this topic.

Refer everyone to Handout B7: K–4 Life Cycles Modified CTS Guide and point out the different outcomes for each CTS Section (I, II, III, IV, and VI) in the left-hand column of the guide. (Optional: you may wish to provide them with Handout A1.6, Anatomy of a Study Guide, to reinforce the purpose of each CTS section for those who may be less familiar with CTS. This handout is on the CD-ROM under Chapter 4, Module A1.) Explain that this is a modified study and they will be using a supplementary reading for Section I and they will not be using Section V for this study since there is no *Atlas* map specific to this topic. If participants have the CTS parent book available, have them turn to page 155 to see the full K–12 topic study guide, "Reproduction, Growth, and Development (Life Cycles)," that was used to develop Handout B7.3.

Does It Have a Life Cycle? (10 minutes)

Show Slide 5. If you have a copy of *Uncovering Student Ideas in Science, Volume 3* (Keeley et al., 2008), hold it up and explain that this probe came from this book of CTS-developed probes available through NSTA. Point out that Chapter 4 in the CTS parent book, pages 80–83, describes how to create these probes. Give participants 5 to 10 minutes to examine the probe and discuss their answers in small groups. Listen for possible content issues to address during the debrief. Provide participants with the explanation sheet (Handout B7.4: Does it Have a Life Cycle Content Explanation) and raise and discuss clarifications or misunderstandings they have about life cycles.

Show Slide 6. Explain that they are about to examine authentic responses that came from two classes in two different schools—one from third grade and one from fourth grade. The responses were transcribed exactly as the students wrote them. The topic of life cycles was taught in the earlier grades. Pass out Handout B7.5: Does It Have a Life Cycle? Student Responses. Give participants 10 to 15 minutes to scan through the responses and discuss the points on Slide 6. The purpose here is not to spend a lot of time examining each student response but to help participants see that even though this is a topic typically taught, students have a variety of different ideas about organisms' life cycles. Remind them they are formatively examining ideas, not passing judgment on the quality of the student responses, grammar, or spelling. Explain that if you were the teacher of the students in one of these classes, there would be tremendous value in doing a CTS to uncover students' commonly held ideas and the roots of the student learning problem.

What Else Do We Want to Know? (5 minutes)

After looking at the student work, participants may have additional questions they want to explore related to the topic of life cycles. Ask them to revisit their K-W-L and post any new things they want to know more about in the second column, particularly as they relate to students' misconceptions, curriculum, or instructional materials. Invite them to add additional sticky notes to the wall chart.

Topic Exploration Using CTS Readings (35 minutes)

CTS Individual Readings (35 minutes)

> **Facilitator Note**
>
> One major difference is that this guide focuses less on the readings explaining the reproductive process. However, if that is an area taught in your group's K–4 curriculum, you may choose to include it in this study.

Show Slide 7. Explain that participants will now further "explore" the topic of life cycles by doing a CTS. Refer participants again to Handout B7.3, clarifying again that this is a modified guide for K–4, taken from the original K–12 guide on page 155 of the CTS parent book.

You might also wish to point out the difference between the *Benchmarks* and the *National Science Education Standards* in this study. *Benchmarks* focus more on humans, in particular the development and maturation stage of the human life cycle. The *National Science Education Standards* focus more on life cycles of all organisms. The *Benchmarks* content may be covered in the health curriculum, and there may be learning goals related to human reproduction that could be considered "controversial" in some schools. Participants may use their own judgment in deciding how much of the *Human Development* part of the study is applicable to their context and which ideas related to human development are applicable to plants and other animals. Also review the purposes for each section in the study guide if participants have had little or no experience using a CTS guide.

Explain that since this is a K–4 version of CTS, participants will not divide up or "jigsaw" the readings. Instead each person will read and take notes on all the sections. However, since some people take longer to read and process than others, and it is important that all sections are read by the group within the time allocated, have participants form small groups of four to five and divide up the responsibility of making sure all the readings are covered by having each person start with a different section (you may want to divide the readings in Section IVB in half). That way, the table groups will cover all the sections even though all participants may not finish reading all the sections. Provide 25 to 30 minutes after giving instructions for participants to read silently and take notes on key findings that relate to the life cycles topic. Let them know that a 10-minute break is built into this time. Remind participants that not everything they read will be related to building an understanding of teaching and learning related to life cycles. It is important to extract the information that is most relevant to the topic. Sometimes it is hard to draw a boundary between one topic and another since they are so interconnected. Let participants know that in this study, although heredity ideas are related, they can skip these to focus more on the fertilization, birth, growth, development, continuity, aging, and death aspects of the cyclic nature of life. Optional: You may choose to select a few questions from Facilitator's Resource B7.2: Questions and Discussion Points to post on a chart or slide for participants to focus their study.

While participants are reading, examine the sticky notes on the *What We Want to Know About Life Cycles* wall chart. Cluster similar notes and make a list on the chart of questions that can be answered from the CTS readings, discussions, or content expertise in the room. As these questions get answered through the remaining course of the session, point them out on the chart.

Topic Development (20 minutes)

CTS Small Group Discussions (20 minutes)

After table groups have completed their readings, have participants take turns in their small groups summarizing and discussing findings from each of the readings. Remind them to be sure to monitor time so that everyone has a chance to share and that all sections have been summarized according to the purpose of the section they read. Also point out the *CTS Reminders* chart so that they remember to speak from the CTS evidence, not their own opinions. Circulate among small groups and answer questions and make sure groups are on task. Use the information on Facilitator Resource B7.2: Questions and Discussion Points to help you probe groups when necessary. Optional: If the group is small, this can be facilitated as a whole group discussion, using questions and key discussion points from Facilitator Resource B7.2.

Topic Synthesis (15 minutes)

Synthesis of Key Take-Home Points (15 minutes)

Show Slide 8. In the same small groups (or mix participants up into new small groups of the same size), ask participants to generate a list of the key take-home points they want to remember and use from this CTS and post them on a piece of chart paper. Encourage participants to think about how the findings from the different readings can inform their own curriculum, instruction, and assessment. Remind participants that they should be able to cite where any of their key points came from in the CTS resources.

Have groups post their charts where everyone in the room can see them (or have participants get up and walk around examining each chart). Ask for a few volunteers to share a key point that will inform their future work. If there is time, ask the group some questions from Facilitator Resource B7.2: Questions and Discussion Points to further reinforce their learning. Tell them they may refer to the *Key Points* charts you just developed for the next activity—a chance to apply what they learned from the CTS.

Application of CTS Findings (30 minutes)

Comparing and Contrasting Two Instructional Units (15 minutes)

Show Slide 9. Ask participants to refer to Handouts B7.6 and B7.7. Explain that these are adapted summaries of actual, published curriculum materials. Each of the units is considered a high-quality instructional unit and is a commonly used elementary kit-based program. Each of the units has a different emphasis and includes strong components in inquiry and other topics in addition to life cycles. They are both considered units that achieve a variety of curricular and instructional purposes. Our task is not to examine which one is "better," or critique their quality, but rather examine them closely for how well they reflect the CTS findings for this topic.

Have participants quickly scan through the two examples of life cycle curricular units and discuss the extent to which they provide opportunity to learn the "big ideas" related to life cycles that emerged from the CTS. Have a group discussion, encouraging participants to back up their reasons with evidence from the CTS findings. If there is time, have them look back on Handout B7.5 and discuss how each curriculum unit might address the commonly held ideas held by the third- and fourth-graders. It is important to point out that one of these units is not "better" than the other. They have different instructional goals. However, if the purpose of selecting materials is to focus on developing big ideas related to life cycles of organisms, not one particular type of organism, or to integrate other ideas about characteristics of organisms, then one unit may serve a better purpose than the other.

Implications for Local Curriculum and Instructional Materials (15 minutes)

Show Slide 10. Ask participants to think for a minute about the two questions on the slide that relate to how they might change local curriculum based on their CTS results. Provide two to three minutes for an individual quick write and then have participants pair up and share their ideas with a partner. If possible, have participants pair up with someone from their own schools, districts, or grade levels. Provide time for small group discussions.

 ## Reflection (20 minutes)

Vignette Wrap-Up (10 minutes)

Have participants do a paired reading of Vignette 3 on pages 96–97 in the CTS parent book or provide a handout of these pages. (*Note:* In a paired reading, each person takes turns reading a paragraph aloud and then briefly sharing a summary comment before going to the next paragraph.) After finishing the reading, ask the whole group to share their thoughts on how the vignette helps them understand the value of doing a CTS.

Individual Reflection (5 minutes)

Show Slide 11. Provide five minutes for participants to revisit their K-W-L chart and complete an individual reflection.

Group Reflection (5 minutes)

Show Slide 12. Give the group a minute to think how they would fill in the blanks based on what they learned from this session. Ask for five to six volunteers to read their sentences aloud to the whole group, filling in the blanks.

Wrap-Up and Evaluation (5 minutes)

Evaluation

Ask everyone to complete an evaluation of the session.

Additional Notes

1. The CTS supplementary Web site provides some examples of videos from the Annenberg Essential Science series (Harvard-Smithsonian Center for Astrophysics, 2003) that show children discussing their life cycle ideas that can be used to supplement this session.

2. For participants who are interested in using CTS to further examine other instructional materials, point out the section in Chapter 4 of the CTS parent book, pages 65–75, that describes CTS tools to use for the selection and implementation of curriculum materials.

ADDITIONAL SUGGESTIONS FOR DESIGNING AND LEADING FULL TOPIC STUDIES

The previous seven modules presented various examples of ways to design and facilitate a full topic study. Each example included standard features, such as incorporating stages in the CTS Learning Cycle, as well as unique nuances, such as engagement and elicitation strategies and ways to develop and debrief the group's understanding after small group study and discussion. These modules can be used as they are or adapted to fit your own context. They can also be used to introduce participants to curriculum topic study. There may be times when a facilitator needs to use a topic situated in the work participants are doing, such as a content institute, a particular topic with which students are struggling, for curriculum selection, or other purposes. You are encouraged to modify any of the modules to fit your specific needs and facilitation style. The following guidelines are provided for facilitators who wish to design their own sessions using a topic other than the ones provided in Modules B1 through B7.

DEVELOPING YOUR OWN HALF-DAY TOPIC STUDY

You will notice that each of the full topic study modules (B1–B7) share key components. You will use these same components to develop your own CTS full topic study for a selected topic and grade level. The following steps can be used to guide the development of your own CTS topic session.

1. Select a topic that is relevant to the group you are working with. Find the CTS guide that best matches the topic. Consult the CTS Grain Size chart located in the Chapter 2 folder on the CD-ROM to determine whether your topic will be fairly narrow and specific or wide and comprehensive. Develop goals for your session.

2. Decide whether the CTS will be K–12, two grade spans (elementary and middle or middle and high school), or a single grade span.

3. Conduct your own CTS on your chosen topic prior to designing your session. By doing the CTS yourself first, you will best learn which sections and readings to include and focus on in your session. Consider creating your own summary of the major findings from your CTS to refer to for planning and when conducting your session. (Examples of summaries can be found for other topics in the Chapter 3 folder on the CD-ROM.)

4. Examine the CTS guide and decide if you are going to use all the readings or pare some of them down. Your decision will also be based on the availability of CTS resources during your session. Refer to Chapter 3 of this *Leader's Guide* for ways to distribute the CTS resources.

5. Decide whether to use the CTS guide as is or create a customized version for a specific grade level, such as the one used for Module B7, the K–4 Life Cycles Study Guide (Handout B7.3). Use the Blank Create Your Own CTS Template 4 in the Chapter 5 folder, Developing Your Own CTS, on the CD-ROM to create your guide. Follow the format used for the study guides in Chapter 7 of the CTS parent book. Add any supplementary resources from the CTS Web site you wish to include.

6. Decide whether to create a CTS Reading Assignments Handout, such as the ones in Modules B1–B6. Consider how to break down the reading assignments in the best way for your group. Sample Blank Reading Assignment Templates are included in the Chapter 5 folder, Developing Your Own CTS, on the CD-ROM. Template 1 is based on the B1 Module and is designed to be used with CTS topics in the "Inquiry and the Nature of Science and Technology" topic studies on pages 229–254 in the CTS parent book. Template 2 is based on Module B3, in which a topic is broken down into concept categories for study. The template is designed for a K–8 study and can be modified for K–12 or other groupings. Template 3 is based on Module B5 and is designed to be used with K–12 audiences. Make sure to fill in your topics, sections, and page numbers that are marked with an XX on the templates.

7. Refer to the CTS Learning Cycle Guide on pages 40–47 in the CTS parent book to organize the different phases of the CTS process. Refer to Modules B1–B7 for different examples of ways to engage participants in each section of the CTS Learning Cycle.

8. Decide whether to select any supplementary materials from the CTS Web site database of CTS supplements to enhance the session, such as videos, assessment probes, and content readings. Embed these within the appropriate CTS Learning Cycle section. If any of the vignettes in Chapter 5 of the CTS parent book match your topic, also consider how to embed one of these in your session. (See example in Module B7.)

9. Review Chapter 3 of this *Leader's Guide* and choose the grouping, reading, and processing strategies you will use within each of the CTS Learning Cycle components.

10. Design your PowerPoint slides and any additional handouts. See those used for Modules B1–B7 as examples to guide you.

11. If possible, try out your CTS design with a small group of colleagues. Gather feedback and make any needed modifications. Good luck designing and implementing your first CTS full topic study session design!

> **Facilitator Note**
>
> Several of the modules in this chapter use assessment probes to examine students' ideas related to the topic and identify misconceptions from the CTS Section IV readings that are evident in teachers' own students. The Chapter 5 folder, Developing Your Own CTS, on the CD-ROM has a crosswalk that links each of the CTS topics to assessment probes that can be found in the *Uncovering Student Ideas in Science* series, Volumes 1–4 (Keeley et al., 2005–2009). Check the CTS Web site for updates to the crosswalk as new volumes of the series are published.

COMBINING TOPICS

Each of the sample Modules B1–B7 as well as the "Design Your Own Topic Study" illustrates how a full CTS can be designed using one topic. Another option to consider is to combine two topic studies when doing a full CTS. This can be accomplished by splitting a group into two with one half doing one topic and the other half doing the other topic. Or conduct each topic study separately in a full-day session and integrate both topics as a final synthesis. During the Synthesis stage of the CTS Learning Cycle, facilitators encourage participants to make connections across both topics. The CTS Web site will feature an example of an integrated topic study. The following are examples of the many combinations of integrated topics that can be considered for combination in a full-day session:

- Cells and Systems
- Solar System and Models
- Biological Evolution and the Nature of Science
- Properties of Matter and Rocks and Minerals
- Fossil Evidence and Constancy, Equilibrium, and Change
- Human Body Systems and Systems
- Physical Properties and Change and Observation, Measurement, and Tools
- Plant Life and Life Processes and Needs of Organisms
- Plate Tectonics and Evidence and Explanation
- Energy and Science and Technology in Society
- Describing Position and Motion and Graphs and Graphing
- Laws of Motion and Technological Design
- Water in the Earth System and Environmental Impacts of Science and Technology
- Energy and Weather and Climate
- Visible Light, Color, and Vision and Stars and Galaxies
- Forces and Mathematics in Science and Technology
- Health and Disease and Medical Science and Technology

CAUTIONS AND NEXT STEPS

This chapter presented various ways to design and facilitate a full topic study. These topic studies are best used after participants have had an introductory CTS session. (See Chapter 4 of this book for the introductory modules.) Facilitators are encouraged to include components from the modules in Chapter 4 if your group is unfamiliar with CTS. Look through the introductory slides and handouts in Modules A1 and A2 and incorporate some of these materials if you are doing a full topic study with CTS first-timers.

The next chapter provides examples of ways facilitators can extend a full topic study to include specific content, curriculum, instruction, and assessment applications.

6

Using CTS in a Content, Curricular, Instructional, or Assessment Context

INTRODUCTION TO CTS CONTEXT APPLICATIONS

Chapters 4 and 5 of this *Leader's Guide* described how to introduce science educators to the curriculum topic study (CTS) tools and processes and described different approaches to designing and facilitating full topic studies. This chapter will address various ways leaders can help participants apply the knowledge gained from doing a CTS in the contextual applications of content, curriculum, instruction, and assessment. The application you select will depend on the purpose of your professional development session or program, the needs of your participants, and their familiarity with CTS. The applications in this chapter closely align with the descriptions on pages 53–90 of Chapter 4, "Utilizing Curriculum Topic Study for Different Contexts," in *Science Curriculum Topic Study: Bridging the Gap Between Standards and Practice* (Keeley, 2005), the CTS parent book. To utilize the suggestions, tools, and designs in this chapter, it is recommended that you have the CTS parent book on hand, as several references will be made to it throughout this chapter. The box that follows shows an outline of the context applications included in this chapter.

CONTEXT APPLICATIONS OUTLINE

A. CTS and science content knowledge

    1. Adult content knowledge

    2. K–12 content knowledge

B. CTS and science curriculum

    1. Curriculum coherence and articulation

    2. Curriculum selection

    3. Curriculum implementation

C. CTS and science instruction

    1. Identifying strategies, phenomena, representations, and contexts

    2. Reviewing and modifying lessons

    3. Instructional design

    4. Strengthening inquiry, technological design, or the nature of science

D. CTS and science assessment

    1. Formative assessment probes

    2. Performance tasks

Within each of the four applications are specific tools and strategies leaders can use to either embed the context application into a full topic study, or use the context application as the focus of a professional development session for participants who are already familiar with using CTS. For example, participants might engage in a full topic study on the Earth, moon, and sun system using the session design provided in Chapter 5, Module B3, pages 125–131 of this *Leader's Guide*. The facilitator might choose to extend this module into a full-day session by having participants use the suggestions and tools for reviewing and modifying instruction described on pages 198–199. Participants would then apply what they learned in the full Earth, Moon, and Sun System topic study to improving an existing lesson related to one of the major concepts in the Earth, Moon, and Sun System topic study.

Alternatively, the contextual application might be the focus of the professional development when working with participants who are familiar with the CTS tools and processes. For example, a facilitator might design a professional development session on how to develop formative assessment probes. Using the design provided on pages 204–215 in this chapter, participants are introduced to the CTS formative assessment probe design process. They then choose their own topics, conduct their own topic study, and apply their results to the development of an assessment probe that can be used in their own classrooms.

This chapter differs from Chapters 4 and 5 in that, with the exception of the Designing Formative Assessment Probes Module, it does not provide full-session scripts accompanied by PowerPoint presentation slides and handouts. Instead it provides suggestions for ways to improve or enhance content knowledge, curriculum, instruction, and assessment by using CTS with a variety of strategies and tools provided in this chapter and in the Chapter 6 folder of the CD-ROM at the back of this book. It is up to the discretion of facilitators to decide how to best use the suggestions and tools in this chapter based on the design of their professional development program, purpose of their work, experience of their participants, and adult and student learning needs.

Applications and the CTS Web Site

As you examine the material in this chapter and think about ways to use the resources that accompany Chapter 4 of the CTS parent book in your work, remember to check the CTS Web site at www.curriculumtopicstudy.org. As new tools and designs are developed to use with the CTS applications, by CTS leaders as well as CTS project staff, these will be shared on the CTS Web site.

CTS AND SCIENCE CONTENT KNOWLEDGE

Introduction to CTS Content Knowledge

As discussed in Chapter 3 of this *Leader's Guide*, one major goal of professional development is to continuously build understanding of science content in order to support quality teaching and student learning. Teachers of science strive to improve their content knowledge throughout their careers. Teachers with a college degree in a particular area of science may seek to improve their knowledge in other areas of science that are less familiar. Teachers without degrees in science may seek to improve their subject matter knowledge of the topics they teach at their grade level as well as gain new knowledge beyond their grade level to deepen their understanding. When we refer to developing and enhancing teachers' content knowledge with CTS, we include the subject matter knowledge all teachers should have as a foundation to be considered science literate. Adult science literacy also includes the knowledge all students are expected to have at the end of their K–12 education. It includes the science knowledge that may exceed what some K–12 teachers teach. Adult science content literacy helps teachers at all grade levels understand science as a unique discipline used to make sense of the natural world. It helps all teachers see where K–12 student learning is headed from one grade to the next as well as after completion of a K–12 education. Content knowledge includes a solid understanding of the curricular content taught by teachers at the grade levels they teach. This is the content that makes up the key ideas and skills students at that grade level learn, as well as the curricular content that comes before and after a teacher's assigned grade level. It is important that teachers know what comes before their grade level so that they can effectively build on earlier key ideas. It is also important to know the next level of content in order to build the foundation students will need at the next grade level as well as to address the content learning needs of students who have met the standards at their grade level and are ready to advance to higher levels of learning.

When teachers have strong knowledge and experiential background in science content, they are able to be more versatile in quickly and effectively responding to students' ideas and learning needs. A teacher with sufficient content knowledge knows the best question to ask next to push student thinking and can guide learning down the most appropriate path. CTS can help strengthen teachers' content knowledge both directly and indirectly.

How Suggestions for Enhancing Content Knowledge Through CTS Are Organized

This section provides suggestions, tools, and strategies for enhancing teachers' content knowledge using CTS. All of the resources and handouts for this section can be found on the CD-ROM at the back of this book by going to the Chapter 6 folder and opening the subfolder labeled "Content Knowledge 6.1–6.11." The following box provides an outline of the suggestions included in this chapter for using CTS to improve content knowledge. Table 6.1 provides a description of the features of these suggestions and the examples provided in the text and on the CD-ROM that facilitators can use to plan inclusion of these suggestions in their CTS professional development.

OUTLINE OF SUGGESTIONS FOR ENHANCING SCIENCE CONTENT KNOWLEDGE

A. Adult science content knowledge

  1. Ways of accessing content on demand

  2. Identifying and recording content for adult literacy

  3. Concept card mapping

  4. K-W-L

  5. Content vignettes

B. Knowledge of K–12 content

  1. Inventory of standards content statements

  2. Unpacking concepts and scientific processes

  3. Ideas before terminology

  4. Using CTS to examine the hierarchal structure of content knowledge in a topic

---

Table 6.1 Features of Tools and Strategies Used for Enhancing Content Knowledge

Strategy or Suggestion	Grade Levels	CTS Topic(s) Examples	CD-ROM Handouts	When to Use This Strategy or Suggestion
Ways of accessing content on demand	K–12 and adult content	Not topic specific	6.1	Use to introduce CTS as a new strategy that can enhance adult and K–12 content knowledge on demand.
Identifying and recording content for adult literacy	Adult content	Nuclear Energy	6.2	Use once participants are familiar with using Sections IA and IB in the CTS guides.

(Continued)

Table 6.1 (Continued)

Strategy or Suggestion	Grade Levels	CTS Topic(s) Examples	CD-ROM Handouts	When to Use This Strategy or Suggestion
Concept card mapping	Adult content	Rocks and Minerals	6.3	Use with teachers who are already familiar with concept mapping.
K-W-L	Adult content	Gravity	6.4	Use once participants are familiar with using Sections IA and IB in the CTS guides.
Content vignettes	Adult content	Gravity	Pages 91–92 in CTS parent book	Use to illustrate how a teacher might use CTS Sections IA and IB.
Inventory of standards content statements	K–12, Grade 4 example	Weathering and Erosion	6.5	Use with Sections II and III for teachers who are teaching unfamiliar content.
Unpacking concepts and scientific processes	K–12; elementary, middle, and high school examples	States of Matter, DNA, Evidence and Explanation, Water Cycle	6.6	Use to deepen teachers' content understanding of the standards and the specific ideas and skills they target.
Ideas before terminology	K–12; K–2, 3–5, 6–8 examples	Forces, Gravity, Water Cycle, Solar System	None	Use to reveal conceptual understanding before introducing terminology. Especially useful at the K–5 level.
Hierarchal structure of content knowledge	K–12; K–2, 3–5, 6–8, 9–12 examples	Life Cycles; Earth, Moon, and Sun System; Landforms; Particulate Nature of Matter	6.7 6.8 6.9 6.10 6.11	Use when teachers have had considerable experience with CTS and have a fairly strong science content background.

How CTS Reveals Gaps in Content Knowledge

There is a saying, "If you don't know what you don't know, you won't know to look for it." This condition of being "unconscious" is not unusual. Take the teacher who has been teaching the same science unit for many years. She may feel competent and comfortable with what she is doing, yet not know that there is content she could learn that could improve her teaching. Sometimes teachers study a topic using CTS and gain indirect learning with respect to the content. They realize what they do not know about the topic and use CTS as an opportunity to identify important content goals for future professional development and learning. This may lead them to sign up for a workshop, summer institute, or university course to fill that identified gap.

CTS is not a replacement for formal science coursework. However, the CTS process does build and enhance the subject matter knowledge deemed necessary to be considered a science literate person. Chapter 4 in the CTS parent book contains a section on "CTS and Science Content Knowledge" on pages 55–61. Facilitators of CTS are encouraged to read and become familiar with this section.

Content Knowledge and Science Literacy

A science literate person is one who has a basic understanding of the science needed to be a productive and informed participant in today's society. A science literate person is not necessarily one who has majored in science in college or works in a science-related career. Science literacy applies to all adults, regardless of their post-secondary education or career choice. It applies to teachers of science who have majored in a science discipline as well as generalists, such as elementary teachers, who teach all subject areas, including science. It is important when using CTS with teachers to reinforce what we mean by *science literate* and emphasize that this also refers to the knowledge all K–12 teachers are expected to have, regardless of what they teach. According to American Association for the Advancement of Science (AAAS; 1989, p. xvii), a science literate person is one who

- is aware that science, mathematics, and technology are interdependent human enterprises with strengths and limitations;
- understands key concepts and principles of science;
- is familiar with the natural world and recognizes both its diversity and unity; and
- uses scientific knowledge and scientific ways of thinking for individual and social purposes.

While CTS Sections I, III, V, and VI specifically target concepts, key ideas, and skills, a careful study of all sections of a CTS guide can contribute to teachers' subject matter knowledge—both the content they should know to be considered science literate adults and an understanding of the content specific to K–12 learning goals. Figure 4.2 in the CTS parent book shows the connections between the sections of a CTS guide and teachers' subject matter knowledge. In Table 6.2, this chart is expanded to show how leaders can use CTS to enhance teachers' content knowledge.

After examining the section, "CTS and Science Content Knowledge," on pages 55–61 in the CTS parent book, the following strategies and suggestions can be used by facilitators in their CTS sessions or professional development sessions to improve teachers' content knowledge. First, we will describe suggestions and strategies for enhancing adult content knowledge. Then we will follow that with suggestions and strategies for improving knowledge of the content ideas and skills taught at different grade levels.

Table 6.2 Using CTS to Enhance Teachers' Content Knowledge

CTS Section and Resources	Ways to Enhance Teachers' Content Knowledge
Section I—*Science for All Americans*	<ul><li>Identify and discuss the big, culminating ideas.</li><li>Identify examples that illustrate and explain the key ideas.</li><li>Identify relevant terminology and clarify definitions.</li><li>Find examples of ways the content integrates across science, mathematics, and technology.</li></ul>
Section I—*Science Matters*	<ul><li>Provide elegant explanations, real examples, and analogies for some difficult science concepts.</li><li>Make connections between concepts and illustrate how the sciences are connected.</li><li>Recognize bigger themes in science (such as energy) and how they connect to different content areas.</li><li>Identify basic facts, vocabulary, scientific principles, laws, or generalizations.</li></ul>
Section II—*Benchmarks for Science Literacy* and *National Science Education Standards* essays	<ul><li>Examine and discuss the *Benchmarks* K–12 overview essays to get a sense of what the "big ideas" are and why they are important.</li><li>Note places in the essays that need further content clarification such as suggestions for learning experiences where a term or phenomenon is unfamiliar.</li></ul>
Section III—*Benchmarks for Science Literacy* and *National Science Education Standards* essays	<ul><li>Unpack and clarify key ideas in the learning goal statements.</li><li>Discuss the "boundaries" of a learning goal—what content it includes and what it does not include (exceeds the learning goal).</li><li>Identify and clarify key terminology used in the learning goal statements.</li></ul>
Section IV—Research Chapter 15 from *Benchmarks for Science Literacy* and *Making Sense of Secondary Science*	<ul><li>Clarify learning goals and adult content (I and III) before teachers examine the research on learning.</li><li>Clarify any statements in the research that are not understood.</li><li>Check for similar commonly held ideas among the teachers.</li></ul>
Section V—*Atlas of Science Literacy*	<ul><li>Note the titles of the conceptual strands and clarify what they mean.</li><li>Note key terminology in the key ideas.</li><li>Analyze how one idea leads to another and how content understanding builds over time.</li><li>Analyze connections between ideas in different topics.</li></ul>
Section VI—State standards or district curriculum	<ul><li>Unpack and clarify key ideas in the learning goal statements.</li><li>Discuss the "boundaries" of a learning goal—what content it includes and what it does not include (exceeds the learning goal).</li><li>Identify and clarify key terminology used in the learning goal statements.</li><li>Ensure the performance verb does not mask the intent and understanding of the content goal.</li></ul>

SOURCE: Keeley, P. (2005). *Science Curriculum Topic Study: Bridging the Gap Between Standards and Practice.* Thousand Oaks, CA: Corwin.

Suggestions and Strategies for Using CTS to Enhance Adult Science Content Knowledge

Typically many teachers seeking to improve their content knowledge will take a university course, participate in a summer content immersion institute, or gain research experience to learn more about a particular area of science. These learning experiences are geared toward adult learners and provide significant, interesting content deemed important by scientists for understanding the natural world. It is important for teachers to realize that the material they are learning is useful for enhancing their own content knowledge, although it may not be directly applicable to their classrooms. Often teachers want to bring their own experience to their students. Rather than finding a way of modifying how they teach (as a result of their experience), they may actually try to recreate the same content or research experience for their students. They fail to see that the immersion or research experience was for them as adult learners, and that they can translate their new understanding in the classroom in subtle ways (such as an improved understanding of the "real" scientific method and what scientists actually do). This section addresses how to use CTS to enhance one's own adult understanding of science in order to be more knowledgeable about the content and process of science.

As teachers participate in content immersion experiences, research opportunities, and courses, CTS can help them translate what they are learning at the university or research level to what all adults are expected to know about science after their K–12 education. Using CTS Section I during a content immersion for adult learners can help teachers better understand the content being presented to them by seeing how it builds on and connects to fundamental ideas for adult science literacy that are the capstone of a K–12 education and endure throughout one's lifetime.

A strategy teachers frequently use to gain or refresh their own adult content knowledge before teaching a new topic is to look in their textbooks, teacher guides, or the Internet to get the information they need. While textbooks include the same content encountered by the students, they often only superficially cover the content, with an emphasis on terminology. Furthermore, they do not provide the rich descriptions, examples, and analogies included in science trade books that are written to make science interesting and accessible to the adult public.

CTS provides a different alternative to developing and enhancing adult content knowledge. Using the readings identified on a CTS guide to study a topic can considerably increase teachers' subject matter knowledge of the topics they teach as well as enhance the basic knowledge teachers need to be science literate adults. The following are some suggestions and strategies for building adult content learning into your CTS professional development designs. All handouts in this section can be accessed by going to the Chapter 6 folder on the CD-ROM at the back of this book and opening the subfolder labeled "Content Knowledge."

Eliciting Teachers' Current Ways of Improving Content Knowledge on Demand

The following activity helps teachers recognize CTS as a valuable alternative to the typical ways they might brush up on their content knowledge prior to teaching a topic. When introducing CTS to teachers as a systematic way to improve teacher content knowledge, consider beginning your session by finding out what teachers currently do to enhance their content knowledge in situations where they do not have the time to take a course or experience a professional development session or institute prior to teaching a unit. Post a chart in the room such as the example provided on Handout 6.1: Enhancing Content Knowledge. Give participants the following scenario:

> Imagine you are teaching a new topic next month. You have some familiarity with the topic but it has been awhile since you had any formal coursework in that topic. You don't have time to take a course and content institutes are not offered until the summer. Before you prepare your instructional unit, you need to strengthen or refresh your own understanding of the content, including understanding the important ideas that a science literate adult should know about the topic. Put a check mark on the chart to indicate the strategy you use the most to improve your own adult content knowledge before teaching a unit.

Give teachers colored dots or have them go up and place a check mark next to the strategy they use most often in this situation. Typically you will see no or few checks next to the CTS strategy of reading and analyzing *Science for All Americans, Science Matters,* or national standards documents. Explain to teachers that the CTS session they are about to experience will provide them with a new strategy that they can use to improve their content knowledge before teaching a topic or working with other teachers to support their content knowledge, such as a mentor coaching a novice teacher. This strategy also shows why it is important for science teachers to have access to the professional resources used in CTS to support their content learning. Prior to CTS, many teachers never thought of using *Science for All Americans* and other science literacy resources to support their own learning.

Identifying and Recording Adult Science Literacy Content Information From a CTS

Figure 4.2 in the CTS parent book shows the content information teachers record from CTS Sections I through IV that can enhance their adult science literacy. Handout 6.2 is an expanded version of this chart, focusing on CTS Section I. Use this handout when teachers are specifically focused on improving their own content knowledge using the CTS Section I readings. It can be embedded within the exploration and concept development phase of a full topic study. It can be used with both Section IA and Section IB readings, or facilitators can modify the handout if using only one of the Section I resources. Facilitators should build in time to debrief the readings and discuss the content recorded on the handout. Time for reflection should also be included—encourage participants to reflect on new knowledge gained from the readings and how it will enhance their teaching.

For example, a group of middle school teachers may be assigned to teach the topic of nuclear energy in their district curriculum. Nuclear energy is one of the topics addressed in the teachers' district's curriculum. Having a minimal background in chemistry and physics, the teachers engage in a CTS on nuclear energy to uncover the knowledge expected of a science literate adult. Using Section I in the CTS guide, "Nuclear Energy," the facilitator guides small groups in recording content statements from their reading of *Science for All Americans.* As the small groups share their findings with the whole group, the facilitator lists several of the major content statements, encouraging clarifying discussion of each concept and the terminology as it is listed. The facilitator guides a discussion around the importance of this background knowledge for the teachers and how it can help them be better prepared for teaching the topic. The following are examples of statements a group might generate from CTS Section IA around the topic of nuclear energy. Note that these are basic ideas for science literacy that come directly from the readings and do not get into the more complex science related to nuclear energy that a chemist or physicist would need to know in order to have specialized knowledge in their field.

- Reactions in the nuclei of atoms (nuclear reactions) involve far greater energy changes than reactions between the outer electron structures of atoms (chemical reactions).
- Fission is when very heavy nuclei, such as those of uranium and plutonium, split.
- Fusion is when very light nuclei, such hydrogen and helium, combine into somewhat heavier ones.
- Fission and fusion release large amounts of energy.
- Fission of some nuclei occurs spontaneously, producing extra neutrons that induce fission in more nuclei, and so on, thus giving rise to a chain reaction.
- Fusion occurs when nuclei collide at very great speeds such as in very high temperatures such as those produced inside a star or by a fission explosion.
- Nuclear reactors use the energy from fission to boil water for steam, which drives electric generators.
- The waste products of fission are highly radioactive and can remain so for thousands of years. This is one of the drawbacks of nuclear energy.
- Controlled nuclear fusion reactions are potentially a greater source of energy, but the technology is not yet feasible to generate energy in this way.

The above example shows how CTS can be used to help teachers identify, discuss, and seek understanding of the important ideas and terminology every adult should know about a topic, such as nuclear energy. For example, the teachers now recognize the tremendous amount of energy derived from nuclear energy and how a nuclear reaction differs from a chemical reaction. CTS also helps to explain two types of nuclear reactions using terminology a science literate adult should be familiar with—fission and fusion. It describes how we get electrical energy from nuclear reactors and what the potential consequence is. It also describes the potential of fusion to generate large amounts of energy but explains that our technology is not yet developed to the point where fusion is used to generate energy. While not all of these ideas may be taught at the middle school level, the background will help the teachers anticipate connections and respond to unanticipated questions that may arise during the teaching and learning process.

Concept Card Mapping

Concept card maps are used to show how learners recognize relationships between concepts and ideas described in all or part of a CTS Section I reading. The strategy is used most effectively with teachers who are already familiar with concept mapping and want to improve their own adult knowledge of the topic. A concept card mapping activity used with CTS Section I takes approximately 40 to 60 minutes, depending on whether you map all or part of the topic. To prepare for concept card mapping, the facilitator reviews the CTS Section I reading, selects key concepts from the reading that go with the topic or subtopic, and places them on cards. Each card contains a single concept. The number of cards varies according to the CTS topic or subtopic selected. Always be sure to provide some blank cards for teachers to write in additional concepts that may arise during their mapping. Handout 6.3 in the Chapter 6 folder on the CD-ROM has an example of concept cards used for mapping "Igneous Rocks," a subtopic of the CTS guide, "Rocks and Minerals." The concepts were selected by the facilitator from the CTS Section IB Science Matters reading on the part of "The Rock Cycle" (Hazen & Trefil, 1991, 2009). Facilitators can make up their own cards for the topics or subtopics they use in their professional development. The following directions describe how the facilitator uses the concept mapping cards with CTS.

Materials:

- Packet of concept cards (no more than fifteen cards)
- Blank sheet of paper or chart paper
- Glue stick

Directions:

1. Distribute a packet of cards to groups of two to three teachers.

2. Give the teachers five minutes to sort through the cards, placing related cards in groups. Encourage participants to discuss and clarify the terminology on each card.

3. Lay the groups of cards out on a sheet of paper. Encourage teachers to arrange them according to any relationships between cards that make sense to them. They may add additional concepts to the blank cards to form relationships. They should discuss the relationships in their small groups and leave space between the cards to draw in connecting lines that describe the relationship.

4. When the groups agree on the arrangement of the cards and their relationships, have them glue them to their papers or charts.

5. Draw in connecting arrowhead lines between related concepts and insert an accompanying phrase that describes the relationship between the two concepts. For example:

VOLCANISM   —————————————————▶   IGNEOUS ROCKS

is a process that leads to the formation of

6. After completing their concept maps, have the participants read the section from which the mapping cards were taken in the CTS Section I reading.

7. Have groups revisit their maps and discuss the relationships based on their reading. Groups should discuss any changes they might make after reading CTS Section I.

8. Debrief the maps with the whole group, asking for examples of key relationships they discovered through their reading. Provide additional content clarification as needed.

9. If time permits, groups can make a new map or write in new connections using different color ink, based on new knowledge they gained from the CTS reading.

This strategy can also be done using sticky notes and chart paper, or the cards can be laid out on a table with strips of paper for writing relationships between the cards. After the CTS Section I reading and discussion, participants can rearrange their sticky notes or table cards as needed or make new connecting statements. Figures 6.1 and 6.2 show before-and-after examples of concept card mapping of igneous rocks using CTS Section IB.

Another way to do concept mapping with a broad topic is to divide it into several subtopics and create concept cards for each of the subtopics. Small groups can be assigned a subtopic. After the mapping is completed, all groups share their maps and discuss the concepts as a whole topic. For example, the topic rocks and minerals can be broken down into four subtopics: the rock cycle, igneous rocks, sedimentary rocks, and metamorphic rocks. Each group can be assigned one of the four subtopics.

Figure 6.1   Concept Map Before CTS

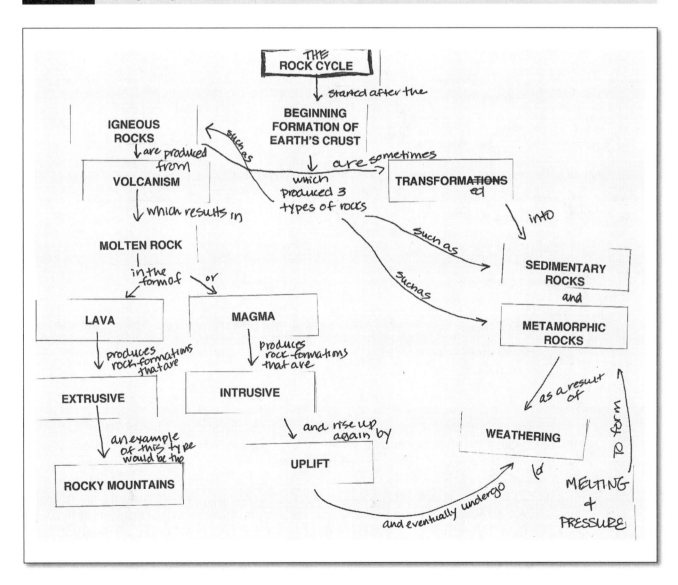

Figure 6.2 Concept Map After CTS

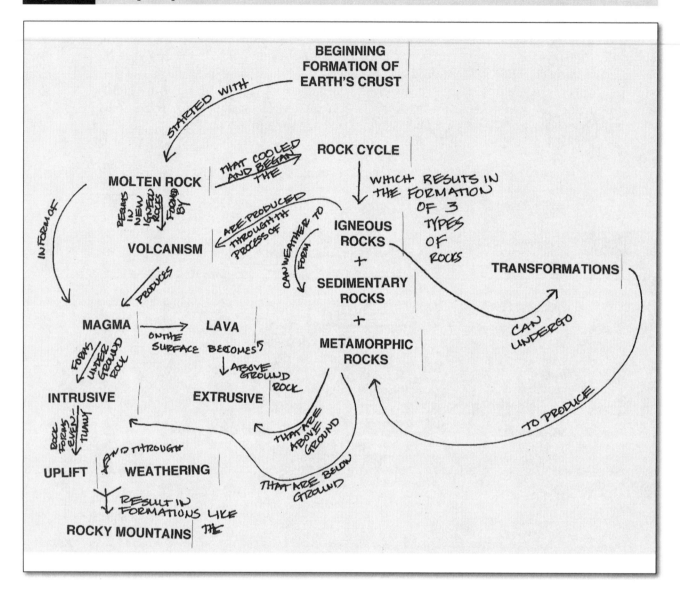

K-W-L

K-W-L is a strategy used to list what participants think they already know (K) about the basic ideas all adults should know about a CTS topic, what additional content knowledge they want to gain (W), and what new content they learned after doing the CTS (L). Template 6.4 for using K-W-L with CTS is provided in the Chapter 6 folder on the CD-ROM. Facilitators should ask participants to fill out the first two columns before they begin the CTS Section I readings. The last column is filled in after completing and discussing the CTS readings. Facilitators might point out Figure 4.4 in the CTS parent book, which shows an example of how a teacher using the CTS guide, "Gravity," might use K-W-L with CTS to enhance her content knowledge.

Content Vignettes

The CTS parent book contains several images from practice useful for introducing teachers to the various ways CTS can be used in their practice. On page 91, "Vignette #1: A High School Integrated Science Teacher Uses CTS to Understand Gravity-Related Content" shows how a teacher used CTS to improve her content knowledge. When introducing CTS as a way to improve upon or enhance one's content knowledge, consider using this vignette as a "mini-case" with the following questions:

1. What is the teacher's content problem? Why did she decide to use CTS?

2. What new content understanding did she gain from reading Section IA, *Science for All Americans*?

3. What did she gain by reading Section IB, *Science Matters*?

4. How did she connect what she read in Sections III and V to Sections IA and IB?

5. Summarize how CTS helped this teacher improve her content knowledge in preparation for teaching the topic of gravity.

Improving Grade Level (K–12) Content Knowledge

While teachers gain new understanding and an appreciation of science content through university courses, research experiences, and summer content immersions, a drawback of these experiences is that they often fail to make a connection to what teachers teach, particularly at the elementary and middle level. Scientists deeply know their content and have a wonderful ability to excite and involve teachers in learning about and doing science. What they often do not have is the specialized knowledge educators have that makes the content accessible to K–12 students. In other words, they lack the pedagogical content knowledge of "school science" needed to help teachers translate their experience into knowledge and activities that are developmentally appropriate, standards-based, and effective in promoting learning with students at their grade levels. For example, a third-grade teacher might participate in a weeklong institute on nanotechnology and learn a lot about this current, relevant area of science, yet not know how to take what she learned and incorporate it into her third-grade curriculum in a way that will build a foundation for her students when they encounter nanotechnology ideas at greater levels of sophistication as they progress through the grade levels. Nanotechnology is a sophisticated topic that requires developmental readiness and careful conceptual building toward particle ideas and understandings of scale. Knowing what nanotechnology is and learning about it in an adult context is very different from what one would teach third-graders. However, incorporating the CTS guides, "Particulate Nature of Matter (Atoms and Molecules)" and "Scale," into the teacher's learning experiences can help her plan backwards to identify the content that is related but appropriate to teach at a third-grade level to build a foundation for introducing progressively more sophisticated content.

For example, the topic study of scale reveals that students in Grades 3–5 are fascinated by extremes and that this interest can be exploited to develop a sense of scale. Furthermore they can be challenged to think about measuring things that are very hard to measure on account of being very small. However, it is not until Grades 6–8, when students' familiarity with very small numbers and powers of ten improves, that extremes of scale become more meaningful (AAAS, 1993).

In addition to the basic science any adult is expected to know in order to be considered a science literate person, as described in CTS Section I, Identify Adult Science Literacy, often there are statements in the K–12 science content standards and grade level essays that may be unfamiliar to teachers who lack a strong science background or who are strong in one science discipline but not another. Additionally, statements in the standards can be used to "unpack" content and identify the key ideas, skills, and appropriate terminology to use at different grade levels. It is important to take the time to clarify the content knowledge needed to understand and effectively teach the K–12 learning goals articulated in the standards. The following are suggestions and strategies used to improve teachers' knowledge of the content of the standards taught at their grade level. Handouts that accompany these suggestions and strategies can be found in the Chapter 6 folder, in the subfolder labeled "Content Knowledge," on the CD-ROM.

Inventory of Standards Content Statements

This is a strategy that can be used with CTS Sections II and III to improve teachers' content knowledge of the science at the grade levels they teach. This strategy is used to identify important content and related content ideas taught at the specific grade levels of the teachers' students. Table 6.3 shows an example of this strategy. The standards statements from the CTS Section III learning goals related to the topic are recorded in the left column as well as statements from the Section II essays. These essays often contain content statements embedded within the description of instructional considerations. For example, the fourth statement on Table 6.3 comes from the *Benchmarks* Grade 3–5 "Processes that Shape the Earth" essay suggestion: "Students should now observe elementary processes of the rock cycle—erosion, transport, and deposit" (AAAS, 1993, p. 72). Although the essay is describing an instructional implication, it also states three processes of the rock cycle that are important for teachers to know.

After participants have listed the content statements in the left column, they are asked to generate questions in the right column related to the content statements. These questions then are discussed in small groups and with the facilitator or content expert in the session. Discussion and clarification of the teacher-raised questions enhances their knowledge of the content they will be teaching and prepares them for questions that might arise later during CTS as well as during classroom instruction and investigation. A template for using this strategy is provided on Handout 6.5 in the Chapter 6 folder on the CD-ROM.

Table 6.3	Example of an Inventory of Standards Content Statements and Questions—Weathering and Erosion CTS (Grade 4)

CTS Standards Content Statements	Questions
Waves, wind, water, and ice shape and reshape the Earth's land surface by eroding rock and soil in some areas and depositing them in other areas, sometimes in seasonal layers (Section III—AAAS, 1993, p. 72).	Why are some layers seasonal? How long does it take for these processes to occur? Is there a difference in how materials are deposited on land, water, or both? How can you tell whether it was waves, wind, water, or ice that shaped a surface? How can you tell if glaciers were once in an area?
Smaller rocks come from the breakage and weathering of bedrock and larger rocks (Section III—AAAS, 1993, p. 72).	What is the difference between erosion and weathering? Do some types of rocks break and weather more easily than others?

CTS Standards Content Statements	Questions
Soil is made partly from weathered rock (Section III—AAAS, 1993, p. 72).	How does the weathered rock form soil?
The surface of the Earth changes. Some changes are due to slow processes, such as weathering and erosion, and some changes are due to rapid processes such as landslides, volcanic eruptions, and earthquakes (Section III—NRC, 1996, p. 134).	What are some examples of surfaces that were changed by slow processes? What length of time constitutes a slow process versus a rapid process?
Elementary processes of the rock cycle include erosion, transport, and deposit (Section II—AAAS, 1993, p. 72).	Do these processes occur in this order? Are they always cyclical? Where does weathering fit in?
Forces on Earth materials can make wrinkles, folds, and faults (Section II—AAAS, 1993, p. 72).	What are some examples of these effects of forces?
Erosion can be simulated in a small tray of soil or a stream table and compared to photographs of similar, but larger scale changes (Section II—NRC, 1996, p. 13).	How do you do this? What should I look for? To what large-scale changes can I compare it? What exactly is a stream table?

Unpacking Concepts and Scientific Processes

Concepts or scientific processes listed by themselves do not describe the key ideas or skills one must know to understand the concept and use the scientific process. For example, kinetic molecular theory is a concept that helps us understand the arrangement and behavior of particulate matter, but what exactly are the particles doing and how do you describe their arrangement? Scientific explanations are an essential feature of inquiry, but what does it mean to provide a scientific explanation in science? What exactly do students need to know about DNA? What do students need to know first before using a word like evaporation? CTS can be used to answer these K–12 content questions through unpacking a concept or skill in the standards by

1. identifying the CTS guide that includes the core concepts and scientific processes;

2. carefully examining the bulleted learning goals in CTS Sections III and V and identifying the concept or scientific process to which they are related; and

3. breaking down the learning goals into key idea or skill statements related to the concepts and scientific processes.

Table 6.4 shows examples of concepts, processes, key ideas, and skills at different grade levels that have been unpacked using CTS Section III or V. This strategy can be used for single concepts or processes within a CTS topic or used with all of the concepts or processes that make up an entire CTS topic. Use this strategy after participants have completed and discussed the readings from CTS Sections III and V. Handout 6.6: Unpacking Experimental Design, on the CD-ROM, shows an example of unpacking content in the

Table 6.4 Examples of Unpacking Grade-Level Concepts, Key Ideas, and Skills

Concept or Process	Grade Level	CTS Guide and Key Ideas and Skills for Science Literacy (From CTS Section III or V)
Kinetic molecular theory	6–8	States of matter • Atoms and molecules are perpetually in motion. • Since increased temperature means greater average energy of motion, most substances expand when heated. • In solids, atoms are closely locked in position and can only vibrate. • In liquids, atoms and molecules have higher energy than solids, are more loosely connected, and can slide past one another. • Some molecules of a liquid may gain enough energy to escape into a gas. • In gases, the atoms or molecules have more energy than solids and liquids and are free of one another except during occasional collisions.
The DNA molecule	9–12	DNA • In all organisms, the instructions for specifying the characteristics of an organism are carried in DNA. • DNA is a large polymer. • DNA is made up of four kinds of subunits (A, G, C, and T). • Chemical and structural properties of DNA explain how genetic information is encoded in genes and replicated by use of a template. • Each DNA molecule in a cell forms a single chromosome. • Changes in DNA occur through mutation.
Constructing scientific explanations	5–8	Evidence and explanation • Explanations are based on observations. • Explanations are different from descriptions. • Some explanations are stated in terms of the relationship between two or more variables. • Explanations emphasize evidence; have logically consistent arguments; and use scientific principles, models, and theories.
Evaporation	3–5	Water Cycle • Materials can exist in different states. • Some common materials, such as water, can be changed from one state to another by heating. • When liquid water disappears, it turns into a gas in the air. • This invisible gas in the air is called water vapor.

learning goals by identifying the key ideas and skills in each concept or process. This handout provides facilitators with an example of unpacking an entire K–12 topic related to experimental design.

Ideas Before Terminology

Examining content statements in CTS Section III, particularly at the elementary levels, often reveals how conceptual understanding is developed before introducing terminology. Developing ideas before giving students the vocabulary word provides an opportunity for them to construct understanding first and then be able to associate this understanding with the appropriate scientific terminology. Often when words are introduced first, the emphasis is on memorizing definitions at the expense of conceptual understanding. Encourage participants to look for examples in CTS Section III, where an idea in a learning goal is described before the scientific term is included. Often this was done intentionally by the standards developers.

Table 6.5 shows examples of learning goals that describe conceptual understanding before introducing the scientific terminology. When using Sections III, V, and VI, encourage teachers to look carefully at the learning goals for evidence of emphasis on developing the idea before including the terminology. As teachers discuss the conceptual ideas articulated in the content of the learning goal statement, ask them to identify the terminology that could be introduced after students have developed a conceptual understanding of the idea. Facilitators can also encourage participants to look at the Section III learning goals beyond their grade levels to identify when it is appropriate to introduce technical terminology.

Table 6.5 Ideas Before Terminology

Conceptual Idea (From Section III Learning Goals)	Grade Span	Terminology (Introduced After Developing the Concept)
The way to change how something is moving is to give it a push or a pull (AAAS, 1993, p. 89).	K–2	Force
Things near the Earth fall to the ground unless something holds them up (AAAS, 1993, p. 94).	K–2	Gravity
When liquid water disappears, it turns into a gas (vapor) in the air and can reappear as a liquid when cooled, or as a solid if cooled below the freezing point of water (AAAS, 1993, p. 68).	3–5	Evaporation Condensation
Large chunks of rock orbit the sun. Some of those that the Earth meets in its yearly orbit around the sun glow and disintegrate from friction as they plunge through the atmosphere and sometimes impact the ground (AAAS, 1993, p. 64).	6–8	Asteroids Meteors Meteorites Meteoroids

Using CTS to Examine the Hierarchal Structure of Content Knowledge in a Topic

Leaders of CTS can use this option to help teachers think about the different levels of content knowledge in a topic ranging from discrete facts to big ideas to overarching unifying concepts. When teachers are engaged in examining and building a hierarchy of content knowledge, they often come to the important realization that the content emphasized in their curriculum or textbooks is often at the lower level of a content hierarchy (e.g., facts, terminology) and fails to develop important conceptual ideas and generalizations. The intellectual act of constructing a hierarchy helps teachers examine whether the focus of their curricular topic is on lower-level factual knowledge, what the key ideas and underlying concepts are in a learning goal, how they can build from the concepts and key ideas to construct understandings of bigger ideas, and how to integrate ideas using unifying principles that cut across the disciplinary content areas of science. The hierarchy is one of the more challenging applications of CTS. This application is most successful when teachers have had experience using CTS and understand the difference between facts, key ideas, concepts, big ideas, and unifying themes and principles. For groups who the facilitator feels are ready to try constructing a hierarchy of knowledge, this strategy proves to be one of the most intellectually challenging and invigorating applications of K–12 content knowledge. The end result is that it increases teachers' ability to modify curriculum and design instruction that promotes higher-level use of knowledge and the development of major scientific principles and ideas, generalizations, and connections across areas of science. However, facilitators need to be warned that this strategy is not an easy one to use if teachers do not have a strong understanding of the content they teach.

Pages 57–61 of the CTS parent book describe the structure of knowledge in a topic. It is important for leaders to read and become familiar with this section before using this application in a CTS session. Handout 6.7 is a revised graphic of the one shown on page 60 of the CTS parent book for facilitators to use when working with groups using this application. Embedding the "Hierarchal Structure of Content Knowledge" in a CTS session helps participants recognize the range of knowledge that a curricular topic can address, from the low-level facts that often end up being memorized and later forgotten to the really big ideas found in the unifying concepts like models, organization, evolution, systems, continuity, and change. These bigger ideas are often overlooked when unpacking standards and teaching specific ideas.

Before using this application with a CTS topic, make sure the operational definitions of the terms used on Handout 6.7 have been discussed and clarified with the group. For example, there are many different definitions of what a concept is. For our purposes, we agree to define a concept as a one word (or made up of a few words) mental construct that frames an idea. Note that the hierarchy example in the CTS parent book takes a curricular topic and breaks it down using the elements described in the hierarchy. Breaking down a curricular topic can be quite cumbersome if participants do not have experience using a hierarchy of knowledge. A simpler way to start is with a concept that is part of a topic, rather than with a full topic. Once participants grasp the process using a single concept, they can then go on to construct a hierarchy of a full topic that includes several related concepts. The strategy described here will address the hierarchy at the concept level within a topic, rather than at the topic level as shown in the CTS parent book.

After participants complete a topic study and discuss the CTS findings for the topic, explain to participants that they will now use their CTS content ideas from Sections I, III, V, and VI to take a single concept in the topic and break it down into subconcepts (for some concepts that tend to be quite specific you can skip the subconcept level), specific ideas, and facts, terminology, and formulas. This deconstruction process is a way of unpacking a concept. However, explain how the opposite direction is also important and is often the direction that is neglected when working with concepts. To go "up," explain how they will construct a hierarchy of bigger ideas that builds on a concept. When they are done, they will see the structure of knowledge unfold from the top level of "really big unifying ideas" to the lower level of factual, discrete information.

Refer participants to Handout 6.8 on the CD-ROM, which scaffolds the process of building a hierarchy starting with a concept (rather than a topic as described in the CTS parent book). Walk them through an example using one of the Chapter 6 examples provided on the CD-ROM or provide a copy of Handout 6.9 to see four different grade level examples of a concept hierarchy. Build in time to discuss each of the steps of the scaffold and examine how they link to the examples provided.

When participants have grasped the notion of a hierarchy, understand what the different levels are, and have had an opportunity to examine an example, they are ready to work in small groups to try out the scaffold with the CTS they completed. As a group, you might start by brainstorming a list of major concepts that were addressed in Section III of their CTS and then have groups select a concept to build a hierarchy around. Provide a copy of Handout 6.10 on the CD-ROM to guide their discussions in small groups as they work on their hierarchy. Handout 6.11 provides a template for creating a hierarchy chart. It is important to emphasize while there is no one right way of doing this, it is the exercise in thinking about the structure of the knowledge that will inform how they teach the content that makes this a valuable learning experience. Provide time for participants to share their examples and reflect on the impact this application will have on their teaching. An extension of this activity is to write hierarchal essential questions to go with each of the levels identified.

CTS AND SCIENCE CURRICULUM

Another context application for which CTS is widely used is curriculum. Curriculum is broadly defined as the way content is designed and delivered. It includes the structure, organization, balance, and presentation of the content in the classroom (National Research Council [NRC], 1996). Leaders can use the CTS tools and process for a variety of curricular applications, including facilitating groups charged with designing a K–12 coherent scope and sequence, selection of curriculum materials that are standards- and research-based, and support for curriculum implementation.

This section provides suggestions, tools, and strategies for using CTS in a curricular context. All of the resources and handouts for this section can be found on the CD-ROM by going to the Chapter 6 folder and opening the subfolder labeled "Curriculum 6.12–6.21." The following box provides an outline of the suggestions included in this section for using CTS for curriculum-related work. Table 6.6 provides a description of the features of the tools, strategies, and examples provided in the text and CD-ROM used to help facilitators plan how to use CTS to support standards- and research-based curriculum.

OUTLINE OF SUGGESTIONS FOR CTS CURRICULAR APPLICATIONS

A. Curriculum coherence and articulation

   1. The Three Little Pigs metaphor

   2. Combining topics for broader study

   3. Choosing curricular priorities

   4. Creating clarification guides for curriculum topics

   5. Clarifying a state standard

   6. Developing crosswalks to state standards

B. Curriculum selection

   1. CTS summary for curriculum materials review

   2. Summary review of curriculum materials

C. Curriculum implementation

   1. Creating customized CTS guides for curriculum

   2. Developing CTS curricular storylines

Table 6.6 Features of Tools and Strategies Used in a Curricular Context

Strategy or Suggestion	CTS Topic(s) Used as Examples	CD-ROM Handouts	Purpose of This Strategy or Suggestion
Three Little Pigs metaphor	Not topic specific	6.12 6.13	To introduce CTS as a tool curriculum committees can use to develop a strong curriculum
Combining topics for broader study	All topics covered	none	To undertake a comprehensive study of all the content likely to be included in a K–12 science curriculum
Choosing curricular priorities	Mechanism for Inheritance (Genetics)	6.14	To unburden the curriculum by deciding what to leave in and what can be left out
Creating clarification guides for curricular topics	Rocks and Minerals	6.15	To provide a summary of CTS findings teachers can use to clarify a topic included in their curriculum
Clarifying a state standard	States of Matter	6.16 6.17	To bring greater clarity to the meaning and intent of a state standard

Strategy or Suggestion	CTS Topic(s) Used as Examples	CD-ROM Handouts	Purpose of This Strategy or Suggestion
Developing crosswalks to state standards	All topics	6.18 and 6.18.1	To provide a tool states can use to link their standards to a CTS topic
CTS summary for curriculum materials review	Used with any topic	6.19	Template used to summarize CTS findings used to review materials
Summary review of curriculum materials	Used with any topic	6.20	Template used to summarize the match between curriculum materials and the findings from CTS in Handout 6.19
Creating customized guides for curriculum	Used with any topic	6.21	Template for customizing a CTS guide to match the topic of the curriculum unit

Curriculum Coherence and Articulation

"A coherent curriculum is one that holds together, that makes sense as a whole; and its parts, whatever they are, are unified and connected by that sense of the whole" (Beane, 1995, p. 3). This involves carefully thinking through the flow of ideas, how they interconnect, and how the process and content of science are intertwined. Studying curricular topics and using the tools provided in this *Leader's Guide* can help curriculum committees make better decisions when grappling with the design and organization of a coherent K–12 science curriculum.

CTS does not recommend a particular approach to designing and organizing a K–12 curriculum. There are a variety of ways to do this, and school districts often have their own formats for putting together the curriculum. The leader's work involves deciding how to best use CTS to inform the work of a curriculum committee. What CTS does provide is the information committees need to think through and make sound curriculum decisions. For groups interested in designs for organizing curriculum that utilize the *Benchmarks for Science Literacy* (AAAS, 1993) or a coherent set of learning goals, we recommend the book *Designs for Science Literacy* (AAAS, 2001), one of the science literacy tools developed by Project 2061. *Designs for Science Literacy* addresses the critical issues involved in assembling sound instructional materials into a coherent K–12 whole and proposes ways to choose and configure curriculum so it aligns with learning goals.

Suggestions for Using CTS to Develop and Articulate a Coherent K–12 Curriculum

Leaders who facilitate committees charged with developing or examining K–12 curriculum to align with standards should first become familiar with pages 63–65 in the CTS parent book. Figure 4.8 in that book addresses five major considerations for curriculum when using CTS to study a topic. As committees make decisions about curriculum, use the questions in Figure 4.9 of the CTS parent book to guide or reflect on decisions. Other strategies and activities leaders can use when working with K–12 curriculum committees include the following.

Three Little Pigs Metaphor

Use the adaptation of the Three Little Pigs story on Handout 6.12 on the CD-ROM to make a case for why curriculum committees would benefit from using the CTS tools to "build" a stronger curriculum. This activity is best used to engage curriculum committees in thinking about a different way to go about the process of developing, revising, or revisiting their K–12 scope and sequence curriculum. This process involves using CTS to ground the committee in the relevant content, appropriate ways to sequence learning goals, alignment of curriculum with standards, developmental appropriateness, and more (see Figure 4.8 on page 64 of the CTS parent book).

The Three Little Pigs metaphor clearly shows why tools, such as CTS, can strengthen the work teachers do in areas like curriculum. Ask for three volunteers to read the story aloud. Designate each volunteer to be one of the three pigs. As they read the story aloud, have the audience read silently along with them. When finished, distribute Handout 6.13 and have small groups or pairs respond to each of the questions (each person responds individually to the first question and places the sticky note on a chart that visually represents a bar graph of Pigs 1, 2, and 3). After placing a sticky note on the wall graph, individuals meet in small groups to begin the discussion of the questions on the handout. Allow about 35 minutes for them to work through the task and then debrief ways they might think about using CTS to avoid the pitfalls of Pigs 1 and 2 and use some of the strategies of Pig 3 in the context of their own curriculum work.

Combining Topics for Broader Study

Dividing the 147 topic study guides up among a curriculum committee to study in order to inform their work is certainly not practical. Many committees look at the disciplinary content through the traditional organizers of life science; physical science; Earth and space science; inquiry, technology, and nature of science; science and society; and unifying themes; therefore, the listing below of twenty-eight topic studies, taken together, should cover the entire K–12 science curriculum. Curriculum committees can divide these up to study the K–12 content for a particular area of science before assigning curriculum topics to grade levels. This study gives the curriculum committee a more connected view of the content within and across disciplinary boundaries. The following are suggestions for condensing the vast number of topics in the CTS parent book into a manageable number of topics that a curriculum committee can undertake over the span of their work. For example, when a committee is working on the life science portion of their curriculum, they may choose to divide the five topics listed under the life science category among their members and use the study results to inform their curricular decisions. This is more efficient than studying all thirty-nine life science topics listed under the four CTS categories of Diversity of Life, Ecology, Biological Structure and Function, and Life's Continuity and Change.

Life Science

- Characteristics of Living Things
- Ecosystems

- Health and Disease
- Biological Evolution
- Mechanism of Inheritance (Genetics)

Physical Science

- Chemical Properties and Change
- Physical Properties and Change
- Particulate Nature of Matter (Atoms and Molecules)
- Energy
- Forces
- Motion

> **Facilitator Note**
>
> The physical science category can be further broken down into two categories: (1) Matter, and (2) Energy, Force, and Motion.

Earth and Space Science

- Processes That Change the Surface of the Earth
- Weather and Climate
- Earth History
- Earth's Natural Resources
- Structure of the Solid Earth
- Earth, Moon, and Sun System
- The Universe

Inquiry, Technology, and Nature of Science

- Science as Inquiry
- Technology
- The Nature of Scientific Thought and Development

Science and Society

- Historical Episodes in Science
- Personal and Community Health
- Science and Technology in Society

Unifying Themes

- Models
- Systems
- Scale
- Constancy, Equilibrium, and Change

Furthermore, leaders can develop customized CTS guides to study a broad area of curriculum at a particular grade level. For example, Table 6.7 shows how middle and high school topic readings from all the matter topic studies were combined into one study CTS guide, "Matter." The curriculum committee used this guide to study a matter strand for their Grade 7–12 scope and sequence. To prepare a combined guide, use the blank template for creating your own CTS found in the Chapter 5 folder, Templates for Developing Your Own CTS, on the CD-ROM. Make a list of all the readings from multiple topics for the grade level(s) that make up a particular strand that your committee wants to study. Combine repeated readings into one and list them on the template like the example shown in Table 6.7.

Table 6.7 Example of a Broad, Combined Topic for Middle and High School Curriculum

Section and Outcome	Selected Sources and Readings for Study and Reflection; Read and Examine <u>Related Parts</u> of:
I. Identify Adult Content Knowledge	**IA:** *Science for All Americans* • Chapter 4, *Structure of Matter*, pages 46–48 • Chapter 4, *Energy Transformations*, pages 49–52 • Chapter 10, *Understanding Fire*, pages 153–155 • Chapter 10, *Splitting the Atom*, pages 155–157 **IB:** *Science Matters—Achieving Scientific Literacy* • Chapter 4, *The Atom*, pages 54–64 • Chapter 6, *Chemical Bonding*, pages 75–93 • Chapter 7, *Atomic Architecture*, pages 94–109 • Chapter 8, *Nuclear Physics*, pages 110–123
II. Consider Instructional Implications	**IIA:** *Benchmarks for Science Literacy* • 4D *Structure of Matter* general essay page 75; grade span essays, pages 77–80 • 4E *Energy Transformations* general essay pages 81–82; grade span essays, pages 84–86 • 10F *Understanding Fire* general essay page 249; grade span essays, pages 250–251 • 10G *Splitting the Atom* general essay page 252; grade span essays, pages 252–253 **IIB:** *National Science Education Standards* • Grades 5–8, Standard B essay, pages 149, 154 • Grades 9–12, Standard B essay, page 177
III. Identify Concepts and Specific Ideas	**IIIA:** *Benchmarks for Science Literacy* • 4D, *Structure of Matter*, pages 78–80 • 4E, *Energy Transformations*, pages 85–86 • 10F, *Understanding Fire*, pages 250–251 • 10G, *Splitting the Atom*, pages 252–253 **IIIB:** *National Science Education Standards* • Grades 5–8, Standard B, *Properties and Changes of Properties in Matter*, page 154; *Transfer of Energy*, page 155 • Grades 9–12, Standard B, *Structure of Atoms*, page 178; *Structure and Properties of Matter*, pages 178–179; *Chemical Reactions*, page 179; *Conservation of Energy and the Increase in Disorder*, page 180; and *Interactions of Energy and Matter*, pages 180–181
IV. Examine Research on Student Learning	**IIIA:** *Benchmarks for Science Literacy* • 4D *Structure of Matter*, pages 336–337; *Heat and Temperature*, page 337; *Heat Transfer*, pages 337–338 **IVB:** *Making Sense of Secondary Science—Research Into Children's Ideas* • Chapter 8, *Materials*, pages 73–78 • Chapter 9, *Solids, Liquids, and Gases*, pages 79–84 • Chapter 10, *Chemical Change*, pages 85–91 • Chapter 11, *Particles*, pages 92–97 • Chapter 12, *Water*, pages 98–103 • Chapter 13, *Air*, pages 104–111

Standards- and Research-Based Study of a Middle/High School Curricular Strand: Matter
(*Note:* Energy ideas are included if they support understanding matter ideas)

Section and Outcome	Selected Sources and Readings for Study and Reflection; Read and Examine <u>Related Parts</u> of:
V. Examine Coherency and Articulation	**V: *Atlas of Science Literacy*** • *Atoms and Molecules*, pages 54–55 • *Conservation of Matter*, pages 56–57 • *States of Matter*, pages 58–59 • *Chemical Reactions*, pages 60–61
VI. Clarify State Standards and District Curriculum	**VIA: *State Standards:*** Link Sections I–V to learning goals and information from your state standards or frameworks that are informed by the results of the topic study **VIB: *District Curriculum Guide:*** Link Sections I–V to learning goals and information from your district curriculum guide that are informed by the results of the topic study
Visit www.curriculumtopicstudy.org for updates or supplementary readings, Web sites, and videos.	

Choosing Curricular Priorities

One of the difficulties curriculum committees often face when assembling a curriculum is deciding what is most important to leave in and what can be taken out. These decisions are often based on personal biases rather than a careful study of the concepts, key ideas, and skills that form a science literacy core. The following quote from *Designs for Science Literacy* (AAAS, 2001) describes the dilemma educators face with an overstuffed, mile-wide, inch-deep curriculum:

> Time in school for teaching and learning is not limitless. Yet many textbooks and course syllabi seem to assume otherwise. They include a great abundance of topics, many of which are treated in superficial detail and employ technical language that far exceeds most students' understanding. And even as new content is added to the curriculum, little is ever subtracted—students are being asked to learn with greater depth. Rarely is more time made available for accomplishing this. Coverage almost always wins out over student understanding, quantity takes precedence over quality. (AAAS, 2001, p. 211)

CTS provides a process that encourages justification of curricular decisions, based on standards and research, for reaching consensus on which concepts and ideas can be eliminated, which ones are essential to learning, and which ones could be included if science literacy goals are met and there is time for additional subject matter. The process also eliminates unnecessary redundancy and helps identify essential technical terminology in order to concentrate learning first on conceptual understanding and avoid the overemphasis on specialized vocabulary. Through a careful examination of CTS Sections III and V, in concert with examining local or state standards from Section VI, informed decisions can be made about which key ideas and skills are clearly emphasized in the standards and where the boundaries should be drawn in order to eliminate unnecessary instruction.

For example, the *Benchmarks for Science Literacy* clearly show that by Grade 8, students should have a conceptual understanding of the atom as a basic building block of matter, but that there is no need, as evidenced in the learning goals in both sets of national standards, to have middle school students learn about the parts of an atom and what they do. Even though the parts of an atom may be in the eighth-grade textbook, this can wait until high school, giving middle school teachers time to focus in more depth on the content that is most essential for students to understand at this grade level. An exception would be if a state decided, contrary to what is proposed in the national standards, to include parts of an atom in their middle school learning goals. In this case, even though this idea is included at the middle school level, it is an indication to teachers that although they must teach this, they should probably place more emphasis, given the limited time teachers have for teaching, on the learning goals that are most central to the middle school curriculum.

The results from CTS Sections III, V, and VI form the center of the graphic shown in Figure 6.3 and used in Handout 6.14: Nested Priorities for Science Curriculum. Curriculum committees can use this organizational chart to examine their current curriculum or curricular suggestions and make decisions about what to leave in or take out. The central content core is essential and focuses on a fewer number of key ideas. As one moves outward from the center of the diagram, the content becomes less essential, and if included, widens the

Figure 6.3 Nested Priorities for Science Curriculum

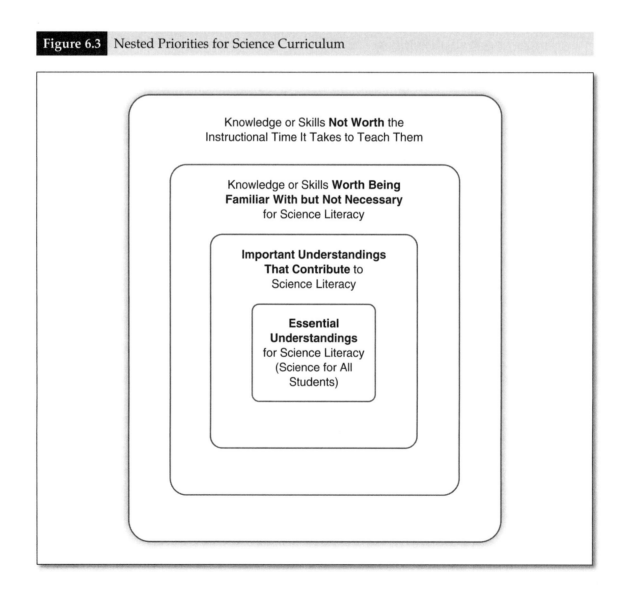

Knowledge or Skills **Not Worth** the Instructional Time It Takes to Teach Them

Knowledge or Skills **Worth Being Familiar With but Not Necessary** for Science Literacy

Important Understandings That Contribute to Science Literacy

Essential Understandings for Science Literacy (Science for All Students)

curriculum. Keep in mind the wider a curriculum gets, the less opportunity students will have for in-depth learning. The four levels are described as follows:

Level 1: Essential Understandings for Science Literacy. These are the key ideas that come directly from the standards, including appropriate terminology used in the learning goals. They are identified by using CTS Sections III, V, and VI. These are the important ideas that are central to the curriculum that all students are expected to learn at the grade level indicated.

Level 2: Important Understandings That Contribute to Science Literacy. These are ideas that can support learning the key ideas and are reasonable for students at the indicated grade level to learn without compromising the time needed to develop in-depth understanding.

Level 3: Knowledge or Skills Worth Being Familiar With but Not Necessary for Science Literacy. These often include ideas that are included in curriculum materials and activities that may be interesting and possible for students to learn if they have achieved the key ideas and teaching these ideas does not replace the time needed to learn essential content. If they are eliminated from the curriculum, it does no harm. These may also be knowledge and skills that exceed science literacy and are targeted for students who are ready to move beyond the standards for achieving science literacy for all.

Level 4: Knowledge or Skills Not Worth the Instructional Time It Takes to Teach Them. These are ideas that exceed the readiness and ability of students at a specific grade level to comprehend, based on their prior knowledge and experience. They often include ideas that will be taught at a later grade span. It would take too much instructional time to be worth the effort to teach these ideas and may confuse students. They should be eliminated from the curriculum for that grade span, but could be considered at a later grade.

Table 6.8 shows how a curriculum committee might use this graphic to make decisions about what to cover in the curriculum. The example shows the decisions made regarding the topic of genetics taught in seventh grade after teachers have done a CTS using the CTS guide, "The Mechanism of Inheritance (Genetics)."

Table 6.8 Curriculum Priorities for Teaching Genetics in Grade 7

Priority	Examples
Essential understandings for science literacy	• In asexually reproducing organisms, all the genes come from a single parent. In sexually reproducing organisms, typically half the genes come from one parent. • Fertilized eggs carry genetic information from both parents. As these cells multiply to form a complete organism, the same genetic information is copied in each cell. • Every organism requires a set of instructions for specifying its traits. Heredity is the passage of these instructions from one generation to another. • Hereditary information is contained in genes, located in the chromosomes of each cell. • Each gene carries a single unit of information.

(Continued)

Table 6.8 (Continued)

Priority	Examples
Essential understandings for science literacy (continued)	• An inherited trait of an individual can be determined by one or by many genes, and a single gene can influence more than one trait. • A human cell contains many thousands of different genes. • The characteristics of an organism can be described in terms of a combination of traits. Some traits are inherited and others result from interactions with the environment. • New varieties of cultivated plants and domestic animals resulted from selective breeding for particular traits.
Important understandings that contribute to science literacy	• Genes make up DNA, the molecule that stores information. • Genes come in pairs. • Within cells, DNA is organized in structures called chromosomes. • There are variations in populations of species.
Knowledge or skills *worth being familiar with but not necessary* for science literacy	• Using Punnett Squares to predict or show results of basic genetic crosses. • Dominant and recessive traits and blending. • History of genetics—Mendel. • Sex-linked traits. • Common genetic disorders.
Knowledge or skills *not worth* the instructional time it takes to teach them at this grade level	• Specific details of mitosis and meiosis. • Structure of a DNA molecule. • Mechanism of DNA replication. • Types of mutations. • Mechanism of cloning and genetic engineering. • Specific causes of genetic disorders. • Alleles. • Genome mapping.

SOURCE: Key ideas from *Benchmarks for Science Literacy* (AAAS, 1993) and *National Science Education Standards* (NRC, 1996)

When using the chart with curriculum committees, make a large chart out of Handout 6.14. Before studying a curricular topic, have participants list concepts, ideas, or skills that are typically taught in relation to the topic identified. Have them write each one on a sticky note and place it on the priorities chart according to where they think it belongs. After studying a topic using CTS, revisit the chart and have participants suggest any changes in the placement of their sticky notes. Invite them to add any additional concepts, ideas, or skills that should be included in the two central boxes as essential or important understandings.

Creating Clarification Guides for Curriculum Topics

After a curriculum committee has conducted a CTS, committee members can put together a summary of their findings to include in their district curriculum guides. These summaries help to clarify the content, learning goals, instructional considerations, possible misconceptions that may surface, and links to state standards. An example of a K–12 curricular clarification for the K–12 topic of rocks and minerals put together by a Maine school district is shown on page 67 in the CTS parent book.

The clarification was included in the district curriculum binders all teachers received and provided a K–12 big picture summary of the specific ideas that underlie the state standards, research on learning that could be used to inform teaching, and implications for instruction at different grade levels. A copy of the template curriculum committees can use to electronically enter their CTS findings is included on Handout 6.15 on the CD-ROM.

Clarifying a State Standard

As curriculum committees work to align their curriculum with new or revised state standards, or improve the existing alignment, they often wrestle with the interpretation of a learning goal from their local or state standards. State standards are written at various levels of specificity, from broad to very specific. Unlike the national standards, they are usually written with the inclusion of a performance verb. Sometimes the addition of a performance verb, particularly if the verb was arbitrarily chosen to ensure a range of levels of performance across standards rather than the cognitive complexity of the idea being assessed, can change or mask the specific science idea or key ideas that make up a learning goal. While almost all states have based their standards on the *Benchmarks for Science Literacy* or the *National Science Education Standards*, they are seldom written word for word in their original language. Examining the national goal statement from which a state or local standard was derived can often reveal much more specific and detailed information about the intent and meaning of a learning goal. The steps facilitators can use to guide participants through the CTS process of clarifying a learning goal include the following:

1. Identify a learning goal from your local or state standards (CTS Section VI).

2. Identify the CTS guide related to the learning goal.

3. Use CTS Section III to identify a similar learning goal from the national standards.

4. Break the national learning goal into its component parts, called "key ideas" (sometimes a learning goal consists of only one key idea).

5. Use CTS Section IV (or the narrative that precedes the *Atlas of Science Literacy* [Vol. 2] maps from Section V) to identify learning research related to the learning goal (not the entire topic).

6. Use CTS Section II to identify instructional implications for teaching the learning goal (not the entire topic).

7. Use CTS Section V to identify connections to other ideas as well as important prerequisite understandings that should be developed before introducing the learning goal.

Handout 6.16 on the CD-ROM shows an example of a goal clarification for a middle school state standard: "Explain the relationship of the motion of atoms and molecules to the states of matter for gases, liquids, and solids" (Maine Department of Education, 2007). What exactly are students expected to explain? Using the CTS guide, "States of Matter," the committee was able to use CTS Section III to trace the state standard goal statement back to the national standard from the *Benchmarks for Science Literacy* from which it was derived. After breaking down the Maine standard into the key ideas students should be able to explain, further clarification from CTS Section IV was provided to grade level teachers on what the research has found to be difficult for students, as well as commonly held ideas. CTS Section II was used to describe considerations to take into account when planning instruction around these key ideas. Use of the *Atlas of Science Literacy* in CTS Section V points out important connections to the key ideas that will help support students' ability to explain the key ideas that make up the standard. Overall the process of clarifying a single state standard, using a CTS guide, led to a greater understanding of the content, curricular, and instructional meaning and intent of the learning goal than what one could derive merely from looking at the goal statement. A template that can be used for this process of clarifying a state standard is provided on Handout 6.17 on the CD-ROM.

Developing Crosswalks to State Standards

Leaders who facilitate implementation of state standards can use the process described above to clarify a particular state learning goal. In addition, leaders who work extensively with a set of state standards might consider developing a crosswalk between their state standards and the CTS guides. The crosswalk provides an easy-to-use tool for teachers or other users of state standards to quickly cross-reference a CTS guide to a state standard that relates to the CTS topic. The CTS guide can then be used to clarify the standard. In addition, leaders who are considering the revision of their local or state standards can develop a crosswalk for committees to evaluate their existing standards and make modifications based on improved interpretation and research on learning. Handouts 6.18 and 6.18.1 on the CD-ROM provide two different examples of crosswalks developed to cross-reference state standards with the CTS guides: the 1995 State of Arizona Science Standards and the 2007 Maine Learning Results. In the Arizona example (Handout 6.18), a table was created that listed each specific standard in the Arizona standards and matched it to a CTS guide that could be used to clarify the standard. In the Maine example (Handout 6.18.1), the CTS guides were entered right into the standards document in capital letters to match the descriptors (specific performance objectives that are expected of students). The CTS guides can be used to clarify the meaning and intent of a descriptor.

Using CTS for Curriculum Selection

In many districts, the process of selecting curriculum materials has been simply to pick something popular with a few teachers or, worse, have all teachers use their own materials with little coordination. Increasingly districts are engaging in more thoughtful analysis of the curriculum and its alignment with standards and research as both a professional development strategy and a means to improve curriculum selection (Loucks-Horsley, Stiles, Mundry, Love, & Hewson, 2010). CTS can be a useful tool for selecting curriculum materials that are based on standards and reflect the research on learning. Curriculum selection leaders and committee members should first read pages 65–69 in the CTS parent book. This section describes how CTS can be used for curriculum selection. The following tools can be used by curriculum selection committees to deepen their understanding of the standards and research on learning that should inform curriculum materials and evaluate the extent to which curriculum materials reflect CTS findings.

CTS Summary for Curriculum Materials Review

If the goal of a curriculum selection committee is to select standards-based and research-informed materials, then doing a CTS prior to examining and selecting new curriculum can make selection committees more aware of what to look for in the material. Publishers often claim that their materials are "standards-based" and developed with the latest research in mind. However, upon close examination, the alignment is often at a topical level or may miss the key ideas at the grade level specified in the standards. For example, national standards and most state standards do not address details of cell organelles at the middle school level. The emphasis in the middle school standards is on different types of cells that perform different types of functions. In looking at materials that address cell ideas at the middle school level, a curriculum selection committee would be looking for this key idea in the standards. If the materials devote more emphasis on parts of the cell and details about cell organelles, but little about different types of cells, this would be considered aligned beyond the grade level of the standards and a selection committee might decide the material does not meet their grade level standards.

Handout 6.19 provides a template selection leaders and committee members can use to summarize findings from a CTS that are then used as a lens through which to rate the material according to the extent to which there is evidence that matches the CTS results. To create a CTS summary guide for curricular review, the committee first chooses the CTS guide that best matches the topic of the curriculum materials or unit that will be examined. For example, on page 70 in the CTS parent book, an elementary committee is examining a third-grade unit on light. The committee does a CTS and records the following information in a concise format that can be used at a glance to examine the curriculum material:

1. *Concepts for Teacher Background Information:* Use CTS Section I and the K–12 overview essay in Section IIA to identify content background material that should be provided for teachers using the curriculum. Record the major concepts identified in these CTS sections. (*Note:* A concept is a word or short phrase that provides a mental construct for the key ideas. For example, the concept *nature of light* provides a mental construct for the key idea that "light travels in a straight line until it strikes an object or material.")

2. *Students' Content Knowledge:* Use CTS Sections III, V, and VI to list specific ideas that should be included in the learning objectives of the curricular unit.

3. *Instructional Implications:* Use CTS Sections II and IV to identify instructional strategies, appropriate learning contexts, relevant phenomena, or developmental considerations that should inform the lessons that make up the unit.

4. *Student Difficulties and Misconceptions:* Use CTS Sections II and IV to identify any difficulties students may have learning the ideas in the topic or potential misconceptions students may develop or bring to their learning. This information will be used to look for explicit mention of these in the teacher's guide or evidence of activities designed to address these difficulties and misconceptions.

5. *Prerequisite Knowledge:* Use CTS Sections III and V to examine ideas that are developed in the prior grade level or precursor ideas within the grade level that should be taught before introducing other ideas. This information will be used to look for evidence in the teacher's guide that alerts teachers to the prior knowledge students will need or activities that consider and build upon prerequisite skills and knowledge.

6. *Connections to Other Topics:* Use CTS Section V to look for connections that can be made to other curricular topics. Look for evidence of these connections in the activities or descriptions of extensions in the teacher's guide that connect to other lessons or units. (*Note:* Occasionally Section II will provide useful information about connections.)

Once this review has been completed and recorded by the committee, it will be used as an overview-at-a-glance of the things a curriculum committee should look for in the unit. The committee then uses this information to rate a material on how well it addresses the standards and research on learning that are revealed through the CTS process. However, this review process only looks at how well it reflects the CTS findings. It does not evaluate the pedagogical design of the materials or the likeliness of instructional effectiveness.

Summary Review of Curriculum Materials

Now that you have the CTS information to guide the review of the material you selected, the next step is to examine the material by looking at the extent to which it addresses the information revealed through the CTS. Handout 6.20 on the CD-ROM provides a template for reviewing the curriculum materials using evidence from the summary completed in Handout 6.19. A description of this tool and an example of how it was used with a CTS and a curriculum unit on light is shown in Figures 4.11 and 4.12 in the CTS parent book. There are five ratings, ranging from 1 to 5, with a 1 being no evidence of the CTS findings in the material and a 5 including strong evidence of the CTS information in the materials as well as additional appropriate features that build from the standards and research that may exceed the basic threshold recommendations of CTS but still contribute to student learning in a meaningful way. For example, in developing the students' content knowledge about light, there is strong evidence that the idea that light travels in a straight line until it meets an object is addressed in the material. However, the material builds upon this by connecting the idea of how light travels to how we see an object. Even though the idea of how we see an object by reflected or emitted light entering the eye is a Grade 6–8 learning goal that exceeds the grade level standards, it is so well developed in the material by building on the key idea of how light travels that it was given a 5 rating. Both 4 and 5 indicate strong evidence of CTS. The distinction is the 5 rating can go further if it is done in a conceptually appropriate way.

Selection committee members go through each of the sections of Handout 6.20, discussing where they see evidence of the CTS findings in the curriculum materials. The group comes to a consensus on the rating and summarizes their reasons for the rating. The discussions among curriculum selection members significantly increase their knowledge of standards and research and increase their ability to look carefully at curriculum materials and not be misled by superficial features.

Doing a CTS on the topic of the materials, prior to conducting an analysis of curriculum materials, provides a lens through which to make evidence-based decisions regarding the likelihood that the materials address standards and research on learning. However, it does not ensure that students will learn the science from the materials. Further analysis on instructional effectiveness, including an examination of the pedagogy, would be suggested if the committee is looking at the instructional quality of the material. For groups interested in taking the analysis further, we recommend that some leaders may want to use CTS with the Project 2061 Curriculum Materials Analysis Process that rigorously examines content alignment and instructional quality of the materials. This process is not described in this *Leader's Guide* or the CTS parent book, but can be accessed at www.project2061.org/publications/textbook/hsbio/report/analysis.htm.

Curriculum Implementation

Supporting teachers to learn science content and pedagogical content knowledge that is directly connected to their curriculum materials increases the likelihood of changes in classroom teaching (Loucks-Horsley et al., 2010). Many districts use teacher leaders and instructional coaches to support implementation of new curriculum materials. One of the ways leaders can help teachers understand the content in their curriculum and know why it is presented and sequenced in such a way to promote learning (which requires the teachers to present material in its entirety rather than skip over or pick and choose only the activities they like or feel comfortable with) is to conduct a CTS with the teachers prior to examining the curriculum materials and experiencing the activities. By doing a CTS first, the teachers develop a common understanding of the content, reasons why certain instructional practices and activities are used, common difficulties and potential misconceptions to anticipate, and connections between lessons and content. The common understandings gained through the CTS process deepen the teachers' understanding of the intent of the curriculum and the importance of maintaining the fidelity of implementation. CTS leaders are encouraged to read pages 69–74 in the CTS parent book. This section provides a background on using CTS to support curriculum implementation. The following are suggested CTS tools and strategies that support curriculum implementation.

Creating Customized CTS Guides for Curricular Units

To use CTS in the context of supporting curriculum implementation, leaders first must identify the CTS guides that most closely match the curricular unit or create a customized guide that combines elements from two or more guides to address the curricular unit. Figure 4.13 on page 73 in the CTS parent book shows an example of a customized guide for Grade 4 that combined readings from the CTS guides, "Magnetism" and "Electrical Charge and Energy" to create a guide that addressed the Full Option Science System (FOSS) "Electricity and Magnetism" unit for Grade 4. Notice the readings focus only on Grade 4 and combine readings from both topics. Handout 6.21 on the CD-ROM provides a custom template for creating your own topic-specific guides for curriculum units that target a specific grade level rather than doing a full K–12 CTS. If there is a major inquiry or technological design skill being developed, your customized guide can include links to readings for that skill as well.

Once you select or create your own CTS guide for the unit being implemented, the leader should first do the CTS, making note of findings that specifically relate to the curriculum unit being implemented. Following the suggestions in Chapter 5 of this *Leader's Guide* for conducting a full CTS with groups, lead participants through the CTS, discussing the results as a group. For the application stage of a full CTS study, refer participants to the curriculum materials. Guide teachers through each of the lessons, encouraging them to make links to the CTS findings from each CTS study section that supports the materials or to refer to the CTS findings when teachers have questions about the materials. By doing a CTS first, the teachers have a lens through which to view their materials and focus their instruction.

Developing CTS Curricular Storylines

CTS conceptual storylines provide a one-page graphic for teachers to see at a glance the content in a curricular unit. The storylines visually organize the intended learning from each activity by subconcepts that are unpacked from the major concepts that make up the unit. Conceptual storylines can be created by the CTS leader and used to support curriculum implementation or can be created collaboratively with teacher leaders supporting the implementation of new curriculum materials. Figures 4.14 and 4.15 in the CTS parent book show the development of a CTS curricular conceptual storyline for the Great Explorations in Math and Science (GEMS) unit, The Real Reasons for Seasons (Gould, Willard, & Pompea, 2000). The "Hierarchal Structure of Content Knowledge" described on pages 57–63 of the CTS parent book and on pages 178–179 of this chapter is used to unpack concepts for the storyline as well as identify big ideas and unifying themes. The storylines help classroom teachers plan their instruction while following the flow of the unit. They also identify the CTS guides if teachers wish to explore further by doing their own CTS on the topic of the unit.

The Valle Imperial Project has developed draft storylines, which can be accessed at www.vipscience.com/storyline_links.htm. These draft storylines are not the same as CTS storylines but can be used as a starting point to develop a CTS conceptual storyline that adds additional information from CTS. Leaders interested in developing or facilitating the development of CTS conceptual storylines for curriculum can use the following steps to construct their own CTS storyline.

PART 1: GROUNDWORK

1. Browse through the unit and become familiar with the content and lessons included in the material. You will keep referring back to the unit as you develop the background information and storyline using CTS.

2. Select a CTS guide that seems to be the primary topic of the unit. For example, the *Real Reasons for Seasons* GEMS guide seemed to best match the CTS guide, "Seasons."

3. Examine CTS Sections III and V to identify major concepts that make up the topic at the intended grade level and are included in the unit. For example, after examining CTS Sections III and V, motion, seasons, models, solar energy, and cycles seemed to be concepts most related to both the CTS guide, "Seasons," and the curricular unit. Record the concepts. These will be your starting point for unpacking the content.

4. Look for subconcepts included in the unit that may be a subset of the major concepts you identified in Step 3. List these subconcepts. These will be used to group the lessons. In Figure 4.14 in the CTS parent book, you will see several subconcepts listed that make up the concepts described above them on the chart. For example, the concepts of Earth's revolution and Earth's rotation are subconcepts of the major concept, motion.

5. List learning goals from your state standards (CTS Section VI) that relate to the concepts or subconcepts. In the example shown on page 74 of the CTS parent book,

there were two Maine standards that related to the subconcept of Earth's tilted axis and seasonal change and Earth's revolution and Earth's rotation.

6. Examine CTS Sections II, III, IV, and V for specific ideas from the learning goals (CTS Section III) as well as specific ideas that can be synthesized from the essays (CTS Section II overview essay), research on learning (CTS Section IV), and connections to and from the maps (CTS Section V). List any specific ideas gleaned from these CTS sections that relate to the lessons in the curriculum material. You might look at related CTS guides connected to the content. For example, in this unit, the CTS guide "Models" helped in describing specific ideas about models related to seasons.

7. Examine the material as a whole and look for big ideas in CTS Section I and the overview essay in CTS Section II that describe the major, overarching ideas developed by the material. For example, the big idea that "models are representations that help us understand the real thing better" is a big idea that comes from several of the modeling lessons in the unit.

8. Select a unifying theme that represents a powerful idea that cuts across disciplines and can be conveyed by the unit. For example, in the *Real Reasons for Seasons*, a powerful, unifying theme is one of constancy and change. Even though some things like the intensity of sunlight and position of the Earth change throughout the year, the cycle of seasons is always the same.

The preceding describes the CTS groundwork that is done in concert with examining the lessons in the material. This information is now used to create a one-page conceptual storyline to help a teacher understand the conceptual flow and learning objectives of the lessons. The following describes how to construct the storyline using a format similar to the one on page 75 in the CTS parent book:

PART 2: DEVELOPMENT OF THE FLOW CHART

1. Start with the title of the unit and the grade level(s).

2. Describe the big ideas and major concepts identified from the groundwork.

3. Examine the activities. Create boxes that describe the name of the activity or lesson and a brief description of what happens in the lesson. Refer to the specific ideas from CTS when developing the description. Don't list the specific idea from CTS but rather try to use it to concisely describe the lesson so that it connects to a CTS idea. However, keep in mind that some lessons may not connect directly to CTS findings but are needed in order to maintain coherency.

4. Group these lessons or activities by related subconcepts. Sometimes there may be only one lesson to address a single subconcept, two or more lessons addressing one subconcept, or a cluster of lessons addressing related subconcepts. For example, Activities 5 and 6 are related. Together they address the related subconcepts of temperature variation and photoperiod. Because each of these lessons is tied together, they are grouped together, even though the first one addresses variation and the second addresses photoperiods. Activity 6 depends on Activity 5.

5. Draw arrows to indicate when a new lesson or cluster of lessons targets a new subconcept(s). For example, Lesson 1 is about seasonal change. The next cluster of lessons that are part of Activity 2: Sun-Earth Survey are not about seasonal change but build from the seasonal change idea. Thus an arrow points to this cluster of three lessons that collectively address Earth's motion in relation to the sun, Earth's shape, and distance from the sun. This cluster of lessons will then lead to a scale modeling activity involving two lessons that will help students understand the vast distance between the Earth and the sun. The flow chart continues by grouping related lessons that develop a subconcept and contribute to the next set of lessons that focus on a new subconcept. Collectively, all the lessons will build an understanding of the major concepts listed at the top of the chart.

6. Review your chart to see if the conceptual flow makes sense and that the activities align with the subconcepts.

7. At the bottom of the chart, list specific skills that are used to develop content understanding through the activities. This is particularly important for inquiry-based materials.

8. List the major CTS guide that was used to examine the curriculum material and inform the structure of knowledge.

9. List any related CTS guides to which the user of the curriculum might refer in order to learn more about the concepts and skills.

10. List the learning goals from your state standards that are aligned with the material.

11. Optional: List the specific ideas identified during the groundwork stage.

When completed, the CTS curriculum storyline provides a conceptual roadmap for teachers to follow. In addition, listing the CTS guides related to the unit provides information for the curriculum user to use CTS to further their understanding of the important ideas that are developed throughout the curriculum.

These are just a few of the many suggestions for ways to use CTS to support the multifaceted nature of curriculum. As you become more familiar with the use of CTS and strategies for leading a topic study, you may find additional ways to support curriculum. Check the CTS Web site occasionally for examples of shared ways CTS leaders have used CTS in a curriculum context as well as new CTS curriculum tools developed by the CTS project.

CTS AND SCIENCE INSTRUCTION

Many professional development programs provide opportunities for teachers to develop their own lessons or modify existing lessons to improve their alignment and instructional quality. Pages 74–79 in the CTS parent book describe ways CTS can be used in an instructional context to target important content-related ideas and skills. Leaders should take the time to read through this section in the CTS parent book. Table 6.9 shows the instructional components, content examples, and CD-ROM handouts described in this section.

| Table 6.9 | Instructional Components, Content Examples, and CD-ROM Handouts |

Instructional Component	CTS Topic Examples	CD-ROM Handouts
Identifying appropriate instructional strategies, phenomena, representations, or contexts	Food Chains and Food Webs, Stars and Galaxies, Properties of Matter	6.22
Reviewing and modifying lessons	Scale Size and Distance in the Universe, Models	6.23, 6.24, 6.25, 6.26, 6.27
Developing a standards- and research-based lesson	Density	6.28, 6.29
Strengthening inquiry, technological design, and nature of science	Controlling Variables, Technological Design, the Nature of Scientific Thought and Development	None

These suggestions and strategies can be used by leaders to help teachers improve instruction that targets important learning goals in a curricular unit. They are described as follows.

Identifying Appropriate Instructional Strategies, Phenomena, Representations, or Contexts

A careful study of CTS Section II reveals implications for instruction that connect to important learning goals included in CTS Sections III and VI. As teachers study CTS Section II, encourage them to identify examples of effective strategies, phenomena, representations, or contexts described in the essays. *Phenomena* are real-world objects, systems, and events that provide evidence of key ideas in science. A phenomenon may be experienced firsthand or vicariously. *Representations* are diagrams, models, animations, simulations, and analogies that can help clarify key ideas in science (Maine Mathematics and Science Alliance, n.d., ¶¶ 3–4).

For example, in the CTS guide, "Food Chains and Food Webs," the Section IIA reading for Grades 6–8 provides examples of phenomena that may be helpful in teaching about land and sea food webs. The essay describes how students should first investigate local food webs that they can study directly. After they have had firsthand experiences, then the vicarious use of films of food webs in other ecosystems can supplement their firsthand experiences, but they should not substitute for them. In addition, a study of CTS Section IV points out the problems some students have with the arrow notation used in trophic relationships. "Hence they failed to understand the underlying principles of the relationship and to complete the activities correctly" (Driver, Squires, Rushworth, & Wood-Robinson, 1994, p. 61). The research helps alert teachers to the use of a representation, such as the food web diagrams in textbooks, used in their instruction.

In the CTS guide, "Stars and Galaxies," a Grade 3–5 key idea is that although some stars are larger than our closest star, the sun, they are so far away that they look like tiny points of light. The Section IIA essay describes use of a phenomenon that can provide an explanation for this idea. Students should observe how a large light source in the dark at a great distance looks like a small light source that is much closer (AAAS, 1993). This is an example of an instructional strategy using a light source such as a flashlight that can be used to develop an understanding of the key idea.

The context in which students learn ideas is also important. For example, Section IIB of the Grades K–4 "Properties of Matter" topic study reveals that students should learn about properties of matter in the context of observing and manipulating common objects and materials in their environment. This points out the importance of designing instruction that uses familiar items. Handout 6.22 provides a template teachers can use to link key ideas in a curricular topic using CTS Section III to instructional considerations from CTS Section II, such as teaching strategies, appropriate phenomena, representations, and effective contexts. Teachers can then refer to the handout during their instructional planning to make decisions informed by CTS.

For more information on phenomena and representations used with the CTS topics, leaders can visit the Web site for PRISMS: Phenomena and Representations for the Instruction of Science in Middle Schools at www.prisms.mmsa.org.

This Web site, shown in Figure 6.4, uses CTS topics to develop clarifications used to analyze the content alignment to key ideas and instructional quality of Web-based phenomena and representations. The analysis process is explained on the site. Leaders might find this a useful source of supplementary material to use with the CTS guides.

Figure 6.4 PRISMS Web Site

Reviewing and Modifying Lessons

Page 76 in the CTS parent book provides suggestions for reviewing and modifying existing lessons. These lessons may come from teachers' instructional materials or may be teacher developed. The CTS process for reviewing and modifying existing lessons strengthens the lesson's alignment to learning goals and improves its instructional quality. This application can be used with any CTS full topic session in which improvement of lessons is used as the context in which teachers apply their CTS findings. It is also useful in helping preservice and novice teachers understand how to improve a lesson's alignment to standards, instructional quality, and address commonly held ideas noted in the research on learning. Handout 6.23 on the CD-ROM provides a scaffold leaders can use with teachers to guide them through the CTS lesson review and modification process. Handout 6.24 provides a worksheet for participants to use as they record their CTS findings that relate to the lesson and track modifications they can make to strengthen the lesson, based on their CTS findings.

An example of the lesson modification process is included on the CD-ROM. Handout 6.25 is a copy of an original lesson found on a NASA Web site. A middle school teacher modified this lesson to address the concepts of scale size and distance used to describe relationships between Earth, moon, a nearby planet (Mars), and the sun. She felt this lesson would provide the first step in understanding size and distance relationships to near-Earth objects before venturing further out in the solar system and beyond. However, the original lesson lacked a pedagogical model and conceptual development that took into account students' existing ideas about scale and models and moved them toward a more accurate conceptual model. After conducting a CTS on related topic studies, using the results to modify the lesson, and adding instructional strategies that reflect the wisdom of the practitioner, the teacher modified the lesson to develop the example shown in Handout 6.26. Handout 6.27 shows how the teacher used the CTS results to record and track suggestions for modifying the lesson. The following suggestions can be used to introduce the CTS lesson review and modification process:

1. Choose a lesson that matches the CTS topic of your session that has potential but needs improvement. After participants have completed and debriefed their CTS, have them read page 76 in their CTS parent book (or provide a handout of that page) and explain that there are materials that elaborate on this description of lesson review and modification that you will introduce to them. Provide a copy of the scaffold (Handout 6.23) and guide participants through each step of the scaffold using the existing lesson you previously identified and the CTS results. Provide participants with a copy of Handout 6.24 to record their CTS findings and track suggested modifications to the lesson. As a whole group, discuss and record how to best modify the lesson, justifying decisions based on CTS as well as knowledge of effective teaching.

2. Provide copies of Handouts 6.25 and 6.26. Ask participants to examine the two lessons and describe the differences they see in the modified lesson (Handout 6.26). Distribute a copy of Handout 6.27 to show how CTS was used to inform the modifications. Discuss and debrief evidence in the lesson that shows how the teacher who designed the modified lesson used CTS to make modifications. Provide a copy of the scaffold (Handout 6.23) and blank worksheet (Handout 6.24) for participants to use in a follow-up session (or as an assignment) to modify a lesson that is related to the topic of the session or a new topic of their choosing. (This will depend on your purpose for using CTS. If it was in a content institute for a particular topic, everyone should be using related CTS guides. If it was for introducing CTS, participants can choose a topic relevant to their own curriculum.)

3. Obtain materials and model and teach the lesson described in Handout 6.26 (participants do not need a copy of the lesson). After teaching the lesson and giving participants a chance to experience it as learners, compare the lesson they experienced to the original one (Handout 6.25). Ask them to comment on the differences in the lesson. Refer to the scaffold (Handout 6.23) and Handout 6.27 to show how a teacher used the scaffold and CTS results to modify the lesson. Give participants an opportunity to reflect on the connections they see between CTS and the revised lesson. As a follow-up, have participants use the scaffold and the worksheet (Handout 6.24) to modify a lesson of their choice.

Developing a Standards- and Research-Based Lesson—SRB Lesson Design

Exemplary instructional materials take into account the specific concepts and ideas in the national standards (CTS Section III), prerequisite knowledge and skills and connections among ideas in science (CTS Section V), implications for instruction related to specific content (Section II), and the research on learning (CTS Section IV). These considerations inform the targeted learning goals, instructional context, and strategies used in a lesson. However, there are times when a teacher may not have access to exemplary instructional materials, or state standards or the local curriculum require supplementary lessons be taught to address gaps in existing materials. In this case, teachers often have to develop their own lessons.

The CTS standards- and research-based approach to lesson design (SRB lessons) uses a backward design approach that begins with a CTS study in order to more effectively address the alignment of specific science content and content-appropriate pedagogy. While there are other excellent generic lesson design processes, such as *Understanding by Design* (Wiggins & McTighe, 2005), the added value CTS brings to these processes is the clarification of the science content, key ideas in the standards, and research on learning before designing instruction to address learning goals. The steps in the process are described in the CTS parent book on pages 76–77 and Figure 4.16 on page 78. Handout 6.28 on the CD-ROM provides a scaffold, based on Figure 4.16, that can be used to design SRB lessons. This lesson design process can be used as an application after teachers have conducted a full CTS on a science topic.

> **Facilitator Note**
>
> A full module for introducing CTS SRB Lesson Design will be available on the CTS Web site. Select "Additional Designs and Resources" in the Science CTS Leader's Guide section of the Web site.

Handout 6.29 shows an example of an SRB lesson designed by a middle school teacher to address her state standard's density learning goal and improve the content alignment by connecting it to a national standards goal that targets the idea that characteristic properties, such as density, are not dependent on sample size. Leaders can use the scaffold and CTS results to guide teachers in developing similar lessons that align with local and national standards and incorporate effective instructional strategies for the specific content.

To introduce the value of using CTS to guide instruction without going through the process of designing a SRB lesson during your CTS session, you can instead have participants read *Vignette #4: A Middle School Teacher Uses CTS to Understand a Difficult Topic to Teach: Density* on pages 98–99 in the CTS parent book. The vignette describes how a teacher plans to modify her instruction, based on the information she gained from using CTS. This vignette helps teachers see the value of doing a CTS first before they design a lesson.

Strengthening Inquiry, Technological Design, or the Nature of Science

Pages 77–79 in the CTS parent book address ways to strengthen a lesson's connection to inquiry, technological design, or the nature of science. When leading CTS sessions that include an opportunity to develop or modify lessons that incorporate the skills and understandings of inquiry, technological design, or nature of science, leaders can connect a "disciplinary CTS guide" from the life, physical, Earth, space, and science and technology topic studies to one of the twenty-seven guides in the "Inquiry and the Nature of Science and Technology" category to develop lessons that link skills, processes, habits of mind, and content. Note that the reference to *inquiry* in the guiding questions on pages 78–79 that go with each CTS section can be replaced with *technological design* or *nature of science*.

Encourage teachers to use the questions on pages 78–79 of the CTS parent book as they think about ways to strengthen inquiry or other crosscutting standards connections to their lessons. Provide them with one of the following examples to illustrate how CTS can be used to strengthen inquiry.

Example 1: A third-grade teacher decides to have her students design a controlled experiment as part of a force and motion unit to find out if the surface a ball rolls on affects how far it rolls. She selects the CTS guide, "Controlling Variables." She is particularly interested in the research on learning to find out if her students might have difficulty with the concept of variables, including identifying and controlling variables during an investigation. As she uses CTS and the questions for CTS Section IV on page 79, she finds out that controlling variables is difficult for students and that it may be better to start with the notion of a *fair test* with younger students. She modifies the activity to focus on getting her students to talk about "what is fair" as they figure out how to design a way to answer the question, "Does the type of surface a ball rolls on affect how far it can roll?"

Example 2: A sixth-grade teacher decides to include a design challenge to accompany a unit on energy resources that includes wind power. She wants her students to design wind-powered machines capable of lifting a small weight. Her state standards require students to use the process of technological design to produce a product that serves a specific function. She isn't sure what the "process" involves so she turns to CTS and does a study, "Technological Design." Using the questions on page 79 for CTS Section III to guide her study, she learns what the abilities of technological design are and the steps students go through in the design process. She uses this information to connect the skills of technological design to the content students must understand about wind machines in order to inform the design process.

Example 3: A biology teacher wants to design a lesson that will involve students in critiquing articles that are both in favor of and not in favor of the use of stem cells in treating certain diseases. She wants them to use critical reasoning skills to defend or argue against the claims made in the articles. She uses the CTS guide, "Scientific and Logical Reasoning," to learn more about how to create an instructional setting for students to engage in a scientific argument as well as the skills they need. Using the guiding questions for CTS Sections II and III on pages 78–79, she reads that students should have plenty of opportunities to critique claims made in print, radio, and television. She decides to expand the activity to other controversial biology-related topics and have students come up with their own examples from various media that they critique. She creates a chart from the bulleted goals in Section III to provide students with examples of things to consider when critiquing claims.

Overall, the strategies and tools for linking instruction to CTS findings can result in changes that focus instruction more deeply on cognitively appropriate learning goals and provide improved opportunities for students to learn concepts and skills through lessons selected and taught by a teacher who has experienced CTS. The instructional applications can be combined with a full topic workshop to create a full-day session in which teachers select, modify, or design lessons based on the CTS topic they studied.

CTS AND SCIENCE ASSESSMENT

Pages 79–86 of the CTS parent book describe how CTS can be used in an assessment context. The following box provides an outline of the suggestions included in this chapter for using CTS in assessment. All of the resources and handouts for this section can be found on the CD-ROM Chapter 6 subfolder labeled assessment.

OUTLINE OF SUGGESTIONS FOR ENHANCING CTS AND SCIENCE ASSESSMENT

A. Formative assessment

    1. Introduction to CTS assessment probes

    2. Professional development module-developing CTS assessment probes

B. Performance assessment

    1. Introduction to CTS performance assessment

    2. Designing CTS performance assessment tasks

The CTS parent book describes the different purposes of assessment and provides suggestions for using CTS to design two types of assessments: assessment probes that are used for diagnostic and formative assessment, and performance tasks that are used for culminating classroom formative or summative assessments. In addition, we recommend using the AAAS/Project 2061 assessment design process for evaluating or designing summative assessment questions that can be used by classroom teachers to create classroom test items or items for districtwide common assessments. Doing a CTS first on a selected topic provides the groundwork needed to utilize the AAAS assessment analysis utility, which can be found on the Project 2061 Web site at www.project2061.org/research/assessment.htm. Table 6.10 describes the tools and strategies for the assessment context applications included in this chapter. In addition, the module "Developing Formative Assessment Probes" is a full session module that includes a facilitator script and PowerPoint slides.

Table 6.10 Assessment Tools and Strategies

Assessment Application	CTS Example Used	CD-ROM Handouts
Developing formative assessment probes	Conservation of Matter, Sound	Handouts 6.30–6.37
Designing performance assessments	Earth, Moon, and Sun System	Handouts 6.38–6.40 and Templates 1–3

Assessment Probes

An explicit link between a standards-based key idea and a commonly held misconception noted in the learning research can be used to develop a type of formative assessment item called an assessment probe. These assessment probes reveal much more than simply an answer. They provide an opportunity for teachers to probe student thinking in order to uncover ideas that may be hidden from the teacher and surfaced by the student prior to or during the instruction.

CTS assessment probes reveal common misconceptions students bring to their learning and provide data the teacher uses to inform instruction. In addition, the probes promote metacognition, as students think through their ideas and justify them with their "rules" or reasons. "Students enter the study of science with a vast array of preconceptions based on everyday experiences. Teachers will need to engage those ideas if students are to understand science" (Donovan & Bransford, 2005, p. 399). Applying CTS to the development of assessment probes provides teachers with the tools they need to engage in continuous assessment of student learning.

In our work with CTS, we have found significant interest in the development of assessment probes. Page Keeley, developer of the CTS process and coauthor of this *Leader's Guide,* is the primary author of a series of books on formative assessment that include diagnostic and formative assessment probes developed by using the CTS assessment probe design process described in the CTS parent book. Published by National Science Teachers Association, *Uncovering Student Ideas in Science* (Keeley et al., 2005–2009) includes formative assessment probes accompanied by extensive teacher background notes that reflect the CTS process. Handouts 6.30 and 6.31 show the assessment probe described on pages 82–83 in the CTS parent book, along with the accompanying teacher notes that reflect the CTS findings. This probe was published in *Uncovering Student Ideas in Science* (Vol. 1) (Keeley, Eberle, & Farrin, 2005). Table 6.11 lists the information provided in the *Uncovering Student Ideas in Science* teacher notes and shows the link between this valuable teacher information and CTS.

> **Facilitator Note**
>
> Inside the "Tools and Templates for Designing Your Own CTS" folder in Chapter 5 of the CD-ROM is a crosswalk that links each of the formative assessment probes in the series mentioned above to a CTS topic.

Teachers who have used these probes and experienced the impact they have made on their students' learning, their instruction, and the classroom climate have taken the next step to learn how to use CTS to develop their own probes. The following is a workshop module leaders can use to facilitate the CTS assessment design process. From our experience with facilitators of CTS and CTS users, this has been the most popular application of the CTS contexts described in Chapter 4 of the CTS parent book.

Table 6.11 Linking *Uncovering Student Ideas in Science* Teacher Notes to CTS

Uncovering Student Ideas in Science Teacher Notes	*CTS Section Used to Inform Development of the Teacher Notes*
Purpose	Sections II and III
Related concepts	Sections III and V
Explanation	Section I
Curricular and instructional considerations	Sections II, III, and V
Administering the probe	n/a
Related ideas in the standards	Section III
Related research	Section IV
Suggestions for instruction and assessment	Sections II and IV
Related NSTA Science Store publications and NSTA journal articles	Includes several of the CTS resource books as well as supplementary resources from the CTS Web site
Related CTS guides	Lists the CTS guides used to inform the development of the probe

Module

Developing CTS Assessment Probes

DESCRIPTION OF THE MODULE

This module introduces participants to the CTS process of developing assessment probes. This context application module takes participants through a cycle of learn, practice, and apply. During this module participants

- *learn* about a CTS process for developing assessment probes that elicit students' ideas and inform instruction;
- *practice* using CTS to unpack specific ideas in the standards and research related to an assessment topic; and
- *develop* an assessment probe using the CTS assessment probe development process.

GOAL OF A CTS ASSESSMENT PROBE DEVELOPMENT SESSION

The overarching goal of this module is to help participants learn how to use CTS to develop their own assessment probes.

AUDIENCE

This session is designed for K–12 teachers and assessment developers. The examples used in this introductory module target K–8 ideas from the CTS guide, "Conservation of Matter," but are also useful at the high school level.

USE OF A SCAFFOLD

The CTS assessment probe development process involves several sequential steps. A scaffold is provided to guide novices through each of the steps until they become proficient in the process.

GUIDELINES FOR LEADING A SESSION ON CTS ASSESSMENT PROBE DEVELOPMENT

Obtain the Following

- At least one copy of the CTS parent book, *Science Curriculum Topic Study: Bridging the Gap Between Standards and Practice* (Keeley, 2005)
- At least one copy of *Benchmarks for Science Literacy* (AAAS, 1993) and one copy of *Making Sense of Secondary Science* (Driver et al., 1994) for each group of two to three participants if they are

going to develop additional probes after the introduction to the process (other books that can also be used but are optional include *National Science Education Standards* [NRC, 1996] and *Atlas of Science Literacy* [Vol. 1–2] [AAAS, 2001–2007])
- Flip chart easel, pads, and markers
- Paper for participants to take notes
- A copy of any of the books from the *Uncovering Student Ideas in Science* series (Keeley et al., 2005–2009) (optional)
- Copy of the crosswalk from the Tools and Templates section of the Chapter 5 folder on the CD-ROM (optional)

Duplicate the Following

See CD-ROM Chapter 6 subfolder assessment for all handouts.

- If participants do not have their own CTS parent books, for each participant, make copies of "CTS and Assessment" on pages 79–83 of that book and the CTS guide, "Conservation of Matter," on page 163.
- If enough resource books are not available, make copies of the selected readings from CTS Sections III and IV from the CTS guide, "Conservation of Matter," on page 163.
- Print copies of Facilitator Handouts 6.32 and 6.33, one copy per group of two to three. Facilitator Handout 6.32 should be copied on yellow paper, Facilitator Handout 6.33 on blue paper.
- Handout 6.34: Assessment Probe Scaffold.
- Handout 6.35: Scaffold Step 4: Conservation of Matter Chart.
- Handout 6.36: Four Types of Probes.
- Handout 6.37: Probe Development Worksheet.

Prepare

- PowerPoint slides for *Designing Assessment Probes* (on CD-ROM) in the Chapter 6 assessment folder. Insert your own graphics and additional information. Optional: Print out copies of the PowerPoint slides for participants.
- Cards for matching activity: Cut out yellow and blue cards from Facilitator Handouts 6.32 and 6.33 and place cards in a plastic zippered bag.

Time

Introducing the process and practicing with the examples provided takes approximately 2 hours. After completing the introduction, if participants are going to develop their own probe using a topic of their choice, add an additional 1 to 1.5 hours. This session can be extended into a full day if participants are working in groups to develop a collection of probes.

DIRECTIONS FOR LEADING CTS ASSESSMENT PROBE DEVELOPMENT

The following directions describe how to lead an introductory CTS probe development session for first-timers using a conservation-of-matter example.

1. Show Slide 1 and 2 and go over goals for the session. Include introductions if participants do not know each other. (3 minutes)

2. Show Slide 3 and give participants an opportunity to discuss the difference between the three types of assessments. Debrief with the large group and explain how this session will

instruct them how to develop their own assessment probes that are used to elicit and identify students' ideas (diagnostic use) and use information about their thinking to inform instruction (formative use). (5 minutes)

3. Show Slide 4. In pairs or groups of three, have participants take turns summarizing each paragraph of the reading to gain background knowledge on formative assessment and the probe development process. Provide time for questions. (15 minutes)

4. Show Slide 5 to describe what a formative assessment probe is. Explain how it is different from a summative assessment item, which is used after instruction to measure the extent to which a student has met a learning goal. Show Slide 6 to explain that there are many terms used to describe the ideas students bring to their learning. During this session, we will refer to students' ideas that are not completely scientific as *misconceptions*. (5 minutes)

5. Show Slide 7 and explain that we are going to look at how a CTS assessment probe is developed. Explain that the development process consists of three steps: (1) identifying a specific idea or ideas from the standards using CTS Section III (or V or VI); (2) identifying findings from the research on student learning that relate to the specific idea (CTS Section IV); and (3) forming from these two components the basis of the assessment probe, which leads to the development of the context of the probe (prompt), the selected responses (distracters and correct answer[s]), and a justification that asks students to explain their thinking and provide reasons for their answer. (5 minutes)

6. Show Slide 8 to give an example of using the CTS guide "Visible Light, Color, and Vision" to identify specific ideas related to light. This slide shows two specific ideas from the *National Science Education Standards* (NRC, 1996) that relate to the idea of light reflection. The two learning goals show that the concept of light reflection is addressed at both the elementary and middle level, with different levels of complexity. The Grade 5–8 goal adds the idea that in order to see an object, light from the object must be emitted or reflected to our eyes. Identifying these ideas is the first component of probe development. We have now targeted a specific idea in the standards. The next component involves looking for a match between these ideas and research on learning that helps us understand how students think about reflection and how light travels. Show Slide 9 to see an example of a match to the research. Now these two components can be put together to develop a probe that can be used across multiple grade levels to find out what kinds of objects students think reflect light and what their reasons are for selecting those objects. Because all of the objects on the list can reflect light, teachers can examine the students' reasoning to see if they recognize the role of light in understanding how we see. (5 minutes)

7. Show Slide 10 to illustrate the third component: choosing the prompt, selected responses, and asking for a justification. If you have a copy of *Uncovering Student Ideas in Science* (Vol. 1) (Keeley et al., 2005), you can mention that this probe came from this book and that all the probes in this book were developed using the CTS process. In this example, a format called a *justified list* was used to see if students were limited by the context they experience either in school or in their everyday encounters with the natural world. They are asked in the prompt to check off the things on the list that can reflect light. Notice mirrors, examples of ordinary things, and colored objects are on the list to match the research findings. In this case, all of the items on the list are correct responses since anything that can be seen by the eye either reflects or emits light. Point out how the last part of the probe asks students to explain their thinking and justify their answer. Reinforce how this probe was developed by connecting the research from CTS Section IV to the specific ideas about reflection in CTS Section III. Show Slide 11 and

provide a few minutes to discuss the questions in small groups and debrief with the larger group. Point out how CTS provides the information and a process to make an assessment link between a key idea and the research on student learning. These types of assessments are important because they help us uncover strongly held preconceptions students bring to their learning that, if not addressed through careful instructional planning, will go unchanged from one grade level to the next. (10 minutes)

8. Show Slide 12 and refer participants to Handout 6.34: Assessment Probe Scaffold. Describe how a scaffold provides the support one needs to undertake a task. Show Slide 13 and refer participants to the first step on the scaffold. For the purpose of learning how to develop a probe, explain that they will practice with the topic conservation of matter. Their first step is to look at the CTS guide "Conservation of Matter" on page 163 in the CTS parent book. Show Slide 14 and point out that they will focus primarily on Sections III and IV when using CTS to develop assessment probes. Section III will help them find the specific ideas that are related to conservation of matter. Since they will not be using the historical episodes for the probes, they will focus the study of Section IIIA only on the 4D benchmarks for Grades K–8. There are no learning goals for conservation of matter after Grade 8 since it is important that by Grade 8, students have achieved understanding of this key idea so they can apply it in high school. However, the probe is useful at the high school level to see if this is indeed the case, as many high school matter-related ideas involve an understanding of conservation of mass. Since this idea in the standards culminates in Grade 8, they will only look at the K–2, 3–5, and 6–8 benchmarks in the *Benchmarks for Science Literacy* in CTS Section IIIA and K–4 and 5–8 *National Science Education Standards* in CTS Section IIIB. Point out Section III on the slide to show what part of the study guide they will focus on. Then point out Section IV and explain how this section will help participants identify misconceptions and learning difficulties that are noted in the research on learning. Point out the readings from Sections IVA and IVB that will be used for assessment probe development. Show that some of the sections in IVB will not be used during this practice example. (7 minutes)

9. Show Slide 15 and refer to Step 2 on the scaffold. Point out the bag of yellow and blue cards, one bag for each small group. Ask participants to take out the yellow cards. Tell them if they were doing their own CTS, these would be the ideas they would identify and record from CTS Section III. The CTS has already been done and the ideas have been placed on cards for this practice activity. Have them spread out the cards and arrange them by grade level. (5 minutes)

10. Show Slide 16 and refer to Step 3 on the scaffold. Ask participants to take out the blue cards and spread them out. If they were doing the CTS, these would be the misconceptions and learning difficulties recorded from the CTS Section IV reading. (3 minutes)

11. Show Slide 17 and refer to Step 4 on the scaffold. Group similar ideas on the yellow cards using the concept categories in the left column on the slide. Repeat with the blue cards using the concept categories on the right column. Explain how this is done in order to make a match between similar clusters of ideas, especially across grade spans. Pass out Handout 6.35 to show how the CTS results are organized for the probe development process, reflecting what participants have done with the cards (or they may have grouped them differently). (*Note:* The cards are used for practicing CTS probe development. The handout illustrates how CTS Sections III and IV are organized when actually doing CTS for the purpose of developing assessment probes.) (8 minutes)

12. Show Slide 18. Explain how Step 4 involves finding a close match between a key idea and a research finding. This match will lead to possible questions that inform the development of the

probe. Explain how the key idea and the research finding shown on the slide both address matter in two different states. This match will lead to the development of an assessment probe that will determine whether students think mass is conserved when ice melts. In addition, the developer of the probe wanted to incorporate the six to eight ideas about conservation of matter in a closed system. (5 minutes)

13. Show Slide 19 and refer to Step 5 on the scaffold. Pass out Handout 6.36 and have participants pair up to read and discuss the four different types of selected response probes. Explain that any of these formats can be used to inform the development of a probe that assesses whether students think mass is conserved when ice melts. Show Slides 19 through 22 to describe each type of probe. (5 minutes)

14. Refer to Step 6 on the scaffold. Looking at the example on Slide 23, point out the prompt, distracters, and correct response that make up the first tier of a two-tiered assessment probe. Point out the justification in the second tier of the probe that asks for an explanation. (3 minutes)

15. Show Slide 24 and summarize the remaining Steps 7 and 8 on the scaffold. Ask if there are any questions about the assessment probe development process before participants work in small groups to develop their own. (5 minutes)

16. Show Slide 25 and explain that the CTS summary on Handout 6.35 can be used to develop other conservation-of-matter probes by finding additional matches between key ideas and the research. Have participants work in small groups to develop a new conservation-of-matter probe by following the probe development process on the scaffold. Provide Handout 6.37 and explain that this worksheet can be used to track and provide a record of their work. (30 minutes)

17. Provide time for participants to share their work either through a gallery walk or brief presentations to the whole group. Provide time for feedback. Feedback should include how closely the probe matches the key idea selected, whether the distracters are likely to reveal misconceptions, and the familiarity of the context and wording used. Remind participants to examine each other's probes with an eye to whether the information revealed will be useful to the teacher. Explain how the next step would involve piloting with students, revising, and then using the probe to gather data on student thinking for the purpose of informing instruction. Ask each small group to briefly discuss with one another how the student data from the probe could be used to adjust their instruction. (10 minutes)

18. Conclude the session with the reflection on Slide 26.

Facilitator Note

Assessment probe development is an iterative process that usually takes several revisions before it is in final, polished form. Make sure participants know they are not expected to produce a final, finished product during the workshop. Rather, it is their idea for a probe that can be used to address a student learning difficulty and the experience of developing their first probe that is important.

EXTENSION

After completing the introduction to CTS assessment probe development using the conservation-of-matter example, this session can be extended into a 3- to 3.5-hour session by having participants develop a probe or set of probes using a topic of their choice. If you are working with a group on a long-term basis, you might consider having teachers come back again after they have refined their probe, used it with students, and collected student work to examine in a subsequent session.

PERFORMANCE ASSESSMENT TASKS

Performance assessment often implies a more formal assessment of students as they engage in a challenging activity or task designed to demonstrate different aspects of their scientific knowledge and skills (NRC, 2001). Performance assessment involves tasks that "use one's knowledge to effectively act or bring to fruition a complex product that reveals one's knowledge and expertise" (Wiggins & McTighe, 2005, p. 346). A well-designed CTS performance task simultaneously provides an opportunity for students to (1) draw upon their existing knowledge and skills, (2) deepen their conceptual understanding of science, (3) further hone their inquiry skills, and (4) demonstrate their content understandings and ability to use science practices.

Performance assessments are more challenging and complex than traditional short answer types of assessments. They require more work by the teacher to create and more effort from the student to complete. Nevertheless, teachers and students frequently prefer these types of assessments because they are engaging, have multiple entry points, require students to *do* science, access higher-level thinking skills, and allow students to demonstrate their understanding in more than one way.

CTS performance assessment tasks are designed by a teacher or teams of teachers for use in the classroom. They are usually embedded within a curricular unit as part of the teaching and learning cycle. CTS performance assessment tasks differ from other types of performance assessments in that they are used by the teacher for both instruction and assessment purposes and closely mirror the curricular and instructional context in which students are learning the ideas. They can be both formative and summative, providing an opportunity for feedback and revision as well as measuring the extent to which students have achieved a learning goal. Other types of performance assessment tasks are used for large scale or districtwide assessment in which results are used for documenting student achievement for accountability purposes. These external summative performance assessments undergo rigorous field-testing and validity and reliability checks. While CTS classroom-based performance assessment tasks are a vital part of classroom assessment, it is important to understand they do not undergo the same level of psychometric analysis used with their high-stakes counterparts.

CTS performance assessment tasks align with and build on the important learning goals targeted by the curriculum. They are strategically placed in a curricular unit to enhance learning and encourage students to pull their science knowledge and skills together, resulting in a product or performance to "show what they know." They can range from a short activity completed within a single classroom period to longer, culminating projects that involve group work and time outside of the classroom. CTS performance tasks can be developed from scratch (with the use of CTS) or they can be adapted from certain types of instructional activities. With modification, a science activity can be adapted for performance assessment if it targets clear knowledge and skill goals and has the potential to provide evidence of the extent to which students can use the knowledge and skills to complete the task.

CTS PERFORMANCE ASSESSMENT
TASK DESIGN AND TEACHER LEARNING

Teacher collaboration is a major feature of CTS professional learning. Collaborative efforts to develop a classroom-based performance assessment task after engaging in the study of a curricular topic is a powerful way to apply CTS learning. At the same time, it results in a standards- and research-based product that can be used in the teachers' classrooms. When teachers work together, using their CTS findings to develop performance assessment tasks, use a peer review process to give each other constructive feedback, and plan together to refine their work, they are engaged in high levels of adult learning.

After the performance tasks have been developed, teacher learning can be extended by having the teachers use their performance tasks in their classrooms and collect authentic achievement data. Teachers may choose to come together again to examine and score the student work. By examining the strengths and weaknesses in the student work, and referring back to the CTS results for "evidence-based" conversations about teaching and learning, the tasks provide ongoing data to help teachers plan for continuous improvement in helping all students achieve high standards of learning.

A CTS PROCESS FOR DEVELOPING PERFORMANCE TASKS

Pages 83–86 in the CTS parent book briefly describe how CTS can be used to develop culminating performance tasks. The example provided shows how a topic study, "Earth, Moon, and Sun System," was used to inform development of a task that would demonstrate students' understanding of the motions of the Earth and moon in relation to the sun in order to differentiate between a lunar eclipse and the new moon phase. Having students compare and contrast a lunar eclipse with a new moon phase helps students give up the common shadow misconception to explain what causes the moon phases. It also shows how a learning goal from the crosscutting unifying theme standard of models was used to explain the difference between the two phenomena. The teacher notes, informed by doing the CTS, provide the background material teachers need to administer the task as well as understand the intent of the task and the instructional considerations that support readiness for students to undertake the task.

Facilitators who lead performance task development using the *Understanding by Design* process (Wiggins & McTighe, 2005) or similar backwards-design models may find CTS to be a powerful tool to use in their performance assessment work. In any design work undertaken by teachers, it is important to provide them with relevant resources to support their work. CTS provides a collective set of resources that are essential for developing standards- and research-based performance tasks in science. Having participants do a CTS first provides a much sharper focus for aligning content, instruction, assessment, and opportunity to learn—all factors that must be considered when designing complex performance tasks that target important learning goals in science.

Performance assessment provides a way for teachers to observe, give feedback on, and document how students *do* science, not just what they know about it. Rich inquiry-based performance tasks can be developed that combine an Earth, space, life, or physical science CTS with the crosscutting curricular topics in science as shown in Table 6.12.

By doing a full study of a CTS content topic and a crosscutting topic before engaging in the design process, teachers deepen their content knowledge and better understand what is reasonable and fair to expect from students. They also have a much fuller view of the likely misconceptions or learning difficulties that might surface during the task and the kinds of instructional experiences and prior knowledge that precede the task.

CTS PERFORMANCE TASK DEVELOPMENT TOOLS

Many grade level teams, schools, and districts are engaged in the work of developing performance assessments ranging from tasks used by teams of teachers for instructional

| Table 6.12 | Performance Task Content |

CTS Performance Assessment Tasks Combine Learning Goals From CTS Topic Categories	
Disciplinary Content From:	*With Crosscutting Content From:*
Diversity of Life Ecology Biological Structure and Function Matter Earth Astronomy Energy, Force, and Motion	Inquiry and the Nature of Science and Technology Implications of Science and Technology Unifying Themes

purposes to districtwide common performance assessment tasks. It is important for leaders to first become familiar with the tools, templates, and processes districts and teachers are using to create performance assessment tasks and then to determine how CTS can best fit into and complement the work being done in their local context.

The following tools and templates are provided for leaders to use (or adapt) in facilitating the development of CTS performance assessment tasks. If participants have access to computers, it is suggested that leaders provide the template electronically. These tools and templates can be found in the Chapter 6 Assessment folder on the CD-ROM:

Handout 6.38: Steps in CTS Performance Task Development

This handout outlines the steps for developing performance tasks. These steps outline the development of CTS performance assessment tasks starting with the CTS groundwork that lays the foundation to inform the development of a content and instructionally aligned task.

Handout 6.39: Criteria for Developing CTS Performance Assessment Tasks

This handout presents established criteria teachers can use as a guide to develop quality, appropriate tasks that provide an opportunity for all learners to demonstrate their knowledge and skills.

Handout 6.40: Guidelines for Drafting Scoring Guides or Rubrics for CTS Performance Tasks

This handout includes criteria for writing a scoring guide or rubric as well as suggestions for writing descriptors.

Template #1: CTS Groundwork

This blank template is used with Part 1 in Handout 6.38. The template provides space for task developers to record their CTS findings used to inform development of the task.

Template #2: CTS Task Development

This blank template is used with Part 2 in Handout 6.38. The template provides space for task developers to record ideas and track their development work as they design their tasks.

Template #3: CTS Performance Task Teacher Notes

This blank template is used with Step 14 in Handout 6.38. The template is used as a guide to develop the teacher notes that provide the background material on the performance task.

INTRODUCING THE CTS PERFORMANCE ASSESSMENT TASK APPLICATION

To introduce teachers to the CTS performance assessment task development process using the example provided in the CTS parent book, the following serves as a guide for CTS leaders. (*Note:* Make copies of pages 83–86, 194, and 269 in the CTS parent book if participants do not have their own copy of CTS.) Leaders are encouraged to make modifications based on the assessment experience of their group and the local context for assessment development. The times given are approximate and depend on the modifications leaders make to fit their unique context. Overall, the introduction to the CTS performance assessment design process takes approximately 2 hours.

1. Ask participants what the purpose of performance tasks is and how they differ from other types of assessments. Discuss their ideas and explain what a CTS performance assessment task is and how these tasks can be used in their classrooms. (*Note:* If working with a school or district, become familiar with the local approach to performance assessment development, such as *Understanding by Design* [Wiggins & McTighe, 2005], and determine how CTS best fits in to that approach.) (5 minutes)

2. Describe the purpose of this session. Will the session be used to introduce participants to the process of CTS performance task development, using the topic and example already provided, or will they be choosing their own topics and developing their own tasks after the introduction? If you are only introducing the process, build in an hour to walk participants through the example. If they are going to develop their own tasks, they will need at least a half-day to develop a draft task as well as conduct their CTS.

3. Ask participants to describe what they would look for in quality, classroom-appropriate performance assessment tasks. Review Handout 6.39: Criteria for Developing CTS Performance Assessment Tasks. Give participants time to discuss and clarify each of the criteria. Provide an opportunity for participants to examine a variety of performance tasks available through their district (optional) or use the example on page 84 in the CTS parent book. Look for evidence of the criteria in the sample task(s) provided. (15 minutes)

4. Distribute Handout 6.38 and explain that the group will now go over each of the steps in developing a CTS performance assessment task using an example created by a middle school teacher in Maine to address the Maine state standards. Use the task on page 84 in the CTS parent book to connect the steps to an actual task. Point out that the process begins by doing "CTS groundwork" to clarify the learning goal(s) being assessed and teaching and learning considerations. After Part 1: CTS

Groundwork is completed, the task development process begins. This is consistent with the backwards planning model that starts with the goal and works backwards rather than starting with a task and trying to match it to a goal. However it differs from most backwards planning processes as it begins with a study of the topic and learning goals, rather than simply choosing the targeted learning goals without doing a study of them.

5. *Step 1*: Using the "Dark Moons" example, explain that the first step is to identify a disciplinary content learning goal from the state standards or local curriculum that they want to assess. Refer participants to the teacher notes on page 85 of the CTS parent book and ask them to find the content goal from the Maine standards that was selected for this example task. Participants should identify "Universe Standard G5: *Describe the motion of moons, planets, stars, solar systems, and galaxies.*" (*Note:* They are looking only at the universe goal right now. The models goal will come later in the process.) State how that very broad and comprehensive goal would be unrealistic to assess in one task. Ask participants what part of the learning goal was selected for assessment. Participants should respond that the motion of the Earth and moon in relation to the sun was selected as the content goal that includes ideas about moon phases and eclipses. (5 minutes)

6. *Step 2*: The next step is to select a CTS guide that can be used to study the goal. Ask which guide would be the best match for this goal. (Response: "Earth, Moon, and Sun System.")

7. *Step 3*: Explain how Section III of the CTS guide is used to clarify the learning goal, specifically looking for key ideas related to motion, phases of the moon, and eclipses. Refer participants to the teacher notes on p. 85 in the CTS parent book to examine the key ideas that were identified to match the learning goal being assessed. Point out the first two bullets that describe the related Earth, moon, and sun ideas from the national standards. Briefly have them discuss how those key ideas help to clarify the state standard that was selected for assessment. (5 minutes)

8. *Steps 4 and 5*: Refer to Sections II and IV on the "Dark Moons: Teacher Notes" on page 86. These two sections are combined in this example. Ask participants why they think it is important to look at instructional implications and the research on learning before developing a task. Have participants look at the first two bullets and ask them to think about and discuss how they think the middle school teacher used them to inform her thinking about the task described on page 84. (10 minutes)

9. *Step 6* (*Note:* This step was not included in the example provided on pages 85–86): Explain how CTS Section V, using the *Atlas of Science Literacy*, helps the task developer consider important prerequisite ideas that students should know beforehand and build from in order to understand the targeted learning goal. If you have *Atlas* (Vol. 1) handy, refer them to the "Solar System" map on page 45. Look for the "Phases of the Moon" strand. Ask participants to identify two to three prerequisite ideas that contribute to understanding moon phases (refer to map ideas 6–8/4F2, 3–5/4A/4, and K–2/4A3). Discuss why it would be important for a task developer to take into account these prerequisite ideas. (*Note:* Eclipse ideas are not included on the *Atlas* map.) (10 minutes)

10. Say that this now concludes the "disciplinary content goal" groundwork necessary to clarify the content and teaching and learning considerations prior to developing the task. It also helps to tightly align the learning goal to the task so that the task invites students to do what the learning goal intends. Remind participants that CTS is sometimes called the "upfront part of backwards design." In other words, it is

important to study a learning goal before working backwards from the goal to align and design instruction or assessment. Mention that Template #1 is provided to record information from Part 1: CTS Groundwork.

11. Point out Part 2: Task Development on Handout 6.38 and explain that this phase of the process involves the actual development of the task. Mention that Template #2 is provided to record information and track progress as you go through the steps in developing the task.

12. *Steps 7 and 8*: Explain how this process can be used to turn some science activities into performance tasks, modify or strengthen existing performance tasks, or come up with an idea to develop a task from scratch. The developer of "Dark Moons" used the CTS Section III idea "Regular and predictable motions explain such phenomena as the day, year, phases of the moon, and eclipses." She decided to focus on eclipses and phases of the moon after looking at the CTS Section IV research and finding that students often confuse phases of the moon with a shadow cast by the Earth on the moon. Since the shadow effect can be used to explain an eclipse, students could be asked to describe and explain each phenomenon and, in the process, could be challenged to give up the shadow explanation for moon phases. Furthermore, the teacher decided to focus on the new moon and lunar eclipse. She reviewed the quality criteria once again, to make sure her task idea was substantively aligned, developmentally appropriate, and so forth. She also considered the cognitive demand of the task and closely matched the cognitive level of the performance verb in her state standards. Ask participants to imagine what kinds of things the teacher was considering as she reviewed these criteria with an idea for the task in mind. Ask participants to share some of their thoughts about the teacher's thinking. (15 minutes)

13. *Step 9*: As Table 6.12 shows, CTS performance tasks also include one or more crosscutting ideas or skills that connect to the disciplinary content and can be addressed in the task. Ask participants what crosscutting idea or skill was identified. (Answer: Use of models.) Refer participants to "Dark Moons: Teacher Notes" to see where additional CTS groundwork was done to clarify the CTS Section VI learning goal on communication. Ask them what part of the learning goal was selected. (Answer: *Make and use . . . three-dimensional models to represent real objects, find locations, and describe relationships.*) (5 minutes)

14. *Step 10*: At this point the teacher then began drafting the task. She started with an engaging question "Did you ever wonder why you sometimes see no moon at all during the moon's monthly cycle of phases?" She then went on to describe what one sees during a lunar eclipse. This phase of development starts with a "hook" or background material students need to engage with and understand the task. Have the participants examine the first paragraph of the "Dark Moons" task on page 84 in the CTS parent book and discuss why it is important to start off a performance task this way. (5 minutes)

15. Task developers often choose a role or audience. Ask participants to list some of the possible roles students can assume in performance tasks (e.g., scientists, journalists, museum display designer, etc.) and audiences (e.g., younger children, museum visitors, parents, etc.). In this case, the students are asked to be "experts" who can distinguish between new moons and lunar eclipses, and the audience is the class. (5 minutes)

16. Now it is time to clearly describe what the task involves, what students are specifically requested to do in order to complete the task (including scaffolding steps if needed), and what the final product or performance will be. Ask participants to

examine the task and discuss what it is the teacher is asking the students to do. Point out that knowing the students, their limitations, and the kind of support they need should be considered when crafting the task. (5 minutes)

17. After the task is drafted, the teacher comes up with an engaging title for the task. In this case, the title is "Dark Moons." Ask participants to practice coming up with a different engaging title for the task. (5 minutes)

18. *Step 11*: After the first draft of the task is written, it is reviewed for accuracy and alignment with the criteria described on Handout 6.39. Modifications are made, and the task is shared with colleagues for feedback. Although we don't have a record of the feedback, further modifications were made after receiving feedback from the developer's colleagues.

19. *Step 12*: Whether the task is used for formative or summative purposes, a scoring guide is developed, which is used to provide feedback or assess the extent to which the student met the standard set for the learning goal(s). A scoring guide or rubric was not included with the "Dark Moons" example. Ask participants what guidelines they use in their schools to draft scoring guides or rubrics for performance assessment. Provide participants with Handout 6.40: Guidelines for Drafting Scoring Guides and Rubrics (changing the language to match the words your teachers use to describe levels of proficiency and adapting it to fit the assessment context in which you work if necessary). Give participants an opportunity to review the guidelines and ask any questions regarding scoring guide or rubric development. (10 minutes)

> **Facilitator Note**
>
> Since different schools, assessment systems, or programs have their own methods and requirements for designing rubrics, this guide will not specify a particular format or way to develop scoring guides or rubrics. You are encouraged to use your own techniques and templates for scoring guide or rubric development.

20. *Step 13*: The last step is to write up teacher notes that would help other teachers use and understand the task. Template #3 is provided as a guide to write the teacher notes. Point out that pages 85–86 show an example of what CTS performance assessment task teacher notes might look like. Discuss with participants the rationale for having teacher notes and how the different sections of the teacher notes provide useful information. Ask if there is other information teachers might add to the teacher notes. (10 minutes)

21. After introducing the task development process, provide time for feedback and reflection. Questions to use might include the following:

   - How is this process different from the way you typically develop performance assessments?
   - What value does CTS add to the development of performance assessments?
   - What other questions do you have about this process?

DEVELOPING THEIR OWN PERFORMANCE TASK

After participants are introduced to the CTS performance task development application, your professional development program might include time for participants to develop their own performance task. Tasks can be developed for a CTS topic that is the focus of your professional development or topics that individuals or teams of teachers can work on together. Distribute the templates and provide time and guidance for participants to develop a draft task using the steps described in the introduction. Depending on the experience of the teachers and whether they develop a task from scratch or adapt an existing activity, you will need a minimum of three hours to produce a first draft. Be sure to provide time for participants to give feedback on each other's tasks using a protocol of your choosing (e.g., tuning protocol, exchanging work and using sticky notes for feedback, etc.). A timely, structured peer review process helps task developers refine their designs and provides an additional opportunity to discuss the link between the CTS results, the task, and teaching and learning. Leaders should encourage participants to give specific, descriptive feedback guided by the criteria for quality performance tasks and results of the CTS findings.

Ideally if participants are part of an ongoing professional development group or program and will be meeting again, encourage them to administer their tasks in their classroom and bring back samples of student work to discuss in small groups at their next meeting. The professional learning can be further extended by having teams examine the student work and do one or all of the following: (1) make adjustments to the performance task or scoring guide or rubric that will improve the use of the assessment; (2) select anchor papers that represent each of the score points on the scoring guide or rubric used for evaluation that will give teachers and students clear examples of what different levels of proficiency look like; or (3) make interpretations that reveal knowledge about their students' learning as well as their teaching and identify further actions for improvement.

By using this CTS context application to design performance assessment tasks, teachers enhance their understanding of the content being assessed as well as the implications for standards- and research-based curriculum and instruction. They understand what it really means for students to demonstrate important ideas and skills in science. While the focus is on development of the performance task, the teachers' conversations and thoughts during the development process keep linking back to CTS. By incorporating CTS as an essential part of the assessment design process, the alignment between content, curriculum, instruction, and assessment becomes much clearer and is addressed more explicitly.

OTHER EXAMPLES OF CONTEXT APPLICATIONS OF CTS

This chapter addressed the content, curricular, instructional, and assessment contexts for applying CTS. There are other contexts described throughout the CTS parent book and this *Leader's Guide*. The remainder of Chapter 4 in the CTS parent book (pages 86–90) describes how CTS is used in the context of preservice and novice teacher support, leadership development, and professional development strategies. The next chapter in this *Leader's Guide* addresses these contexts by providing examples of the ways CTS can be embedded in preservice courses and inservice professional development designed to support science educators at various stages of the teacher continuum ranging from preservice teachers to novice teachers to experienced teachers to teacher leaders and professional developers.

7

Embedding CTS Within Professional Development Strategies

WHY USE CTS WITHIN OTHER PROFESSIONAL DEVELOPMENT STRATEGIES?

A major theme of the preceding chapters is that CTS is a professional development strategy that helps teachers develop their important professional knowledge base in science and supports and helps focus learning on the standards and research to improve curriculum, instruction, and assessment. This last chapter addresses how CTS fits into and enhances many other professional development strategies that are used to improve science teaching and learning. The chapter provides guidelines for using CTS as a component or activity within seven other specific professional development strategies, including Application 1: Study Groups; Application 2: Collaborative Inquiry Into Examining Student Thinking; Application 3: Video Demonstration Lessons; Application 4: Integrated Lesson Study; Application 5: CTS Action Research; Application 6: CTS Seminars; and Application 7: Mentoring and Coaching.

For example, if teachers are engaged in using the professional development strategy of demonstration lessons, a teacher conducts a classroom lesson that other teachers observe. The teachers usually meet before the observation to discuss the goals and intent of the lesson and again following the observation to debrief the experience (Loucks-Horsley, Stiles, Mundry, Love, & Hewson, 2010). A critical element for the success of this strategy is that all of the teachers involved are prepared to observe how the important science ideas are developed in the lesson. Yet some groups have found it difficult to build a consensus of what to look for in demonstration lessons. Embedding CTS in the process of planning for and debriefing the demonstration lesson is an ideal way for the teachers to come to a shared understanding of what concepts or ideas the students should be learning from the lesson and the ways a teacher might probe for and address students' prior knowledge and misconceptions to increase the chances that the scientific ideas are understood.

Likewise, a group of teachers using the professional development strategy of examining student work and thinking may have the same challenge: What evidence of learning should they look for in the students' work? What is the important content for students to understand? Examining student work is a strategy for teacher learning and school improvement that is widely used in schools organized as professional learning communities and in grade-level teams. The purpose is to increase teachers' pedagogical content knowledge by examining students' work carefully and reflecting on what students understand, any misconceptions they have, and what instructional strategies would support further learning (Loucks-Horsley et al., 2010). Productive examination of student thinking requires teachers to clarify and really understand what students should know about a science concept at their particular grade level as well as common misconceptions.

Examining the standards and cognitive research through the CTS process, before looking at student work, provides teachers with several advantages. When teachers use CTS as part of their strategy for examining student work, they might conduct a topic study even before they gather the student work. The results of the study inform the questions or assessment prompts they use to gather the student work. This ensures that their assessments directly address the science ideas that students most need to know, and they can adequately probe for whether their students hold any of the misconceptions discussed in the research. Later, as teachers examine the student work, they can see the extent to which their students hold misconceptions similar to those cited in the research and examine the instructional implications for teaching the topic to inform next steps and how they may intervene to improve student understanding.

CTS AND PROFESSIONAL LEARNING COMMUNITIES

Many of the professional development strategies included in this chapter are ones that are "job-embedded" or that take place in the course of teachers' regular work. Increasingly these strategies are used in schools organized as "professional learning communities" that value ongoing teacher collaboration in the service of learning—for students *and* staff. Professional learning communities come in many shapes and sizes, with some focused on constant assessment and interventions in the classroom, others on intensive improvement of lessons, action research to collect evidence of learning in the classroom, and other strategies (Mundry & Stiles, 2009). One way to encourage the ongoing use of CTS in teachers' work, whether it be planning for the classroom or enhancing their professional knowledge base, is for the continued development of such professional cultures among teachers that use standards and research to inform practice. We have advocated for many years a halt to one-shot, one-size-fits-all professional development and a redirection of resources toward more ongoing teacher learning and organizational arrangements in schools that support teachers to collaborate on practice. The National Science Teachers Association, the National Staff Development Council, and other key education groups now advocate for professional learning that is grounded in practice, builds professional community, and is sustained over time (National Staff Development Council [NSDC], 2001; National Science Teachers Association [NSTA], 2007; Wei, Darling-Hammond, Andree, Richardson, & Orphanos, 2009). We are heartened by the growing recognition and evidence that schools organized as learning communities can make substantial changes in practice and improve learning, and we see CTS as a tool that can make the work of these professional communities more grounded in research and the standards.

> **Facilitator Note**
>
> If you are using Module A1 in this *Leader's Guide* to introduce CTS to professional learning communities, substitute Handout A1.5 with Handout 7.1.0. This version of the Snapshots is the same as A1.5 except the individual teacher asking the question in the snapshot has been replaced by the professional learning community.

DuFour, Eaker, and DuFour (2005) defined several characteristics of professional learning communities. The first is a *focus on learning* where faculty members are committed to meeting clearly defined student learning goals and continuously work to improve results. We contend that it is very difficult for faculty to develop a productive focus on learning unless they have a solid grounding in the standards and research and are able to draw upon their professional knowledge to inform their deliberations and practice. In settings that lack such a foundation, faculty may say they have a "focus on learning," but they may be focusing on the wrong learning goals or strategies. CTS is a basic tool used by professional learning communities to ensure that the faculty are consulting the standards and research to inform what the important content for their focus on learning should be and the research that describes students' commonly held ideas related to that content.

Another characteristic of professional learning communities is that they have a *collaborative culture focused on learning*. Teams plan instructional interventions, gather and examine evidence of student learning, and reflect on what they are learning. For these collaborations to yield the desired results, it is essential that faculty members are able to engage in productive dialogue, including speaking from the evidence that comes from examining the standards and research, not just from their own opinions. Topic studies provide teachers with ample practice using "CTS talk," where teachers use their CTS findings to support their claims about teaching and learning.

Staff in professional learning communities also engage in *collective inquiry*, where they routinely ask themselves how well they are achieving the desired outcomes, and what they need to learn or do next to be more successful. CTS can be particularly helpful in informing those next steps. When groups of teachers see students struggle with ideas in science, CTS can help them to become aware of common misconceptions that their students may hold or developmental issues related to learning and get ideas for how to effectively move students from where they are in their conceptual understanding to where they need to be. They may also need to step back from the specific idea with which students are struggling and trace the development of that idea back through the grades to discover what it is that might be the missing knowledge in the students' understanding. One high school chemistry teacher we worked with used the *Atlas of Science Literacy* to map out what ideas students needed to know to support their understanding of chemical reactions involving gases, an area where her students were having difficulty. She identified several concepts at the sixth through eighth and the third through fifth grade spans that she wasn't sure her students fully understood. She used this information to formatively assess what the students knew about gases, identified barriers that impeded their learning, and then adjusted her instruction, going back to earlier fundamental ideas that students hadn't had sufficient opportunity to develop. These are just a few examples of how CTS can enrich the work, dialogue, and decisions of professional learning communities, opening up new doors into understanding science teaching and learning that teachers may never have encountered without access to the professional resources and process for using them provided through CTS.

This chapter provides guidelines and materials that can be very applicable to making the work of professional learning communities productive and results oriented. For example, they can apply the guidelines for using video-demonstration lessons to study and improve practice, use the design for examining student thinking (CIEST) to inquire into student learning, or engage in the lesson study design to enhance and improve specific lessons. Professional learning communities provide the vehicle through which many of these CTS professional development strategies can be implemented. Several of these CTS strategies can be used outside the context of an ongoing professional learning community by groups of teachers engaging in a similar professional development strategy; for example, content immersions during summer institutes or CTS teacher action research as part of a math-science partnership program involving teachers from several different districts. Regardless of what your professional development structure is, CTS is a versatile process that can be embedded within a variety of structures ranging from a simple one-day workshop or seminar to the long-range, ongoing work of professional learning communities. The following sections illustrate the ways you can use CTS within a variety of professional development applications. Table 7.1 lists the professional development strategies that can be strengthened by embedding within them the CTS process. How you would use CTS within each of these strategies will vary depending on your purposes and the focus of your particular work. Therefore, the guidelines in this chapter also vary widely. For some of the professional development strategies, we provide actual session designs including a facilitator script, PowerPoint slides, and handouts to invite you to replicate what we have done with specific groups. For others, we provide a general discussion and guidelines for using CTS with a professional development strategy and offer tools and protocols that we have used for you to adapt for your own use. (See Chapter 7 folder on the CD-ROM at the back of this book.)

Table 7.1 Enhancing Professional Development Strategies With CTS

Professional Development Strategies (Loucks-Horsley et al., 2010)	Description	How CTS Enhances
Curriculum alignment and selection	Teams of teachers and administrators from appropriate grade levels analyze the content, student learning activities, instructional methods, and teacher background information and assessment strategies within instructional materials to select what materials to use with their students.	Teams of teachers and administrators use the CTS results to identify the accuracy of the content as compared with the CTS resources, how well the materials reflect the learning goals for their grade level suggested in the standards, if the learning strategies are aligned with the instructional implications from the standards, and how well the assessment activities uncover students' prior knowledge and common misconceptions.

Professional Development Strategies (Loucks-Horsley et al., 2010)	Description	How CTS Enhances
Curriculum implementation	Teachers are oriented to new curriculum materials to see how concepts are developed. They learn the content by engaging in selected lessons themselves as learners; they examine how they will use the lessons in the classroom, including reflecting on how their students will learn through the activities; and they practice and get feedback on how to implement the activities in their own classrooms.	Teachers conduct a CTS on topics within the new curriculum to identify how the concepts develop from grade to grade and if there are gaps that need to be filled. They examine the units and the lessons within the curriculum to see how they align with the standards and research, select appropriate phenomena and representations, strengthen the connections among concepts, and identify any modifications needed. Teachers review the assessments to identify how well they uncover common misconceptions and make adjustments as needed.
Partnerships with scientists	Science faculty or other scientists and teachers collaborate to build content knowledge through research projects, immersion in inquiry, or in real world science applications.	CTS provides invaluable support for scientists and teachers working together. They develop a shared understanding of the learning goals for K–12; teachers can increase understanding of the science content in the standards, and they both can explore the areas of difficulty for students and instructional implications, especially bringing in real world science applications.
Study groups	Study groups are collaborative groups of educators who convene regularly to examine specific topics related to teaching and learning.	CTS findings can guide study groups as they identify and examine instructional tools and methods to improve student learning. The study groups can explore student misconceptions, investigate if the students in their schools hold these misconceptions, and if so, develop strategies to address them.
Action research	Teachers conduct classroom-based research to investigate their own teaching and students' learning through observation, data collection and analysis, and reflection on results.	Teachers conduct a topic study to inform their research question, decide what data would constitute evidence of student learning, and interpret findings.

(Continued)

Table 7.1 (Continued)

Professional Development Strategies (Loucks-Horsley et al., 2010)	Description	How CTS Enhances
Case discussions	Groups of teachers review and discuss a case of teaching (written or video) designed to provoke discussion and reflection on practice.	Teachers conduct a CTS on the topic covered in the case to inform their analysis of the case. They explore the extent to which the case reflects the standards and research in the CTS findings and derive learning about what actions they can take in their own practice.
Examining student work and thinking	Groups of teachers examine student responses to prompts or test items to move beyond just knowing if students got the right answer to being able to identify what students know and understand and identify next steps to deepen understanding.	CTS findings provide a shared understanding of what to look for in student work, for example, what key ideas and learning goals are most important for student to know. Teachers use CTS to assess whether students understand key ideas, hold misconceptions, or are lacking understanding of important precursor knowledge.
Lesson study	Groups of teachers meet over long periods of time to develop, try out, and refine lessons. Teachers consult research and instructional materials, discuss effective practice, and observe the lessons to gather data on needed refinements.	Lesson study groups use CTS to make research-based lessons. They conduct CTS on the topics of the lessons or unit and then make modifications to the lessons that reflect the research. They speak from the evidence found in the CTS to inform their lesson design. They use CTS findings to inform the observation of the lesson and suggest adjustments as needed.
Immersion in content and immersion in inquiry	Groups of teachers participate in multiple-day sessions led by content or inquiry experts, to learn science content or the content and skills of inquiry either for teaching at their own grade level or to develop their adult-level science knowledge.	CTS Section I provides guidance on the adult literacy target that can clarify the learning targets for the adult learners in the immersion, as well as an appreciation for the scientific literacy target that the cumulative K–12 educational experiences for students are intended to achieve. CTS Section V can help make the links between the adult literacy targets and those specific grade level targets within a student's K–12 experience.
Coaching	A teacher and an instructional coach work together to enhance classroom practice through observation and feedback.	The coach and the teacher with whom the coach is working conduct a CTS on the topic of the lesson prior to planning the lesson. They incorporate CTS findings by ensuring

Professional Development Strategies (Loucks-Horsley et al., 2010)	Description	How CTS Enhances
	Most coaching involves a pre- and post-conference and classroom observation of a particular teaching technique, grouping strategies, student-teacher interaction, or other topics of interest to the teacher.	that the key ideas and learning goals from the CTS are reflected in the lesson and that assessment strategies uncover misconceptions and ways of addressing them.
Demonstration lessons	A teacher conducts a classroom lesson that is videotaped or other teachers observe. The demonstrating and observing teachers meet before the observation to discuss the goals and intent of the lesson and following the observation to debrief the experience.	The demonstrating and observing teachers conduct a CTS prior to the demonstration to gain a shared idea of what learning goals should be targeted in the lesson and what prior knowledge students need. They identify any common misconceptions that may need to be addressed in instruction. As they debrief the lesson, they explore how the instructional strategies reflect the standards and research found in the CTS.
Mentoring	New teachers are assigned a more experienced teacher to provide them with support and guidance about the school, instructional strategies, and the content they teach.	Mentors teach the CTS process to their mentees so that whenever new teachers have a question about content, they can turn to the CTS resources to find answers. Mentors meet with the new teachers to discuss the CTS findings and how they apply their instruction.
Developing professional developers/leaders	Staff developers and teacher leaders learn to facilitate professional development activities such as study groups, lesson study, mentoring, and so forth or have training to lead specific institutes, courses, and workshops in science education.	Teacher leaders and staff developers conduct a CTS on the topics they are covering in professional development. They use the findings to design their professional development and may include CTS introductory sessions in their work with teachers so the teachers can learn how to access the standards and research any time.
Workshops, institutes, courses, and seminars	Teachers participate in structured classes lasting from a half-day to weeks or months to learn content or instructional strategies.	Leaders of these learning sessions use CTS prior to designing their workshops and courses. They use the findings to build in activities and content that is aligned with standards and research. They lead CTS introductory sessions or full topic studies in their workshops or courses to develop teachers' understanding of the standards and research on the topics they teach.

CTS Professional Development Strategy Application 1: Study Groups

DESCRIPTION OF CTS STUDY GROUPS AS A CTS PROFESSIONAL DEVELOPMENT STRATEGY

Through the professional development strategy of study groups, teachers engage in regular, structured, and collaborative interactions about topics identified by the study group to be of interest to them for improving student learning (Loucks-Horsley et al., 2010). A key feature of study groups is that they provide opportunities to study research, gather information, examine resources, and decide how to apply what they learn through their study to classroom practice. Study groups may form to focus on learning about a specific topic or developing a new skill or to read and discuss books related to education practice. Murphy and Lick (2001) developed the model of whole-faculty study groups, where the purpose is to create study groups that are focused on student learning. They defined the study group as "a small number of individuals joining together to increase their capacity through new learning for the direct benefit of students" (p. 10).

Study groups have been used to lead whole school improvement, to guide the implementation of new curriculum materials, or to simply read and discuss professional literature. In a CTS study group, each of these purposes is tied to studying a science topic. CTS study groups can also form for the purpose of meeting on a regular basis to do curriculum studies on different curricular topics selected by the group. The following are twenty ideas for applying CTS in a study group:

1. Teach each other lessons that show how to use instructional strategies identified in CTS Section II.

2. Examine student work using a protocol such as the CTS CIEST (Collaborative Inquiry into Examining Student Thinking). See Application 2 in this chapter on pages 235–243.

3. Use formative assessment probes, observational data, interviews, audiotapes, and videotapes of student learning to examine students' thinking for commonly held ideas identified in CTS Section IV.

4. Clarify performance indicators or other learning goals from CTS Section VI to ensure consistency in targeted learning goals.

5. Engage in a full CTS or a partial study to examine whether instructional materials used or under consideration for purchase include the teaching of the important ideas reflected in the standards.

6. Share strategies used in the classroom to have students demonstrate their learning of key ideas identified in CTS Sections III, V, and VI.

7. Share and demonstrate elicitation strategies that can be used to uncover students' existing conceptions before instruction and compare students' ideas to those identified in the research from CTS Section IV.

8. Design a lesson or modify an existing one based on CTS findings and the instructional design suggestions described in Chapter 4 of the CTS parent book.

9. Design a common assessment to be used by the study group based on district assessment protocols and analysis of the targeted student learning goal(s) from CTS Sections III, V, and VI.

10. Identify, discuss, and practice strategies that help develop students' inquiry and technological design skills and understandings using the CTS guides from the category, "Inquiry and the Nature of Science and Technology."

11. Identify, discuss, and practice strategies for helping students develop nature of science understandings in the context of life, physical, Earth, or space science content by conducting a joint CTS between a disciplinary CTS guide and "The Nature of Scientific Thought and Development" CTS guide.

12. Investigate ways to increase students' motivation and engagement in science by examining key ideas using CTS Sections III, V, and VI and strategizing ways to make the content more accessible and relevant to students.

13. Videotape a lesson using the CTS video demonstration lesson protocol and share and discuss with the group how to improve the lesson, based on CTS. See Application 3 in this chapter on pages 244–250.

14. Use CTS findings to inform a discussion of a case of science teaching. Use published case studies that describe a decision point or dilemma faced in the classroom or address content that is rife with student misconceptions (e.g., electricity or seasons). Study group members can also write their own "cases" of when they saw students struggle or run into difficulty learning a specific science topic. The study group would explore possible explanations for the difficulties or dilemmas in the CTS findings.

15. View and discuss videotapes of teaching and student learning available through the Annenberg collection at www.learner.org. Use CTS to look for evidence of how the lesson reflects standards and research on learning.

16. Use the Annenberg Essential Science series at www.learner.org to focus a study group around a particular content area such as physical science. Select CTS guides to go with each of the videotapes. Conduct a CTS prior to viewing the video and follow by discussing how the video can inform classroom practice.

17. Examine the technical terminology from CTS Sections I and III that is essential for science literacy in a particular topic and discuss ways to help students develop the language of science.

18. Gain a deeper knowledge of particular science content by teaching each other, using Section I of the CTS guides or reading and discussing adult trade books together. (See www.curriculum topicstudy.org for links to books and resources.)

19. Use the *Atlas of Science Literacy* and CTS Section V to examine connections among ideas and plan for increased curriculum coherence.

20. Work on a collaborative action research project using the CTS action research protocol and write up and publish results. See Application 6 in this chapter on pages 274–279.

GOALS OF A CTS STUDY GROUP

The goals of a CTS study group are to

- engage teachers in accessing new information, resources, and research that address student learning needs,

- deepen teachers' understanding of research and standards related to the science topic of interest to them or that their students have difficulty learning, and
- apply the CTS findings to identify an effective instructional method or material to address a student learning need.

AUDIENCE

Using the CTS study group design is appropriate for any grade level but is best for teachers who can focus on similar content and grade levels.

Facilitator Note

This scaffold is based on the work of Murphy and Lick (2001), modified to focus on embedding CTS within an existing study group. For more information on establishing whole-faculty study groups see Murphy and Lick (2001).

STRUCTURE OF THE CTS STUDY GROUP

The CTS study group uses a protocol consisting of seven steps summarized as follows. You can use these steps to facilitate a CTS study group. A scaffold (Handout 7.1.1) guides study group members through each of the steps. All handouts can be found on the CD-ROM in the Chapter 7 study group folder.

Step 1: Review Group Norms

Clarify the standards of behavior that your study group will use and if you do not already have a norm to "speak from the evidence," add that norm to your list. (Consult tips for establishing other norms for your study group in Chapter 3 of this *Leader's Guide*.)

Step 2: Identify Student Needs in Science

As a group, set a goal for the area(s) of science education in which you would like to make improvements in student learning. Use local or state assessment results to inform your selection of goals for improved science learning. For example, if your study group is composed of teachers who teach the same grade level, you might identify a particular topic for your grade level for which you would like to learn more about ways to enhance learning. If you are a schoolwide team, you may wish to study a combination of topics to see what students should be learning across multiple grades. If you are a districtwide team examining end-of-course studies, you may be interested in better understanding what all twelfth-grade graduates should know. Handout 7.1.2 is an example of a planning template that can be used by study group members to plan what they will do during their study, the CTS guides they will use, the local or state standards they will target, and the data sources they will examine to identify areas for improvement in learning (Murphy & Lick, 2001).

Step 3: Elicit Prior Knowledge

Provide time for the study group members to surface their prior knowledge about the topic and what they know about effective teaching and learning and difficulties related to learning the topic. (You can use a tool such as the K-W-L used in the mock study group design later in this chapter or other elicitation strategies suggested in Chapter 3, Table 3.2 of this *Leader's Guide*.) Use this knowledge as a base to build from as you explore the CTS findings or challenge prior assumptions about teaching and learning the topic(s).

Step 4: Study Research, Standards, and Other Information

Identify the CTS guide(s) most relevant for the goal(s) established in Step 2. If your group needs a primer on how to find the right study guide (or section if you are doing only partial studies), you can use the Introduction to CTS Scaffold activity in Module A1 in Chapter 4 to give them a little practice (substitute Handout 7.1.0 for Handout A1.5 in Module A1). Use readings from the guides to

1. clarify the adult content knowledge associated with this area for science improvement (CTS Section I),

2. clarify the learning targets (CTS Sections III and VI),

3. examine the curricular and instructional implications (CTS Sections II and V), and

4. identify the common student misconceptions (CTS Section IV).

Handout 7.1.3 provides an example of a recording sheet you might use with a study group to record the findings from each CTS section and how they might be used to address the identified student-learning problem. Select at least one reading strategy and a report out strategy from Chapter 3 of this *Leader's Guide* for use by the study group.

Step 5: Investigate Effective Instructional Methods or Materials

In this step, based on what the study group learned from the CTS, it will identify and select some new instructional method(s) or materials or modify ones already in use to address the student learning problem. Examine the methods or materials and discuss how they (1) link to the CTS findings and (2) why teachers would use them with their students. Encourage participants to take time to reflect on the study group session—what they learned and what they will try out with students back in the classroom.

Step 6: Use New or Adapted Instructional Methods and Materials

Discuss how the group will use the new or adapted methods or materials in the classroom and what evidence of student learning they will look for to provide evidence the new methods are enhancing learning. Evidence can consist of student work, common assessment scores, observations, anecdotes from students, videos, or even observations made by study group members who observed the other members teaching a lesson.

Step 7: Collect Evidence of Learning

Collect evidence to bring back to the study group to show how the new or adapted methods or materials worked with the students and what might need to be adjusted in order to be more effective.

Step 8: Reflect on Results and Decide Next Steps for New Topics to Study

Reflect on how CTS informed changes in teaching and learning, including how well the new or adapted methods or materials worked with students and what the next steps are to sustain new practices that are working. Discuss modifications one might make to further improve the instruction. Decide on new topics to investigate by returning to Step 2 of the protocol and repeating the cycle.

OTHER RESOURCE MATERIALS FOR CTS STUDY GROUP LEADERS

CTS study group members may wish to take turns completing a log to document the work of their study group. An example of a log is provided on Handout 7.1.4. Time to meet and lack of incentives can pose barriers to forming and sustaining study groups. Handout 7.1.5 offers suggestions for finding time for study groups to meet and incentives to encourage participation. Handout 7.1.6 provides procedural suggestions for setting up and maintaining a CTS study group.

USING A MOCK STUDY GROUP TO INTRODUCE A CTS STUDY GROUP

The following guidelines provide a full design for leading a simulated or mock CTS study group session. You can use this design to introduce the study group strategy to teachers who may be new to study groups and are interested in forming groups in their school, or you can use it to teach an existing study group how CTS is applied to their study group work. Following their experience in the mock CTS study group, they will be prepared to engage in their own CTS study group on topics of interest to them following the protocol described previously.

Time Required

A full CTS mock study group session takes approximately 1.5 to 2 hours. (Your own CTS study group sessions will vary depending on your goals. Try to allow a minimum of one hour for study group sessions.)

MATERIALS AND PREPARATION

Materials Needed by Facilitator

CTS Parent Book

Science Curriculum Topic Study: Bridging the Gap Between Standards and Practice (Keeley, 2005)

Resource Books

- Atlas of Science Literacy (Vol. 1) (American Association for the Advancement of Science [AAAS], 2001).
- Since summaries of the readings are used, the rest of the resources books are not needed for this session, but it is recommended that the facilitators have one full set of the resource books on hand.

Application 7.1 CTS Mock Study Group PowerPoint Presentation

The PowerPoint presentation is located in the Chapter 7 folder on the CD-ROM. Review it and tailor it to your needs and audience as needed. Insert your date and location on Slide 1, add additional graphics as desired, and add your own contact information on the last slide.

Handouts

The following are resource handouts that can be used when planning study groups. (*Note:* Not all of these handouts are required for the mock study group design in this section, but they should be used when you plan your own CTS study group.) All handouts can be found on the CD-ROM in Chapter 7, study group folder.

- Handout 7.1.2 Action Planning Template
- Handout 7.1.3 Recording Sheet
- Handout 7.1.4 Study Group Log
- Handout 7.1.5 Suggestions for Finding Time
- Handout 7.1.6 Study Group Procedures

Supplies and Equipment

- Computer and LCD projector to show PowerPoint presentation
- Flip chart easel, pads, and markers
- Paper for participants to take notes
- Sticky notes for notes and marking charts

Wall Charts

- Write out the titles of the eight steps of the CTS scaffold on Handout 7.1.1 on chart paper and post in the room.
- Prepare and post a sheet of chart paper labeled *Evaluation*. Divide the sheet into two columns (a T-chart) and label the left column *What I Got* and label the right column, *What I Need or Suggest*.

Facilitator Preparation

- PowerPoint slides for Application 7.1 Study Group (in the Chapter 7 folder on the CD-ROM). Insert your own graphics and additional information as desired. (Optional: Print out copies of the PowerPoint slides for participants.)
- Conduct your own "Experimental Design" CTS prior to leading the session and review the CTS summaries already prepared for this session (print the four summaries from the Chapter 7 folder under Study Group on the CD-ROM).

Materials Needed for Participants

(*Note:* All handouts and the summary notes can be found on the CD-ROM in the Chapter 7 study group folder.)

- Handout 7.1.1 CTS Study Group Scaffold
- Handout 7.1.2 Action Planning Template
- Handout 7.1.3 Recording Sheet
- Handout 7.1.7 Mock Study Group Scenario
- Handout 7.1.8 K-W-L Sheet
- Handout 7.1.9 Instructional Scaffold for Experimentation
- Handout 7.1.10 Linking CTS Results to the Instructional Scaffold
- Copies of CTS summaries for Sections I, II, III, IV (in Chapter 7 study group folder on CD-ROM)
- Table copies of the map *Scientific Investigations* from the *Atlas of Science Literacy* (Vol. 1), pages 18–19
- Copy of PowerPoint slides for Application 7.1: CTS Mock Study Group (optional)

DIRECTIONS FOR LEADING A CTS MOCK STUDY GROUP

Welcome and Introduction (5 minutes)

Show Slide 1. Have participants sit in small groups of four to six people. Show Slide 2 and share that the session goals are to introduce the idea of using curriculum topic study as a component of a study group to increase focus on science content and pedagogy and address particular student learning needs in science. Share what a study group is by reviewing the points on Slide 3 of the PowerPoint. (*Note:* If your group is familiar with study groups, you can skip this.) Remind everyone that as a study group you are going to learn to embed a process called CTS in your study group by going through an experience in a mock study group. Show Slide 4. Point out that each table group is a "mock" study group. Ask everyone to review Handout 7.1.7: Mock Study Group Scenario and clarify that for this session the group is the Verifine School District's Middle School Study Group. This study group's goal is to examine issues related to inquiry-based science teaching and learning. In this study group, teachers meet twice each month to read and discuss articles and resources related to a particular aspect of inquiry they wish to improve in their instructional practice. The study group chooses the focus related to inquiry for each meeting and teachers take turns facilitating each meeting. (Point out Handout 7.1.2 and show how the study group, when planning its work, would use this Action Planning Template. They can use this template if they lead their own study group sessions later.)

Clarify that as the Verifine School District's Study Group, they will go through a set of steps called the CTS study group protocol as they simulate the study group session.

Point to the chart paper that you posted with the eight steps of the CTS study group protocol and refer them to the Handout 7.1.1: CTS Study Group Scaffold, which has the same eight steps listed. Let them know they will start the mock study group with Step 1.

Review Group Norms (5–10 minutes)

Refer to your own group norms that you posted on chart paper or if you don't have group norms, choose some from Chapter 3 of this *Leader's Guide* to guide your group and post them on a piece of chart paper in the room. Ask members to review the list, suggest other norms they would like to add or any they would like to delete or clarify. Point out that one of the norms of using CTS is to "cite the evidence" and that they will add that norm to work on today (if you don't already use it).

Once you are done, place a check mark next to Step 1 on your chart paper and say they have just completed the first step of a CTS study group. Remind everyone to use the norms throughout the rest of the session.

Identify Student Learning Needs in Science (5 minutes)

Point out that for this session, the Verifine School's Study Group decided to examine the difficulty their students are having in designing their own experiments that involve variables. They agree to do a CTS on the topic experimental design, focusing on the research-identified difficulties students have, the learning goals for their grade level, and the K–12 progression of development in the ability to design controlled experiments.

One of the teachers has found an instructional activity that seems to have the potential to help upper elementary and middle school students understand how to design controlled experiments. She will walk the teachers through the activity after they do the CTS readings. While they do the activity, the teachers will think about how the steps in the lesson reflect what they learned in the experimental design topic study.

Check to make sure everyone understands what the student learning needs in science are that the Verifine Study Group is focused on and check to see if there are any questions. Return to your chart with the study group protocol steps listed and check off Step 2.

Elicit Prior Knowledge (5–10 minutes)

Show Slide 5. Ask everyone to pull out the Handout 7.1.8: K-W-L and jot down their response to the questions on the slide: "What do you know about experimental design before you begin the study? What types of instructional opportunities or contexts help students learn skills and knowledge of experimental design? What are some important concepts at your grade level? What difficulties or misconceptions do students have with this topic? What would you like to learn about teaching experimental design?" After participants complete this step, return to the wall chart and check off Step 3 to remind everyone of the steps they are moving through.

Study Research, Standards, and Other Information (20 minutes)

Have everyone put their K-W-L sheet aside and say that ordinarily they would now divide up sections of the CTS guide and do the readings. Since this is a mock study group to learn the CTS study group process, not their own study group, explain that you are giving them the summaries of Sections II, III, and IV, which they will use for their CTS discussion. Explain that you are also giving them a summary of Section I to model the results that a study group member might share as well as their personal reflection after sharing the section. Move the chart pointer to Step 4. Have everyone look at the Section I summary while you as the facilitator use the handout to model how the study group member would report out on the reading and share personal reflections on the reading. (The top part of the summary discusses the findings and the bottom includes a reflection on what was learned.) Go over the summary and then show what your "personal reflections" were. Ask everyone to use this as the model for their own summaries and reflections.

Next, ask someone at each table to take responsibility for one section, either Section II, III, or IV. Pass out the copies of the CTS summaries. Have them use the CTS summary to review findings and record them on Handout 7.1.3: Recording Sheet (or in their own notebooks or on laptops if they prefer) and to write a reflection like the model from the summary of Section I. Allow time at the tables for each person to read and share their personal reflections on what they read. Show Slide 7 and have the group look at the *Atlas* map, *Scientific Investigations,* at their tables guided by the questions on the slide. After a few minutes, ask each group to share at least one thing they noted from the map. Based on the discussion and report out, point out that they have just clarified what students should know and be able to do according to the standards and the instructional implications for teaching experimental design that are suggested in the readings. They will use this now to examine a new instructional tool.

Investigate Effective Instructional Methods or Materials (30 minutes)

To remind everyone where they are in the process, point back to the chart paper of the steps and review the four steps you have already done: (1) Identify group norms including the norm of citing the evidence, (2) clarify a student learning goal to focus their work, (3) identify what they know and want to learn about the topics of their work, and (4) study the topic by reading CTS sections. Point to the poster that you prepared listing these steps, checking off any steps you have not yet checked.

Now you as the facilitator will model Step 5 (point it out on the chart) by leading the study group to investigate an instructional tool called the "Instructional Scaffold for Experimentation" (see Handout 7.1.9), which you brought to the group. Say that in Step 5, they will look for a possible solution to the student learning problem. One of the Verifine Study Group members has identified a possible resource that may help their students address the problems they are having with experimental design.

Show Slide 8 and distribute Handouts 7.1.9 and 7.1.10. Tell the group that they will use what they learned from their CTS readings and discussions to examine the tool, Instructional Scaffold for Experimentation (Handout 7.1.9), and to link what they learned from the CTS findings by using Handout 7.1.10: Linking CTS Results to the Instructional Scaffold.

Facilitator Note

The following activity is also used in Module B1 in Chapter 5 and uses an example of an investigation into seed germination—you can use this example or substitute it with an example of your own choosing. Make sure the study group makes the connections between the experimental design scaffold and its connections to CTS.

Begin to model Step 5 by explaining to the Verifine Study Group that you have brought an instructional tool for their consideration and analysis. Tell them the tool is the Instructional Scaffold for Experimentation (Handout 7.1.9) and that it was adapted from work by Ann Goldsworthy (1997) in the United Kingdom and by the Colorado Department of Education for their Goals 2000 Inquiry Toolkit. Explain that it is an instructional tool, not a methodology that may imply a rigid set of steps in doing science. Be sure participants know that the Instructional Scaffold for Experimentation is intended to help students who struggle with the steps involved in setting up an experiment and controlling variables, but that they need to make sure students understand this does not mean there is a single lock-step method all scientists follow when engaged in investigation. Reinforce this concern by pointing out the findings from CTS, from the readings in Step 3, that alert teachers to the danger of teaching a rigid scientific method. The Instructional Scaffold for Experimentation is a type of representation for modeling the steps in designing an experiment, but like any model or representation, it does not fully capture the process and has some drawbacks. This scaffold guides students through the basic process of experimentation. It does not address more complex types of experiments that involve multiple variables or control groups such as those used for comparison in clinical trials. It also does not capture how scientists often cycle back and forth when testing their ideas.

Explain that germinating (or sprouting) bean seeds is a common activity used to help students understand how living things are dependent on physical factors in their environment. Explain that they will use this as their example for exploring the use of Instructional Scaffold for Experimentation to see how it addresses their students' learning problems and to see how well it connects to their CTS findings.

Explain that the group will work through the Instructional Scaffold for Experimentation using the bean seed germination example. As they go through the sections of the scaffold, encourage the group to make notes on Handout 7.1.10: Linking CTS Results to the Instructional Scaffold of any linkages they see between the steps in the scaffold and the findings from the CTS readings and discussion.

Explain how inquiry often begins with wondering about some object, event, or process in the natural world. This wondering is often coupled with accessing prior knowledge or experiences related to the phenomenon at hand. Show Slide 9 to illustrate an example of a student's "wondering" about seeds that initiates the inquiry and leads to the design of an experiment. Refer them back to Handout 7.1.9, Part I, and the first question: "What am I wondering about?" Show how students would write their beginning question there on the scaffold.

Show Slide 10. Point out that students would then brainstorm things they could change or vary related to their beginning question. Invite the study group to brainstorm their own list of at least four examples. Explain that this part of the scaffold helps students identify the different things that might affect seed germination. In Part A, they list things that could be changed when investigating what seeds need to germinate. Explain that for the purpose of trying out the instructional resource, they will choose four things they could change.

Ask participants to write things that could be changed when investigating what seeds need to germinate. Put each factor on its own colored sticky note (use the same color sticky note for all factors—we will call "Color A"; in the PowerPoint they are blue). Place the sticky notes on the boxes on Handout 7.1.9 as shown by the blue boxes on Slide 11.

Facilitator Note

The germinating bean seeds example is one that most teachers are familiar with. Since this part of the CTS does not involve doing the experiment, it is best to select a familiar example. Depending on the background of your group and the context they teach in, feel free to substitute the germinating bean seeds example with another one that ties to the instructional materials, curricular goals, or grade level of your teachers.

Facilitator Note

Guide participants to include amount of water on their lists. Other examples might include type of seed, type of soil, temperature, amount of time for sprouting, and amount of light.

Facilitator Note

You need not be limited by the four boxes on the handout—use more if desired by extending the layout of the boxes.

Next show Slide 12 and ask participants what kinds of things could be measured or observed during seed germination. After a few minutes, ask for ideas from the group.

Ask participants to place four things they identified that can be measured or observed in each of the four boxes using a different colored sticky note ("Color B") as shown by the yellow boxes on Slide 13. Debrief this part by explaining that students are brainstorming possible things that can be measured or observed. They have not yet refined their experiment to focus on two experimental variables.

Show Slide 14 and refer to the "Identifying Variables" section of the scaffold. Explain that now students are beginning to isolate a manipulated and responding variable. Ask them to imagine that students decided they wanted to know how the amount of water affects germination. Guide participants into moving the Color A sticky note, amount of water, into the box under "I will change." Ask them to imagine the students decided they would count how many seeds in a given sample sprouted. Guide participants to move the Color B sticky note, number of seeds sprouted, into the box under "I will measure or observe." Now have participants pick up and move the other three Color A sticky notes into the three boxes under "I will not change." Explain how the type of soil, type of seeds, and temperature will all stay the same (as well as other factors that may have been identified). The only thing that will change is the amount of water. Have them repeat this with the three remaining Color B sticky notes, showing that they do not need to observe or measure these things.

Pause and ask participants to reflect on their CTS readings and findings and discussion and make the link between what they have done so far with the scaffold and how it reflects the CTS findings. Have them record their ideas on Handout 7.1.10. It is important for them to take the time to reflect on these linkages between the instructional tool and their findings from CTS so that they can better understand how CTS informs the selection of appropriate instructional methods or materials by the study group. Encourage them to cite the research from CTS Section IV that describes the difficulty students have identifying and controlling variables and how the scaffold guides them in developing this skill.

Next, have participants refer to the "Asking a Testable Question" section on page 3 of Handout 7.1.9. Explain that they will now move the two colored notes from the previous steps into these boxes to frame a testable question. They will use the question frame to write their question that will guide the experiment. Show Slide 15 as an example of framing a testable question. Point out that this is a time when they access their prior knowledge and experience to think about what they already know in relation to their question. Show Slide 16 and ask group to link the scaffold (on Handout 7.1.9) to their CTS findings.

> **Facilitator Note**
>
> Guide participants to make sure the number of bean seeds sprouted is included as one of the things that could be measured or observed. Other examples are the size of the sprouts, mass of the sprouts, color of sprouts, and the number of days the seeds take to sprout.

> **Facilitator Note**
>
> Again, you need not be limited by only four boxes—use more if desired by extending the layout of the boxes.

Use New or Adapted Instructional Methods and Materials (5 minutes)

Show Slide 17 and point out Step 6 on the chart. Encourage the mock study group to talk about how they will go back and try the Instructional Scaffold for Experimentation with their own students, or make modifications for younger students based on the idea of a *fair test*, the prerequisite to designing controlled experiments. Ask if there are other things they will do differently when teaching experimental design based on the CTS. Even though this is a mock scenario, ask them to imagine they are going back to the classroom to try this out. Ask them to discuss what kind of student learning evidence they will bring back to the study group next time they meet. Point out Steps 7 and 8 and describe that these steps take place in the classroom (Step 7) and when the study group meets again (Step 8). Remind everyone of the six steps they just went through as a simulated CTS study group and the two other steps that take place after this study group meeting.

Wrap-Up and Reflection (10–15 minutes)

Show Slide 18. Wrap up by having everyone return to the K-W-L sheet and write what they learned. Use one of the report out strategies from Chapter 3 of this *Leader's Guide* to allow everyone to share something they learned, value they see in using CTS with study groups, or something they plan to do differently. Bring the group back together and ask for several responses.

Evaluation (5–10 minutes)

Show Slide 19. Ask everyone to provide feedback by writing a short list of what they got or learned from the session and what suggestions they have for changes in the session for the future. Ask them to put each idea on a separate sticky note and post them on the chart paper labeled *Evaluation* as they leave the room.

CTS Professional Development Strategy Application 2: Collaborative Inquiry Into Examining Student Thinking (CIEST)

DESCRIPTION OF CIEST AS A CTS PROFESSIONAL DEVELOPMENT STRATEGY

CIEST is a collaborative process that combines three elements: curriculum topic study, student work from assessment probes, and a tool called data driven dialogue (Lipton & Wellman, 2004). As shown in Figure 7.1, the overlap of these three elements results in a process that deeply engages teachers in constructing a deeper understanding of students' thinking for the purpose of improving students' opportunities to learn important ideas in science.

CIEST differs from other strategies for looking at student work because the primary focus is not to examine student work for the extent to which a student has achieved a learning goal. Instead, CIEST is designed to examine students' thinking, even before instruction addresses a learning goal. It uses assessment probes that uncover students' ideas before, throughout, or even years after instruction. The assessment probes are designed to reveal students' commonly held ideas noted in the research in CTS Section IV and related to specific ideas in the standards as described in CTS Section III. The process helps teachers think about the actions they can take in their classrooms or curricular contexts to address students' commonly held ideas related to a concept they teach. Teachers work collaboratively to construct meaning around the ideas students bring to their learning and analyze data in order to inform curricular and instructional decisions.

Figure 7.1   Three Elements of CIEST

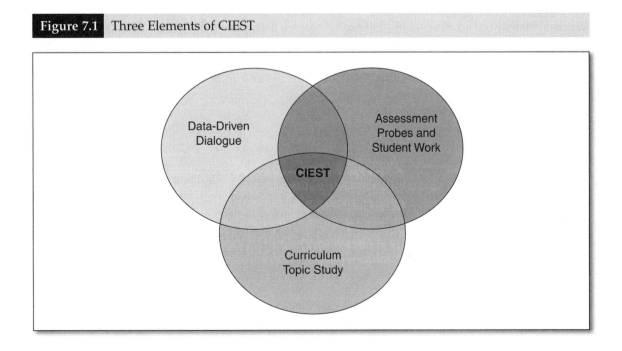

235

Schools are increasingly moving to use ongoing collaborative arrangements such as professional learning communities and study groups, where teachers learn together and make improvements in practice. While research suggests these collegial arrangements can lead to improved outcomes (Lee, Smith, & Croninger, 1995; Marks, Louis, & Printy, 2000; McLaughlin & Talbert, 2001), we caution that success will depend on how well teachers build the norms, tools, and protocols to focus on learning. Collaborative inquiry is a powerful and increasingly popular way teachers are reflecting on their practice and student learning and making adjustments to improve outcomes. They use data and reflective dialogue to come to a shared understanding of student learning problems and agree on potential solutions to test out in the classroom (Love, Stiles, Mundry, & DiRanna, 2008). While teachers certainly examine their own students' work and thinking quite often, the nature of collaborative inquiry changes the focus to also enhancing teacher learning, not just student learning. Teachers become very good at interpreting students' thinking and ideas and using what they learn to inform their instruction. This process also uses an approach, called data-driven dialogue (Lipton & Wellman, 2004; Love et al., 2008) that is a four-phase process in which (1) a team of teachers *predicts* what they will see in data; (2) then they *go visual* by looking at the raw data as a group; (3) they state facts that they *observe* in the data, sticking to just the facts; and (4) then they infer or ask questions of the data (Figure 7.2). This process grounds everyone in the facts and data, before they jump to inferences.

Figure 7.2	Data-Driven Dialogue

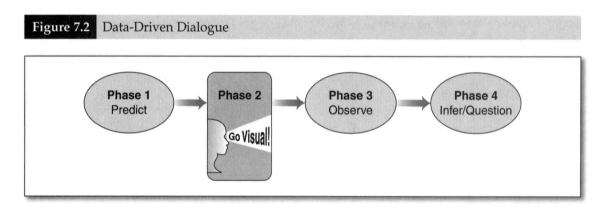

Source: Adapted from Lipton, L. & Wellman, B. (2003). *Data-Driven Dialogue: A Facilitator's Guide to Collaborative Inquiry.* Sherman, CT: Mira Via. In Love, N., Stiles, K. E., Mundry, S., and DiRanna, K. (2008). *A Data Coach's Guide to Improving Learning for All Students: Unleashing the Power of Collaborative Inquiry.* Thousand Oaks, CA: Corwin. Used with permission.

GOALS OF A CIEST SESSION

The goals of this session are to

- learn about a CTS professional development strategy for collaboratively examining student thinking,
- practice using the CIEST protocol with a sample of student work, and
- consider how you might use the CIEST strategy and CTS developed probes to examine student thinking in your work.

AUDIENCE

CIEST is appropriate for any grade level and can be conducted with mixed grade levels looking at samples of student work from different grade levels or with same grade level groups. This session uses an eighth-grade example of student work to introduce participants to the process, making it accessible to both elementary and secondary teachers in the group, and is appropriate for introducing the process. Once the audience has learned the CIEST process, they can gather their own student work and use the process to focus on their own teaching.

STRUCTURE OF A CIEST SESSION

The CIEST Session uses a protocol consisting of ten steps broken into four parts summarized below. A CIEST Scaffold (see Handout 7.2.1 in the Chapter 7 CIEST folder of the CD-ROM) guides participants through each of the steps.

CIEST Protocol

Part One: Groundwork (Prior to Looking at Student Work)

- Establishing group norms
- Examining and completing the probe
- Clarifying the content
- Using CTS Sections III and VI to identify related ideas in the standards
- Sharing assumptions about how students might respond

Part Two: Examining Student Thinking

- Sorting and tallying Tier 1 responses
- Sorting and organizing students' reasoning
- Displaying the data (go visual)

Part Three: Analyzing the Data

- Analysis of data from student work
- Examining commonly held ideas from the research (CTS Section IV)
- Comparing research findings to data from student work
- Examining coherence (CTS Section V) and instructional implications (CTS Section II)

Part Four: Integrating Data From Student Work and CTS Findings

- Drawing inferences, making conclusions, and offering explanations
- Examining new understanding of students, student thinking, curriculum, and instruction
- Group and individual reflection

Time Required

A full introductory CIEST session requires approximately 3 hours.

MATERIALS AND PREPARATION

Materials Needed by Facilitator

CTS Parent Book

Science Curriculum Topic Study: Bridging the Gap Between Standards and Practice (Keeley, 2005), or copies of the CTS guide, "Heat and Temperature," on page 216 of that book.

Resource Books

CTS selected resources or copies made of selected readings from CTS guide (include the *Atlas* [Vol. 2] map on *Energy Transformations,* if available).

State or Local Curriculum Framework or Standards

If you are working with a group from the same district, copy sections of the local curriculum frameworks or standards that address heat or thermal energy for the CTS Section VI readings.

PowerPoint Presentation

The PowerPoint presentation is located in the Chapter 7 folder on the CD-ROM. Review it and tailor it to your needs and audience. Insert your date and location on Slide 1, add additional graphics as desired, and add your own contact information on the last slide. (Adapt slides as described below if you are not using the *Mixing Water* probe.)

Supplies and Equipment

- Computer and LCD projector to show PowerPoint presentation
- Flip chart easel, pads, and markers (each table group will need at least one sheet of chart paper and one to two markers)
- Paper for participants to take notes
- Sticky notes for notes and marking charts

Wall Charts

- Prepare a Bar Graph wall chart of PowerPoint Slide 17 if whole group is using the *Mixing Water* probe. (Substitute your own visual if you are using a different probe for student work.)
- Prepare a wall chart of Figure 7.3.

Facilitator Preparation

- PowerPoint slides for CIEST (in Chapter 7 folder on CD-ROM). Insert your own graphics and additional information as desired. (Optional: Print out copies of the PowerPoint slides for participants.)
- Conduct your own "Heat and Temperature" CTS prior to leading the session.
- Prepare an evaluation form to collect feedback from participants.

Facilitator Note

You can substitute copies of your own student work from two-tiered assessment probes. (Go to NSTA Press for information on the *Uncovering Student Ideas in Science* series—a source of CTS-developed assessment probes. A crosswalk between the CTS topics and the probes from this series can be found in the "Developing Your Own CTS" folder in the Chapter 5 folder of the CD-ROM—or create your own using the assessment design process described in Chapter 4 of the CTS parent book or pp. 204–215 in this book.)

Materials Needed for Participants

(*Note:* All handouts can be found in the Chapter 7 folder CIEST on the CD-ROM.)

- Handout 7.2.1: CIEST Scaffold
- Handout 7.2.2: *Mixing Water* Assessment Probe (or substitute your own assessment probe)
- Handout 7.2.3: *Mixing Water* Cards (cut and placed in plastic bags or envelopes for each small group of three to four participants)
- Handout 7.2.4: *Mixing Water* Explanation

DIRECTIONS FOR FACILITATING CIEST

The following directions describe how to lead an introductory CIEST session for first-timers using the *Mixing Water* (Grade 8) example. If you use a different probe, substitute your probe in the directions and make modifications to Slides 12, 14, 15, 17, 18, and 21.

Introducing CIEST (10 minutes)

Show Slides 1 and 2. Welcome the participants and explain that this session will introduce them to a CTS professional development strategy called CIEST—Collaborative Inquiry into Examining Student Thinking. This process will engage them in using CTS to enhance inquiry into student ideas and thinking. Review the goals on Slide 2.

Show Slide 3. Describe how CIEST is used to examine student thinking for the purpose of informing curriculum and instruction. It differs from other ways of looking at student work since the assessments used are often not summative in nature and the emphasis is on examining thinking rather than determining proficiency.

Show Slide 4 and suggest that there is great benefit to working with others to share insights and engage in collaborative inquiry. Point out that one tool that supports collaborative inquiry is the use of a protocol. Show Slide 5 and give participants a few minutes to activate their prior knowledge about protocols and discuss their use. Show Slides 6 and 7 explaining what a protocol is and why we use protocols in professional development.

Show Slide 8 and describe how the CIEST protocol combines three elements: CTS, assessment probes and student work, and data-driven dialogue. Data-driven dialogue allows us to move away from our opinions and inferences and speak from the student data and CTS findings. We use assessment probes to generate examples of student thinking that often match findings from cognitive research. We use CTS to make our analysis of student work research and standards based. Show Slide 9 to describe the resources and tools we will be using. Explain that the *Mixing Water* assessment probe we will use in this session comes from *Uncovering Student Ideas in Science Volume* 2 (the blue book on the slide).

Refer everyone to Handout 7.2.1: CIEST Scaffold. Explain this will be the step-by-step process for engaging in collaborative inquiry to examine student thinking. Give everyone a minute to look it over.

Step 1: Establish Norms (5 minutes)

Show Slide 10. Explain that the group will start at Step 1 by developing norms for the group. Ask the group to take one minute to brainstorm norms they would like to guide their work together.

Have a cofacilitator or volunteer write the norms on chart paper as they are offered. Then ask the group if there are any additions. Add one more, "sticking to the protocol," if it isn't already there and discuss ways the group can monitor to make sure they stay on track. Post the norms where everyone can see them throughout the session. Remind everyone to "enforce" the norms in their small group work.

Facilitator Note

The assessment probes in these books were developed using the CTS process. However, you can also develop and use your own CTS probes. The process for developing them is described in Chapter 4 of the CTS parent book (facilitators may also want to use the process described in this book on pp. 204–215).

Facilitator Note

If anyone seems confused, offer an example of a norm such as "avoid vague terms and clarify meaning" or "begin and end on time." Ask for one suggestion from each table (going around the room quickly).

Facilitator Note

The facilitator will have to pay close attention to whether the participants follow the steps of the protocol. Ask them to stick to the scaffold and refrain from sharing stories about their own students or making suggestions for how to fix the problem. Remind everyone there will be time for that later, but for now, they should stick to the protocol!

Step 2: Examine and Complete the Probe (10 minutes)

Show Slide 11 and move to Step 2 of the protocol. Pass out Handout 7.2.2: *Mixing Water* Assessment Probe. Ask everyone to complete the probe and jot down the notes requested on the slide. In case you have a mixed group with some participants who may not be comfortable with their science content, let them know they will not need to make their answers public. Show the probe on Slide 12 and be sure everyone understands the task. Allow 5 minutes to work on the probe, using the questions on Slide 11.

Step 3: Probe Clarification and Standards Groundwork (15 minutes)

Show Slide 13. In table groups of two to three people, ask participants to discuss the first three questions on the slide. Depending on the grade levels and background of participants, they will often be able to identify the concept (transfer of energy or heat) but may be unsure of the correct answer or how to explain it using the specific idea that when warmer things are put with cooler ones, there is a transfer of energy. The warmer things get cooler and the cooler things get warmer until they all are the same temperature. Tell participants that it is essential for them to understand the topic before they examine students' ideas about the topic. Distribute Handout 7.2.4: Mixing Water Explanation and allow time to review. Ask if there are any questions about the science content in the probe. Ask for a volunteer to summarize the scientific explanation. (As an alternative to this step, you can do a quick science talk on the topic to give the scientific explanation or ask a "scientist" in the group to explain. See note.)

Refer everyone to the last two points on the slide. Ask them which CTS guide would be helpful for this probe (response: "Heat and Temperature" is the most specific guide. "Energy" could be used but it is a much broader topic). Tell them they will now use CTS Section III to find out the specific ideas in the standards related to the probe. Show Slide 14. Tell them to look at the grade span of the students who responded (Grades 6–8) as well as the grade span that comes before (Grades 3–5). Explain that the latter is important because often students miss precursor ideas when they move from one grade span to the next, leaving gaps in their understanding. If participants have their local or state frameworks or standards available, ask them to identify their related standards. Check to make sure they indicate the following specific ideas in the standards in the following box and that these are posted on a wall chart.

> **Facilitator Note**
>
> This may be a time for you or a knowledgeable partner to distinguish between the terms *heat* and *thermal energy* (commonly called heat energy with younger students). In this probe, when we use the word *heat* we are referring to the movement of energy. Thermal energy is associated with the temperature and is sometimes described as the total internal energy a system, object, or substance has. Thermal energy is transferred as heat.

SPECIFIC IDEAS ABOUT HEAT AND THE TRANSFER OF THERMAL ENERGY

Benchmarks Grades 3–5:

- When warmer things are put with cooler ones, the warm ones lose heat and the cool ones gain it until they are all at the same temperature.
- When warmer things are put with cooler ones, heat is transferred from the warmer ones to the cooler ones (AAAS, 2007, p. 25).

Benchmarks Grades 6–8:

- Heat can be transferred through materials by the collision of atoms or across space by radiation.
- Energy can be transferred from one system to another thermally, when a warmer object is in contact with a cooler one.

- Thermal energy is transferred through a material by the collisions of atoms within the material. Over time, the thermal energy tends to spread out through a material and from one material to another if they are in contact. Thermal energy can also be transferred by means of currents in air, water, or other fluids. In addition, some thermal energy in all materials is transformed into light energy and radiated into the environment by electromagnetic waves; that light energy can be transformed back into thermal energy when the electromagnetic waves strike another material. As a result, a material tends to cool down unless some other form of energy is converted to thermal energy in the material (AAAS, 2007, p. 25).

National Science Education Standards 5–8:

- Heat moves in predictable ways, flowing from warmer objects to cooler ones, until both reach the same temperature.

Spend a few minutes discussing the meaning and intent of each learning goal. Point out how each learning goal addresses ideas related to thermal energy (heat energy) and the transfer of thermal energy as heat.

Step 4: Anticipate Student Thinking (5 minutes)

Show Slide 15. Based on the teachers' knowledge of their own students, and the expectations of the learning goals described in the standards, ask them to share at their tables some assumptions they have about how eighth-graders would respond to this probe. Have each table share one assumption they discussed as a small group.

Break (10 minutes)

Step 5: Organizing Data (35 minutes)

Refer to Step 5 of the scaffold and Slide 16. Explain that they will now analyze the student responses. Point out the bags of student responses that you made from cutting up all the student responses on Handout 7.2.3: Mixing Water Cards. Tell participants to clear a space on their tables where they can all read and sort the student work. Ask them to quickly scan through the work samples and sort them by response. Direct the teachers to follow the prompts on the slide. Ask the table groups to first tally the tier one responses (how many students answered with responses A–E?). Then show the bar graph of student responses on Slide 17.

Remind the groups they are still working on Step 5 and are not drawing any conclusions yet—they are just sticking to the facts they see in the student work. Show Slide 18 and ask them to examine the reasoning used by the students in their tier two responses. Ask them to look at the student work and see if they observe different categories of reasoning. For example, how many students used additive reasoning (added temperatures together); how many used subtraction (subtract cold from the hot); how

> **Facilitator Note**
>
> Put out bags of student response cards (Handout 7.2.3) during the break.

> **Facilitator Note**
>
> If you are using the *Mixing Water* probe with the whole group, they do not need to make a graph since all the graphs would look the same. At this point, you post the wall chart graph made in advance and show Slide 17 to illustrate what their graphs would look like. After participants have been through this introduction to CIEST session, they will be able to make their own graphs when they analyze their own student work. The point of this session is to introduce the CIEST process so they can go through the steps on their own later with their own sets of student work.

many used scientific vocabulary without a clear understanding? How many had the wrong answer but a sound explanation? How many had a correct mathematical answer with no scientific concepts used in their explanation? Let them come up with their own categories of reasoning. Then each group should "go visual" and create a chart that visually shows the different types of reasoning they saw in the student responses.

Step 6: Analyze the Data (20 minutes)

Show Slide 19. Ask participants to examine and analyze their data display. Have groups state facts that are evident from the data on their charts. Look for trends or patterns. Note interesting or surprising findings. Examine correct responses for evidence of the specific idea(s) identified in the CTS groundwork (Step 3). Remind everyone to remember the group norms and also refrain from making inferences, drawing conclusions, or suggesting ways to address the problem at this stage. Engage participants in a "fact-based" discussion. Remind them of the old Jack Webb quote from Dragnet: "Just the facts, ma'am, just the facts."

Ask everyone to post their charts of the student responses on one long wall and invite everyone to visit each chart for a few minutes. After everyone has had a few minutes to look at the charts, gather everyone around and ask participants to report out on what they noticed. For example, you might ask the group, "What kind of thinking is represented in the student responses?" Someone might point out that students primarily use mathematical (e.g., additive) thinking to approach the problem with no use of scientific concepts. Refer to the questions on Slide 19 to guide the analysis and discussion. Stop people from making any inferences—remind them they are still just looking at the facts.

Step 7: Examine Cognitive Research (20 minutes)

Ask participants to return to their seats to move on to Step 7 of the scaffold (Handout 7.2.1). Show Slide 20. Ask everyone to work with a partner and read the selected CTS readings for Section IV and examine the research notes for Section V if available. Explain that commonly held ideas come from the research but there are often common ideas that teachers' own students have that might not be referenced in the research literature. Define an idiosyncratic idea as one of the "outliers" or unusual idea a single student or two might have that are not seen in the rest of the group.

> **Facilitator Note**
>
> There is a new map in *Atlas of Science Literacy* (Vol. 2) that came out after the CTS guides were published. If participants have access to Volume 2 of the *Atlas*, refer them to the *Energy Transformations* map on pages 24–25.

Ask them to jot down notes or highlight sections of the research that are related to students' thinking in the work they analyzed. After about 10 minutes, ask them to share their findings at their tables. Allow 10 minutes for table discussions and then ask for a few people to respond to the question on Slide 21: "How did CTS Section IV, Research on Student Learning, help you understand the eighth-graders' thinking about heat transfer in the *Mixing Water* probe?"

Step 8: Examine K–12 Learning Goals and Instructional Implications (10–15 minutes)

Have participants examine the *Atlas* map in CTS Section V to see what a coherent progression of learning might look like that leads up to and connects with the learning goal targeted by the *Mixing Water* probe. Focus on the Grade 3–5 and 6–8 bands. Point out how Project 2061 changed the language in the new *Atlas* benchmark and the 2009 Benchmarks Online by using the term *thermal energy* instead of *heat energy* because of problems students have distinguishing between the ways the word *heat* is often used. Ask participants to also examine CTS Section II to get a sense of the types of instructional experiences that may help students in Grades 3–8 understand the ideas on the map. If you have a group from the same district and have the local or state standards, examine the learning goals for this topic. Show Slide 22 and provide time for discussion of questions on the slide.

Step 9: Integrate the Data (10 minutes)

Bring the group back together for Step 9. Show Slide 23. Have the table groups create a list of the inferences, explanations, or conclusions they can draw from the data. What additional data would they like to collect? How has their understanding of students, student thinking, curriculum, or instruction changed as a result of the CIEST?

Step 10: Next Steps and Reflection (15 minutes)

Show Slide 24. Ask table groups to discuss what they will do as a result of what they learned. What actions would they like to take as a group? What are the implications of their findings for ensuring all students in their school have opportunities to learn the ideas necessary for science literacy?

Ask participants to step back from the exploration they have just done to think about implications for them as individuals. Ask them to stand up and make eye contact with someone with whom they haven't worked today. Ask them to pair up with that person and discuss the questions on Slide 25. Remind them to share their talk time so that each one can share a few ideas and reflections.

Wrap-Up and Evaluation (10 minutes)

Bring the group back together and with everyone still standing ask for three to four people to share one reflection and then ask them to return to their seats. Show Slide 26, explaining that CIEST is powerful because it brings the classroom directly into professional development, thereby increasing its potential impact on practice. Using Slide 27, point out the professional development value of CIEST and invite participants to think more about how they can use the CIEST protocol in their own work. Thank everyone for their participation and ask them to complete an evaluation form of your own choosing, if applicable.

Optional Extension to Make This a Full-Day Session

Once participants learn to use the ten-step scaffold for CIEST, they will benefit greatly from examining their own students' work and using CIEST to explore science topics they teach. You can use the above session design to introduce CIEST using a common set of student work from the *Mixing Water* probe. After participants have experienced the introduction as a group, they can break into small groups conducting their own CIEST with student work from their own students. (*Note:* they will have to choose an assessment probe ahead of time, administer the probe to their students, and select student work samples to bring to the session.) An optional poster session, sharing each group's analysis and findings, can be included. There is also an advantage to having teachers go through the process by looking at anonymous work from an unknown class or school first, since it helps them learn to use the process and stick to the protocol in a safe and comfortable way, before using it to analyze their own students' work.

CTS Professional Development Strategy Application 3: CTS Video Demonstration Lesson (VDL)

DESCRIPTION OF VIDEO DEMONSTRATION LESSON (VDL) AS A CTS PROFESSIONAL DEVELOPMENT STRATEGY

A CTS Video Demonstration Lesson (VDL) provides a structured opportunity for teachers to collaboratively examine and discuss classroom practice related to specific science content and teaching goals. VDL consists of three elements: (1) CTS, (2) a discussion protocol, and (3) a diagnostic classroom observation tool on the content of a lesson (Saginor, 2008). Combined, these elements guide teachers to identify and discuss evidence of accurate content being taught, see the connections to standards and research on learning, and explore teaching practices that support learning the content. VDL provides an opportunity for teachers to view and discuss a typical example of teaching (not an exemplar), viewed through the lens of the standards and findings from a CTS. The strategy's effectiveness lies in its ability to promote and support a cycle of discussion based on evidence—evidence that provides participants with an understanding of the criterion through which the lesson will be examined prior to viewing and discussing the lesson.

As outlined in the beginning of this chapter, demonstration lessons are one of several professional development strategies that focus squarely on the practice of teaching. They involve teachers preparing and teaching lessons in the classroom while other teachers observe (or they are videotaped for later viewing). Teachers who will observe meet ahead of time to discuss the goals and intent of the observation and then observe the lesson and debrief the experience to derive learning and connections to their own teaching. It is important to stress that demonstration lessons are usually not exemplars or the perfect lesson; rather they should be thought of as examples of practice that can be sources of learning for other teachers (Loucks-Horsley et al., 2010).

Demonstration lessons have been widely used as a professional development strategy because they provide a direct window into the classroom and give teachers opportunities to examine instruction carefully. The strategy is often used when teachers are learning to use new instructional strategies and also for beginning teachers who may benefit from seeing more experienced teachers teach a lesson. While demonstration lessons are often done "live" with a few teachers visiting the demonstrating teacher's classroom, VDLs are easier to use for professional development because they can be viewed anytime. They can also be used more than once, and the school can build its own video library of lessons. Video allows teachers to stop and rewind a section of the lesson to inform their discussion, and they can view over and over from different perspectives or focus areas. For example, one might view the first time to just get a sense of the whole lesson, how it flowed, what the instructional strategies were, and what students were doing. Then one might revisit the tape to analyze how well the lesson aligned with the learning goals and standards for the lesson.

The CTS VDL differs from other types of video lesson analysis in that it focuses primarily on the science content of a lesson and utilizes findings from a topic study to guide participants to examine how the lesson reflected related standards and research.

GOALS OF A VDL

The overarching goal of a VDL is to provide an opportunity for teachers to improve their content and pedagogical content knowledge by examining teaching. The following box describes some of the reasons for using videotaped lessons.

WHY USE VIDEOTAPED SCIENCE LESSONS

- Creates norms and routines for teachers to collaborate and learn together
- Moves teaching from a solitary act to a shared practice
- Encourages teachers to shift focus from observation of teachers to observation of instructional practices and student learning
- Promotes teacher reflection on practice—their own and that of their colleagues
- Increases teachers' awareness of different instructional practices and their impact on student learning
- Provides opportunities for teachers to get feedback from their peers on their use of new practices
- Helps to strengthen and build professional community
- Enhances science content knowledge and pedagogical content knowledge
- Levels the "playing field"—all teachers, experienced or novice, are seen as having something others can learn from

---

As a result of participating in a VDL, teachers will

- become familiar with viewing ethics and a protocol for watching, discussing, and providing feedback on a colleague's (or other teacher's) videotaped lesson;
- use CTS to improve their understanding of effective teaching and learning as it relates to the content of a lesson; and
- recognize the value and importance of making teaching public.

AUDIENCE

CTS VDL is appropriate for teachers at all levels of the teacher professional continuum, from novice to experienced. The VDL was originally developed for use with mentor teachers and novices in the Northern New England Co-Mentoring Network (www.nnecn.org) as a way to collaboratively view and safely discuss teaching practice in a learning community made up of coaches and mentor teachers and novices.

COMPONENTS OF VDL

VDLs require the following components:

- Videotapes of classroom lessons (teachers make videos of a complete or a segment of a lesson or choose commercial videos of classroom lessons)

- Curriculum topic study (teachers complete a CTS of the topic of the lesson in the video)
- Norms (teachers establish and follow viewing ethics and norms)
- Content and instructional indicators to guide viewing
- General indicators for other factors such as classroom management
- Discussion
- Reflection
- VDL protocol

STRUCTURE OF A CTS VDL

CTS VDL is structured around the use of a protocol with six parts described as follows and shown in Table 7.2.

VDL Protocol

Step 1: Review of Related CTS

> **Facilitator Note**
>
> Doing the CTS beforehand helps develop a shared understanding among participants of the content of the lesson, connection to grade level standards, implications for teaching, and possible misconceptions or difficulties students may encounter.

Prior to the observation, participants conduct and discuss the grade-level specific CTS on the topic or subtopic of the lesson they will observe (Option A) or participants review and discuss a summary of the CTS topic that the demonstration lesson teacher or the facilitator prepared on the specific topic of the lesson (Option B).

Step 2: Teacher Introduction to the VDL

The VDL teacher sets the stage for viewing the video by describing the classroom context (grade, discipline, grouping, topic, curriculum materials used, etc.). The teacher will briefly describe what the purpose of the lesson is, what students will be doing, and what students did prior to this lesson. The VDL teachers provide an observation sheet for the lesson with key ideas identified and indicators that describe what they are particularly interested in having the viewers look for in the video segment.

Step 3: Viewing of the Video

Participants observe the VDL, noting specific evidence that the lesson addresses the key ideas from the standards and the instructional indicators on the observation sheet.

> **Facilitator Note**
>
> If you have a small number of participants such as a grade level team, you may only have one group.

Step 4: Small Group Discussions of the VDL

Small groups discuss their observations and point out the evidence the lesson addressed the indicators. Each small group generates one to three comments or "I wonder" questions to report out to the large group.

> **Facilitator Note**
>
> The VDL teacher may respond to questions for clarifications as needed, but keep this very brief so that the focus is on the observers sharing what they observed.

Step 5: Group Share: Observations and Comments

One by one the facilitator will ask the small groups to share their findings with the large group. At this point, it is the role of the VDL teacher to listen and take notes, but not respond to any comments.

Table 7.2 Structure of a CTS Video Demonstration Lesson

Time	The VDL Protocol
15–40 minutes	Review of related CTS (see Options A and B)
5 minutes	Teacher introduction to the VDL
20–30 minutes	Viewing of the video
15 minutes	Small group discussions of the video
5–10 minutes	Group share: observations and comments
5–10 minutes	Demonstration teacher comments and reflection

Step 6: Demonstration Teacher Comments and Reflection

Following the discussion, the VDL teacher clarifies or comments on the points made during the large group discussion referring to the notes taken. The teacher will summarize what he or she has gained from the process and share any personal reflections on how this will benefit his or her teaching.

Time

The CTS VDL takes about 90 minutes if the CTS is done in advance. If the CTS readings are done during the VDL session, build in an additional 30 to 40 minutes.

GUIDELINES FOR INTRODUCING AND LEADING VDLS

Materials Needed by Facilitator and for Participants

- *Science Curriculum Topic Study: Bridging the Gap Between Standards and Practice* (Keeley, 2005)
- Access to all CTS resource books used with Sections I–VI of the CTS guides
- Optional: CTS summary developed ahead of time by the facilitator
- 20–30 minute segment of a videotaped lesson
- TV/VCR, TV/DVD, or LCD/DVD and speakers
- Handouts (all handouts are in the Chapter 7 video demonstration lesson folder on the CD-ROM):

  Handout 7.3.1: VDL Protocol

  Handout 7.3.2: VDL Viewing Ethics and Professional Courtesies

  Handout 7.3.3: VDL Technical Tips

  Handout 7.3.4: VDL Content of the Lesson Indicators

  Handout 7.3.5: VDL Observation Worksheet (blank)

  Handout 7.3.6: Example Observation Sheet—DNA Lesson (optional)

  Handout 7.3.7: CTS Summary from DNA Lesson (optional)

Facilitator Note

The VDL professional development strategy is most useful when the videos come from the participants' classrooms. However, it is possible to modify this strategy by using commercially available videos of teaching. If these videos are used, the facilitator takes on the role of "speaking for" the teacher in the video, omitting the personal comments and reflection that come at the end and summarizing and responding to comments from the facilitator's perspective instead.

Facilitator Preparation

- Collect CTS books and resources for the group to use or have them bring their own if they have them. One set of books per five people is sufficient if they are doing the CTS in a meeting prior to observing the lesson. If there are not enough books, copy the study guide you will be using and the readings from the resource books that go with the study guide or prepare a CTS summary of the topic (see note below).
- Identify teachers to videotape a 20–30 minute segment of their classrooms to model the VDL process, offer to videotape it for them so you can use it in the CTS VDL sessions, or select a commercial video from the Annenberg collection at www.learner.org or other videos. Some teachers may already have videotaped lessons they are willing to share, such as lessons they may have prepared for certification, Presidential Award applications, graduate study, or to gain national board certification. You might tap these people to see if they have some video you can use to get the group started.
- If a teacher-made video is being used and the process is being modeled by the teacher, the facilitator works with the teacher to create the VDL Observation Worksheet that will be used for introducing the VDL. The VDL teacher and the facilitator will review the video in advance and select key ideas and one to two indicators of the content of the lesson (from Handout 7.3.4) to focus on and one general education indicator. Include these on the Observation Worksheet (see Handout 7.3.5) and fill in the rest of the information needed to view the lesson. (See example on Handout 7.3.6 from a high school DNA lesson.)
- If a commercial video is used, the facilitator can represent the teacher in the video to model the process for the teachers who will later be videotaped and become the VDL teachers themselves. In this case, the facilitator will create the Observation Worksheet based on what was observed while watching the video.
- If participants do not have their own copies of the CTS parent book, make copies of the topic study guide from Chapter 5 that they will need. *Option*: The facilitator can complete the CTS on the topic of the lesson and make a CTS summary that is grade-level specific for the topic of the VDL. Participants will read the summary in advance of viewing the lesson instead of doing the CTS themselves. (See Handout 7.3.6 for an example of a summary for a high school DNA lesson in the Chapter 7 folder on the CD-ROM.)
- Make a set of all handouts for the session.

DIRECTIONS FOR INTRODUCING CTS VDL

Introducing VDL (5 minutes)

The first time you do this with your group, introduce VDL as a professional development strategy that uses examples from real classrooms as the subject for teacher learning. Ask your group to suggest what they might gain from viewing and discussing video demonstration lessons. Describe how this introduction will use a prepared video (explain that it is from one of the teachers in your VDL group who volunteered, or say you will be using a commercially available video to kick off the process and that in the future you will ask for some volunteers to videotape a lesson).

Setting Norms: Review of Video Ethics and Professional Courtesies (5 minutes)

Point out that it is essential to create a safe environment for discussion so everyone is comfortable engaging in honest, professional dialogue about teaching and learning. Acknowledge that watching a videotape of one's practice can be uncomfortable and share the importance of establishing norms for reviewing practice. Review Handout 7.3.2: Video Viewing Ethics and Professional Courtesies and discussion norms. Clarify any questions and allow for discussion of the norms the group wants to use. Add any additional norms that the group suggests. Ask the group to agree to follow the guidelines and, during the session, remind everyone to keep the norms in mind. Ask one or two people to remind everyone if the norms are not being followed.

Introduce the Protocol and Observation Worksheet (5–10 minutes)

Introduce Handout 7.3.1: VDL Protocol. Briefly explain the flow and the purpose of each section of the protocol. Clarify questions.

Distribute Handout 7.3.5: VDL Observation Worksheet, which you prepared to address the video you will view. The VDL Observation Worksheet contains the key idea(s) from the lesson, content and instructional indicators connected to CTS, and one or two general instructional indicators.

Explain how the VDL Observation Worksheet has been created specifically for the VDL lesson that will be viewed. Introduce the worksheet and explain each of the components, including how the facilitator and the VDL teacher prepared the worksheet in advance. Provide Handout 7.3.4: Content of the Lesson Indicators so participants can see the indicators that the VDL teacher selected from.

Explain that the group will use the VDL Observation Worksheet to record specific evidence they find as they watch the video demonstration lesson. Divide the group into three smaller groups and assign each group one of the three components on the worksheet. Ask the small groups to focus on and collect evidence on their assigned component throughout the lesson. For example, in a small group of six, two people would look for evidence of key ideas, two people would divide up the content of the lesson indicators, and two people would look for evidence of the general indicator(s). Everyone should be observing for evidence of how the CTS findings are reflected in the demonstration lesson.

Review of the Related CTS (15–40 minutes, depending on whether summaries are used)

Explain how a partial CTS (i.e., reading just Sections II, III, IV, V, and VI for the specific grade level of the lesson only) is done by the group prior to viewing the lesson, or a CTS summary is prepared by the VDL teacher or facilitator and reviewed by the group in advance of viewing the lesson.

Have participants conduct the CTS or spend a few quiet minutes reviewing the CTS summary for this lesson. Pose the following question for small group discussion:

> **Facilitator Note**
>
> If using CTS summaries, two to three VDLs can be viewed in a half-day session. If the teachers will be doing the CTS readings themselves in the session before the VDL, plan on just one VDL for a half-day session.

- Based on your review of the CTS information, what do you think you might look for or find in the lesson in terms of learning goals, specific content, instructional approaches, or research-identified misconceptions or student difficulties?

After small groups have discussed their ideas, have each report out one or two points that summarize their findings from the CTS.

Teacher Introduction to Video Lesson (5 minutes)

Introduce the teacher featured in the VDL. The VDL teacher sets the stage for viewing the videotaped lesson by briefly describing the classroom (grade, discipline, grouping, topic, and any relevant context issues), explaining what learning experiences have come prior to the segment to be

Facilitator Note

If using a commercially available video to introduce VDL, the facilitator will stand in for the teacher in the video and provide similar information on the lesson that will be viewed.

Facilitator Note

Break the video viewing into 10–15 minute sections. Briefly pausing the tape allows participants time to process what they have seen and record their observations up to that point.

viewed, and the key idea(s) (learning goals) of the lesson. The teacher may choose to provide the group with a one-page written summary of these details. The teacher also shares what he or she is particularly interested in the audience looking for in the demonstration lesson, referring to the indicators on the VDL Observation Worksheet.

Viewing of Video (20–30 minutes)

Allow time for small groups to determine who will be observing for evidence of each component on the worksheet. Provide time for participants to familiarize themselves with the indicators or descriptors of evidence assigned to them. Remind participants to use the VDL Observation Worksheet to note specific evidence that supports their observations. Remind everyone to record any connections they see to the CTS findings.

Small Group Discussion of Video (15–20 minutes)

Prompt groups to discuss the observations made during the video. Remind participants to base their discussion on the evidence collected. Ask each table group to generate one to three comments or questions to report out to the larger group. The group may choose to summarize their findings or ask clarifying questions.

Group Sharing of Observations and Comments (5–10 minutes)

Invite the small groups to take turns sharing the comments and questions each group generated about the lesson. During this time, the VDL teacher is not directly addressed, but listens and takes notes regarding comments made or questions posed. The role of the facilitator at this time is to act as a "sounding board," keeping the group's comments and questions focused on evidence.

Teacher Comments, Reflection, and Wrap-Up (5–10 minutes)

Invite the VDL teacher to address any of the comments or questions raised during the large group discussion and then publicly reflect on what he or she has learned from the process.

After this, ask all participants to write a short reflection on what they learned during the process. Ask them to focus on what they learned about the topic studied and the lesson and also from the process of collaborating with other teachers to examine practice. Bring the group back together and ask, "How might this type of work be useful to the teachers in the future?" Ask for a few people to share their reflections. If it does not come up in the teachers' reflections, point out that VDLs can help redefine the solitary view of teaching, moving it toward a more collaborative, open view of teaching, a professional activity open to collective observation, study, and improvement, and grounded in standards and research on learning.

Be sure to publicly acknowledge, thank, and celebrate the VDL teacher(s) for making their teaching public to the group.

Next Steps

After this introduction to VDL, solicit volunteers to become the VDL teachers for future meetings. VDLs can be used in the context of professional learning communities where teachers use common planning time to do curriculum topic studies and share and learn from their videotaped lessons with their colleagues.

CTS Professional Development Strategy Application 4: CTS Integrated Lesson Study

DESCRIPTION OF LESSON STUDY AS A CTS PROFESSIONAL DEVELOPMENT STRATEGY

Lesson study is a form of job-embedded professional development grounded in classroom practice that seeks to gradually improve student learning through structured reflection on teaching. Through an iterative process of developing, observing, and debriefing a research lesson, lesson study group members (most often groups of teachers) come to deepen their understanding of how students learn and how implementation of the research lesson supports student learning. Embedding CTS within lesson studies enhances the traditional lesson study process by providing teams access to resources needed to design standards-based research lessons, to conduct student-focused classroom observations, and to have evidence-based, post-observation debriefings.

This application engages participants with a robust lesson study protocol that seamlessly integrates the CTS resources throughout the entire lesson study cycle. Participants simulate the work of the Pleasant Hills Lesson Study Group and through the simulation, learn how CTS enriches all parts of the lesson study cycle. Further, participants explore how to apply the process to their own lesson study work. This session is designed for those people who already use or know how to use the lesson study process. For those interested in lesson study, but without prior experience in its practice, we encourage you to review the resources listed on Facilitator Resource 7.4.1 in the Chapter 7 lesson study folder on the CD-ROM. Once you have a deeper understanding of lesson study, return to this session to learn how to infuse CTS into the lesson study.

GOALS

The goals of this session are to

- learn how CTS can support and enhance all phases of a lesson study cycle,
- practice using the CTS–Lesson Study Scaffold with a selected lesson to simulate the work of an actual lesson study group, and
- consider how the CTS–Lesson Study Scaffold and supporting tools can enhance your own work with current or future lesson study groups.

AUDIENCE

This session is for lesson study leaders as well as teachers, teacher leaders, professional developers, and teacher educators with experience in lesson study who seek additional tools and resources for applying research and standards on instruction and student learning to the design, implementation, and evaluation of their lesson study work.

STRUCTURE OF CTS INTEGRATED LESSON STUDY

The CTS Integrated Lesson Study session is guided by a protocol with ten steps as outlined below.

Step 1: Establish Norms and Set Goals

In this step, the group clarifies the standards of behavior that will guide its work and identifies shared aspirations for students and the gaps between those aspirations and how students are currently learning.

Step 2: Examine Existing Science Unit and Select a Research Lesson

In this step, the group outlines the sequence of key science concepts, skills, and knowledge in the unit of instruction they will study. They select a lesson within the unit that addresses an important big idea and supports the group's goal for student learning. They clarify the adult content knowledge associated with the lesson by engaging in CTS Section I.

Step 3: Clarify the Learning Goals of the Research Lesson

In this step, the group identifies the CTS guide(s) most relevant to the topic of the research lesson they will study and uses readings from the CTS guides to (1) clarify the learning targets (CTS Sections III and VI), (2) examine the curricular and instructional implications (CTS Sections II and V), and (3) surface common student misconceptions (CTS Section IV).

Step 4: Construct the Instructional Sequence of the Research Lesson

In this step, the group constructs an instructional sequence that describes the learning activities and anticipated student and teacher responses. They use the analysis of the research lesson from Step 3 to suggest modifications to activities, anticipate student responses, and select appropriate instructional approaches.

Step 5: Define Evidence of Learning for Observers to Collect

In this step, the group identifies evidence of student learning that can be collected by observing students or analyzing their work and indicates where in the instructional sequence the evidence will be generated.

Step 6: Implement the Lesson and Collect Evidence of Learning

In this step, a group member teaches the constructed research lesson. Other group members observe the lesson guided by the instructional sequence and evidence of learning created in Steps 4 and 5.

Step 7: Analyze Evidence of Student Learning

In this step, the group members assess the extent to which the learning goals were attained in the research lesson by analyzing and citing evidence to support their conclusions. They use the CTS summaries constructed in Step 3 to support discussions and remain grounded in evidence. They develop a rationale for making any modifications to the research lesson.

Step 8: Revise the Instructional Sequence

Based on the discussion in Step 7, the group revises the research lesson and the instructional sequence, noting how students are likely to respond and any new guidance needed for observers.

Step 9: Repeat Implementation of Lesson (Optional)

In this step, another group member teaches the modified research lesson while others observe as described in Step 6. The group debriefs the lesson as in Step 7, and revises the lesson again if necessary, as in Step 8.

Step 10: Create a Lesson Study Report

In this final step, the group develops a lesson study report that describes the extent to which the learning goals were attained and indicates the evidence to support any conclusions. They identify people within the school and beyond with whom they will share the findings.

GUIDELINES FOR INTRODUCING AND LEADING A SIMULATED CTS INTEGRATED LESSON STUDY

Materials Needed by the Facilitator

CTS Parent Book

Science Curriculum Topic Study: Bridging the Gap Between Standards and Practice (Keeley, 2005)

Resource Books

One copy of each of the CTS resource books. (*Note:* It is important for CTS facilitators to have both volumes of the *Atlas of Science Literacy*; but for this session, only Volume 1 is used.)

CTS Integrated Lesson Study PowerPoint

The PowerPoint presentation is included in the Chapter 7 folder on the CD-ROM. Review it and tailor it to your needs and audience. If your group is new to CTS, consider adding a few slides from Module A1 as noted in the following directions to provide an overview to the CTS parent book and resource books. (See p. 63 in Chapter 4 of this Leader's Guide.) Insert your date and location on Slide 1, add any graphics desired, and add contact information on the last slide.

Facilitator Resources

The facilitator resources can be found on the CD-ROM in the Chapter 7, lesson study folder.

- Facilitator Resource 7.4.1: Lesson Study Background Resources
- Facilitator Resource 7.4.2: CTS Summary: "Water in the Earth System" for CTS Section I
- Facilitator Resource 7.4.3: Completed Pleasant Hills Lesson Study Group Packet

Materials and Supplies

- Blank paper for note taking
- Sticky notes (small and large) (optional)
- Highlighter pens (optional)
- Easel, chart paper, and markers

Wall Chart

Prepare a wall chart for your group norms. Some suggested norms are as follows:

- Focus on and engage with tasks
- Actively learn and support others in their learning
- Monitor "air time"
- Listen and pause before responding

Materials Needed by Participants

All handouts can be found on the CD-ROM in the Chapter 7, lesson study folder.

- One to two copies of the CTS parent book per table. If you do not have enough books, make copies of the "Water in the Earth System" CTS guide on page 190 for each person.

- CTS Resource Books: one set of books for every three to four participants. (*Note: Science Matters* and Volume 2 of the *Atlas of Science Literacy* are optional for this session.) If you wish all participants to have their own readings, make copies of the readings listed on the "Water in the Earth System" CTS guide (p. 190 of CTS parent book).
- Handout 7.4.1: Agenda at a Glance
- Handout 7.4.2: Lesson Study Reflections
- Handout 7.4.3: CTS–Lesson Study Scaffold
- Handout 7.4.4: Pleasant Hills CTS–Lesson Study Packet
- Handout 7.4.5: List of Science CTS Guides
- Handout 7.4.6: Packet of Student Work (4 samples)
- Handout 7.4.7: CTS–Lesson Study Template
- Handout 7.4.8: Step 10: Lesson Study Report for Pleasant Hills Lesson Study Group
- Copy of PowerPoint slides (optional)
- Copies of Facilitator Resources for those people who may replicate this session with others (optional)

DIRECTIONS FOR INTRODUCING CTS INTEGRATED LESSON STUDY

Welcome

Show Slide 1. Welcome the participants and explain that this session will help them learn how to integrate a process for accessing research and standards called curriculum topic study with lesson study to enhance the quality of their research lesson, their classroom observations, and their post-lesson debriefing. Review the agenda (Handout 7.4.1) and explain that the session is designed to provide the group with varied learning activities. Show Slide 2 and ask participants to show by a raise of their hands which item best describes their experience with lesson study. Use this information to help you understand the lesson study background of the group. If participants are unfamiliar with lesson study, define lesson study and share a little background on its purpose. If participants do not know one another, have them introduce themselves to their table group and have them mention their prior experience with lesson study in their introductions.

Overview of Goals and Norms

Review the three goals of the session listed on Slide 3. Clarify that the process they will learn today does not alter the integrity of the lesson study process, but rather gives lesson study users additional tools and resources to support their research lesson, classroom observations, and post-observation debriefings. Show Slide 4 as a means of modeling the use of norms, just as lesson study groups establish norms as a part of their work. Refer to the list of norms posted on chart paper prepared for this session and invite any additions to the list. (If you are working with an established group that already has norms, you can simply remind everyone of the norms you use and that they will be used again today.) Ask the group to abide by the norms for the day.

Facilitator Note

This is material your participants may already know if they are all experienced in lesson study. Therefore, you can cover it lightly or omit it based on your knowledge of the group. It is worth including in a mixed group just to make sure everyone has the same information.

Three Strands of Lesson Study

Ask everyone to take a minute and write down what they know about lesson study. Then ask, "What do you know about curriculum topic study?" Ask everyone to turn to a partner and discuss what he or she knows about lesson study and CTS. Show

Slide 5. Share that when people first hear about lesson study, they often assume the purpose is to develop the "perfect lesson," but lesson study is about much more than a single lesson. It is a form of collaborative professional learning grounded in classroom practice that seeks to gradually improve student learning through structured reflection on teaching.

Show Slide 6 and point out that lesson study experts from Northwest Regional Education Laboratory and Learning Points identified three interdependent strands of lesson study that are important to consider. Many people are familiar with the "process"—the basic steps—but equally important are the "big ideas" and "habits of mind" that are infused in the process. Point out that the group will look at all three strands and explore how CTS can support each of the strands and their interconnections. They will begin by a quick review of each of these strands, then they will engage in a series of explorations that link CTS to them.

For example, the strand of "big ideas" leads study groups to clarify the aspirations they have for their students, explore content in-depth, carefully consider what students already know, their anticipated responses to instruction, and misconceptions held. The "habits of mind" strand addresses the dispositions such as habits of collaboration, a research stance, and self-efficacy used by a study group.

> **Facilitator Note**
>
> For more information on the three interdependent strands of lesson study, visit www.nwrel.org/lessonstudy/ls-guide/materials/handouts/handout_4_2.pdf.

Use Slides 7–12 to review the first interdependent strand: the process. *Note:* The amount of time spent on these slides will vary depending on the experience level the group has with lesson study. Key features in the process are as follows:

- Begin by identifying a broad goal to guide their work.
- Conduct research, examine instructional materials, and develop a detailed, annotated lesson plan.
- Teacher teaches, while other group members observe and document student "learning" (i.e., collecting evidence from what students are doing that support the lesson goals).
- Share observations and reflections.
- Analyze findings and revise lesson.
- Teacher teaches the lesson again (sometimes, but not always).
- Debrief entire cycle and create a report to share learning, both regarding that specific lesson and overarching insights.

Remind participants that it is easy to think of lesson study simply as this process, this protocol, or series of steps. But the two additional strands—big ideas and habits of mind—underscore that how teachers engage in that process is equally important.

Show Slide 13. Point out that the second strand of lesson study is big ideas. Key features for the lesson study group to address in this strand are as follows:

> **Facilitator Note**
>
> Even for those with experience in lesson study, thinking about these two additional strands may be a new way of thinking, so this is a nice "new insight" into lesson study for them.

- *Long- and short-term goals:* Establish long-term goals for students—what teachers want them to become, where they are now; establish short-term goals, often on science content and process.
- *Content focus:* Explore topics within the unit and selected lesson in depth.
- *Student focus:* Consider students' prior knowledge, anticipated responses, misconceptions—and think about how students learn.

> **Facilitator Note**
>
> This really starts to foreshadow how CTS is important as a tool for the lesson study process in that it intensifies the focus in the areas of content, students, and instruction.

- *Instructional focus:* Consider instructional methods that will help students reach learning goals, drawing upon existing lessons, research on "best practices," and identification of effective instructional tools.

Show Slide 14 and point out the third strand of lesson study—habits of mind. The habits of mind that support lesson study are as follows (Northwest Regional Educational Laboratory, 2008):

- *Learning together:* There is a safe environment; group members are clear and respectful and open to new ideas and approaches.
- *Research stance:* Teachers pose questions and problems, research solutions, try new ideas, collect data, and analyze findings. They engage in inquiry, reflection, and critical examination of practice.
- *Self-efficacy:* Teachers are motivated and persistent in improving their craft; they take responsibility for the process and believe they can make a difference.

Facilitator Note

The "habit of mind" strand connects directly with the purposes of CTS in terms of taking a research stance and taking personal responsibility for learning from research-based resources to make improvements in learning.

Summarize this segment by pointing out that lesson study is a well-articulated process that teachers use to gain professional knowledge and to continuously improve the quality of learning experiences they provide for students. Ask if there are questions about the three strands and the importance of addressing all of them in lesson study work. Pause and ask participants to reflect on how these strands match or are different from their prior experience with lesson study. Show Slide 15. Using Handout 7.4.2: Lesson Study Reflections, invite participants to take notes in all three strand areas—process, big ideas, or habits of mind in response to the question, "How can my lesson study work improve?" After everyone has recorded a few thoughts individually, ask them to take a few minutes at table groups to discuss some of their thoughts.

Listen in on the discussions and note ideas that surface. Paraphrase a few of them to the larger group that most directly foreshadow what using CTS with lesson study can offer and indicate that the tools and processes the group will explore can offer assistance for many of the issues raised in the discussions.

Facilitator Note

You will need to modify Slide 16—either insert slides from Module A1 if you choose to do a short introduction to CTS (see Chapter 4 folder on the CD-ROM) or simply make this slide into a quick overview to the CTS process from your own knowledge of the tool. If you insert slides, be aware that you will need to update the slide numbers referenced in the rest of this script so you can use it to lead the session.

Introduce CTS (Optional)

Show Slide 16. If your group has never used CTS, you can introduce them to the CTS parent book using the introductory activities in Module A1 in Chapter 4 of this *Leader's Guide.* (Use the activities "Getting to Know the CTS Parent Book and Study Guides" and "Getting to Know the Resource Books" on pp. 68–69 in Module A1.) You will need to add approximately 20–30 minutes to the schedule if you use these activities. If your group already has a basic understanding of the CTS process, you can skip this step.

Introduce the LS-CTS Scaffold

Show Slide 17 and point out that it lists the ten steps on the CTS–Lesson Study Scaffold. Refer everyone to Handout 7.4.3: CTS–Lesson Study Scaffold to review what is done at each step. Indicate this scaffold offers a framework for how to seamlessly integrate CTS into lesson study to strengthen the process and support the big ideas and habits of mind just discussed. Make it explicit that CTS in this context isn't an "add-on" or something else they have to do, rather it provides access to resources that weave into the regular work of a lesson study group.

Give people a few minutes to look through the scaffold. Indicate that they are going to work through the scaffold in small groups using a worksheet designed to follow the scaffold steps (Handout 7.4.4: Pleasant Hills CTS–Lesson Study Packet) to explore the impact CTS has on the three strands of lesson study. They will do this by simulating the work of a lesson study group.

Applying the Scaffold to a Science Unit

Distribute Handout 7.4.4, which introduces participants to the Pleasant Hills Lesson Study Group. Ask everyone to review the general information on the first page. Explain that the handout follows the same steps as the scaffold the group just reviewed. Show Slides 18–20 to review the information about the study group, including the following:

- Group members
- Group norms
- Group goals—long and short term
- Unit focus

Point out that what is on Handout 7.4.4 represents what the lesson study team has completed in prior sessions consistent with Step 1 and the beginning of Step 2 on the scaffold (Handout 7.4.3). Indicate that during the current session they will simulate the role of members of the Pleasant Hills team by completing Step 2 and moving through the other steps on the scaffold. Together, they will prepare and debrief a research lesson using the CTS–Lesson Study Scaffold and artifacts from the actual work of the Pleasant Hills team. As they move through this process, they will also periodically step out of the simulation role, and reflect on their roles as lesson study group leaders or members. They will return to the initial reflections they made at the beginning of the session to make connections between what they are learning through this simulated lesson study work and their ongoing efforts to support lesson study in their own settings.

Break

Show Slide 21. This is a good place for a short break or adjust this slide to the place in the schedule you will provide a break.

Examine a Typical Science Unit and Lesson

Show Slide 22. Have participants look at the bottom of the first page of Handout 7.4.4: Pleasant Hills CTS–Lesson Study Packet, which lists the flow of topics within the unit on storms for Grade 7. The topics listed are as follows:

- Uneven heating of the Earth's surfaces
- Heat transfer
- Convection currents
- Cloud formation
- Convection in the ocean

Facilitator Note

This will be particularly important for participants new to CTS, as it requires them to carefully look over the guides and be able to find those that are going to be most useful to them. If you have experienced CTS users and less experienced users, ask people to help each other at the tables.

Indicate that these are the main section headings within the unit the Pleasant Hills Lesson Study Group selected to use for their study. Explain that before they dig into the work, they will take a few minutes and practice using the CTS guides. Ask everyone to spend a few minutes in small table groups to review the 147 science CTS guides (Handout 7.4.5) and identify which CTS guides might be useful in deepening their understanding of the topics in the storms unit that the lesson study group will use.

After table groups have had some time to look at the list of 147 science CTS guides, ask them to report out on which guides might be useful. Use this simply as a check for understanding of how to use the CTS guides. Ask for responses from each small group and then show examples on Slide 23. Typical responses might include the following:

- Uneven heating of the Earth's surfaces: "Energy," "Solar Energy," "Processes that Change the Surface of the Earth"
- Heat transfer: "Energy," "Heat and Temperature"
- Convection currents: "Energy," "Heat and Temperature"
- Cloud formation: "Water Cycle," "Water in the Earth System," "Weather and Climate," "Cycling of Matter in Ecosystems," "Air and Atmosphere"
- Convection in the ocean: "Flow of Energy Through Ecosystems"

Debrief this segment by helping participants recognize the range of topics included within the guides and how the organization of the CTS parent book itself puts all the CTS resources within easy access of lesson study groups to inform their work. (If you have display copies of the CTS parent book on each table, invite them to look at a guide in Chapter 6.)

Select a Research Lesson Selected From Within the Unit

As in any lesson study group, ultimately a single lesson must be selected as the focus for the research lesson. Refer back to Handout 7.4.4, page 2, and indicate that the Pleasant Hills Lesson Study Group chose to focus on Lesson 6.1, which addresses the effects of temperature on cloud formation. Show Slide 24. Ask everyone to review the group's reasons for choosing this lesson on Handout 7.4.4 and share any observations or questions. The group's reasons for choosing this lesson were as follows:

- Students have had prior experiences in earlier grades with the water cycle and the movement of water between the Earth and the atmosphere, but few experiences to examine how this cycle relates to cloud formation or weather patterns.
- All students have seen clouds so the lesson is grounded in something familiar to them. They hoped to pique students' curiosity about how clouds form.
- The activity in the lessons is relatively simple and the concepts being explored can be readily observed, making the learning targets more accessible to students.
- The lesson design includes a lot of opportunity for questioning on the part of the teacher that the group hoped would promote student thinking and ownership of their learning.

Make the link between these reasons and the long-term group goals for students established by the lesson study members as indicated on the first page of Handout 7.4.4.

Ask participants to identify which CTS guide will be most useful to inform their research on Lesson 6.1. Anticipate that in a cursory look you could get responses ranging from "Water Cycle" (see CTS parent book, p. 189) to "Water in the Earth System" (see CTS parent book, p. 190) to "Weather and Climate" (see CTS parent book, p. 191) to "Air and Atmosphere" (see CTS parent book, p. 175). Tell your group that these are all potentially right answers since all of those study guides are likely to be useful. Indeed there is even some overlap in the assigned reading listed in those guides. In practice, most lesson study groups access reading from multiple guides since a lesson may address overlapping topics. Show Slide 25 and explain that for their purposes today—to simply familiarize themselves with how the guides can support their lesson study team, they are going to focus on the study guide, "Water in the Earth System" (see CTS parent book, p. 190).

Provide the context for your group by sharing that before moving into carefully analyzing the lesson, its intended learning targets, and the associated classroom activities, the lesson study group decided to spend some time to become more familiar with the content addressed by the unit. Several group members have had little formal preparation in Earth science or weather and felt they needed a better understanding of the concepts before they could look critically and confidently at the student materials. Point them back to Handout 7.4.3 and show that Step 2 of the scaffold provides a way for the group to address these content questions. To do this, the team did readings from CTS Section I. As members of the Pleasant Hills Lesson Study Group, they will do these readings and construct a group summary of the adult content knowledge needed for the topic, "Water in the Earth System."

Show Slide 26. Invite participants to quietly complete the readings in *Science for All Americans* (pp. 42–44), or the copy of this resource you made for them, and *Science Matters* (1991 edition, pp. 196–201) (optional). Ask them to use sticky notes, highlighters, or note taking to document the main points and the page numbers so they can cite them later. After reading as a full group, discuss the key ideas in the reading(s).

> **Facilitator Note**
>
> Refer to Facilitator Resource 7.4.2 on the CD-ROM for a summary of the key ideas that should be raised. Prompt the group to bring out any central ideas that are not surfaced in the group discussion.

Ask everyone to turn to page 2, item 3, on Handout 7.4.4 and construct a group summary of the readings to which they can later refer as they continue their lesson study work. Remind them to look back at their notes and record the source and page numbers for their readings so that they can easily refer back to the original text for clarification if necessary.

Make it clear that all too often teachers jump ahead into lesson design without first taking the time to make sure that they, as adults, fully understand the content themselves or consider the big ideas that the student learning targets are working toward developing by the twelfth grade. Understanding these ideas will help them better support and assess student learning when they actually begin teaching the lesson.

Clarify the Learning Goals of the Research Lesson

With a firm grounding on what scientifically literate adults should understand about this content, our lesson study group is now prepared to more carefully examine the learning targets associated with the research lesson. According to the instructional materials, the selected lesson asks students to "explore the effect of water temperature on evaporation and condensation." To inform their analysis of the lesson and be able to assess how well the activities meet the intended learning targets and support student learning, the Pleasant Hills Lesson Study Group decides to explore the standards and research on student learning using the CTS.

Show Slide 27. Have participants work through Step 3 on Handout 7.4.4. Using their CTS guide, "Water in the Earth System," from the CTS parent book or the copy you made of this page, small groups will research the following questions:

- What science concepts, skills, and knowledge should students learn from this lesson? (Read CTS Sections III and VI.)

- What prior knowledge and skills are necessary to learn the content of the lesson? (Read CTS Section V.)
- What commonly held ideas related to the learning goals do students have? (Read CTS Section IV.)
- What instructional strategies or contexts are important to help students learn this content? (Read CTS Section II.)

Show Slide 28 and ask each table group to assign one to two individuals to each of the four prompts and assigned readings depending on the size of your groups. If you have five-person groups, assign two people to split up the reading for Prompt 1, which has two sections. All participants should individually summarize their findings on page 3 of Handout 7.4.4.

After each person at the table has completed the individual task, invite table groups to share their individual findings with one another to build a common base of knowledge across all prompts for all table members and record their group discussion on Handout 7.4.4. Ask everyone to put their summaries aside for now and tell them they will return to them in a few minutes.

Show Slide 29 and indicate that they are now going to step out of the lesson study simulation for a moment to think about their actual lesson study work in their home school or district—make this shift explicit so people can step away from the details of the simulation and look at it from the perspective of how it will relate and support their work. Invite participants to return to Handout 7.4.2: Lesson Study Reflections used earlier. Ask them to think now about the work they just completed in this particular phase of the lesson study cycle—designing the research lesson. How could the CTS process they just completed support or enhance that aspect of their lesson study work? Ask them to record thoughts or insights.

Ask participants to share some of their reflections with the group and show Slide 30 to highlight some of the points and build on participants' ideas. Try to draw out the following ideas:

- CTS provides participants access to content knowledge they may need and a mechanism to identify what they want to learn more about to deepen their content knowledge.
- CTS provides participants access to the standards, clarifying the required learning goals and better preparing the team to focus the lesson on what the standards and research suggest students need to know or be able to do.
- CTS gives a team access to research on how students think about science ideas, which allows the team to better anticipate student responses in the lesson.
- CTS orients teams to the prerequisite knowledge students need to master the intended learning goals.

Developing the Instructional Sequence

Indicate that your lesson study group is now prepared to develop their research lesson by constructing the instructional sequence (Step 4 on Handout 7.4.4). To begin developing the research lesson, the team will look carefully at the intended outcomes of their selected lesson (as indicated in the instructional materials) and look at the desired student outcomes as suggested in the standards and research. Step 4 of the scaffold (Handout 7.4.3) leads the lesson study team to make adjustments to their existing instructional materials to ensure the activities in the lesson address student needs.

Show Slide 31. Point out that the Pleasant Hills Lesson Study Group began by looking at the stated student objectives associated with the lesson they selected. The objectives are listed on Handout 7.4.4. According to the instructional materials,

Students will model and describe how water evaporates and condenses and how these processes play a part in cloud formation.

Show Slide 32. Based on the stated student objective, ask the group what key ideas from the standards that they summarized in Step 3 are best addressed through this lesson. Invite table groups to look back at the relevant standards for the topic that they surfaced in Step 3 and select those that would best fit the objective of the lesson if it were designed to intentionally address them and if the teacher provided sufficient scaffolding and questioning to guide student thinking. Give table groups time to refer back to their prior work and have some discussion about which standards could be addressed. Ask for responses from each group. Anticipate that the following key ideas from the standards are most likely to be suggested (Slide 33):

- Water evaporates from the surface of the Earth, rises and cools, condenses into rain or snow, and falls again to the surface.
- The cycling of water in and out of the atmosphere—the water cycle—plays an important role in cloud formation.
- Clouds form by the condensation of water vapor.
- The sun is a major source of energy for phenomena on the Earth's surface, such as the water cycle.
- Oceans have a major effect on climate because water in the oceans holds a large amount of heat.

If other key ideas from the standards are raised, ask participants to explain the link between the objective as defined in the instructional materials and the key ideas and justify why they think it is a strong link.

With these learning targets from the standards clearly defined, it is time for the lesson study team to develop the instructional sequence to ensure that students will be engaged in the intended concepts and the team would be able to gather appropriate evidence to assess student understanding as indicated in Steps 4 and 5 of the CTS–Lesson Study Scaffold.

Indicate that the current instructional sequence as shown on pages 4 and 5 of Handout 7.4.4 simply outlines the first two activities of the existing instructional materials. Those activities or procedures are briefly described in the first column of the five-column chart. Make clear that this is a critical juncture for the lesson study team to apply their insights from the readings on standards and research to their instructional approach and assessment strategies.

Indicate that it is now time for each table group to draw upon their research and apply it to the development of the research lesson. Show Slide 34. Each table group will complete Step 4 and Step 5 on Handout 7.4.4 on Activity 1 of the lesson. To do this, ask them to look at the first activity listed in the chart on page 4 of the Handout 7.4.4 and consider whether the listed activity is well aligned with one or more of the five intended learning goals listed just above the chart on the handout. What modifications would they suggest to improve the alignment between the activity and the learning goals? Ask the groups to discuss the question and make note of any suggested modifications in the space on the chart under Activity 1 or on sticky notes. The team will then identify the student actions and responses and the teacher responses to the students that they anticipate seeing and record them on the chart in the second and third columns, respectively.

> **Facilitator Note**
>
> In practice, a lesson study group may list more activities within the lesson depending on the focus of their research lesson and how much data they need to collect to inform their research question. Since their purpose here is to learn how to use CTS to support lesson study, we will simply practice this step of the scaffold with the first activity.
>
> Without experience in the lesson, it is difficult to complete the instructional sequence for Activity 2. If you are working with a group that is very familiar with this lesson, you can use both Activity 1 and Activity 2 to work on the instructional sequence. If not, just use Activity 1.

Show Slide 35. To complete Step 5, the team will consider what evidence of learning observers would look for to assess the extent to which students understood the learning target, and list the evidence in the fourth column on the chart on Handout 7.4.4. A summary of their task is as follows:

- Describe modifications to the listed activities needed to better align with the learning targets (CTS Sections III, VI). Point to the list of content-focused learning goals above the chart on Handout 7.4.4.
- Describe anticipated student responses (CTS Section IV). Point to second column.
- Describe planned teacher responses and actions (CTS Sections II, V). Point to third column.
- Identify appropriate evidence of learning (CTS Sections III, V). Point to fourth column.

Remind them that it is important that they refer to pages 2–3 in Handout 7.4.4, where they recorded their CTS work, and to draw upon this work as they continue their lesson study. Ask them to record their instructional sequence on chart paper. Indicate that when everyone is done, they will do a gallery walk and look to see how different lesson study groups drew upon their CTS work to inform their instructional sequence and observation points.

Offer an example of what a group might enter into the chart to help make the task more accessible to everyone. Once the task is clear, invite table groups to work collectively on their task. Remind participants that for the purposes of this session, they are working on just one activity out of the entire lesson. In practice, an actual lesson study team would complete this work for the entire instructional sequence of the research lesson.

As groups complete their task, prepare them for the gallery walk. Show Slide 36. As they go around and look at the work of other teams, ask them to look for evidence that the resulting instructional sequence and observation points incorporated the CTS findings. Debrief by asking participants to share their observations from the gallery walk as well as the decisions they made in their own table group. Focus the sharing on how their sequence incorporated what they learned through their CTS work. Highlight those components specifically to make them apparent to participants. Draw out the following ideas (Slide 37):

- CTS informs revisions of the activities to more directly target the learning goals. Modifications include removing things that are not necessary or inserting things that were missing.
- CTS gives lesson study groups access to anticipated student responses from the research on student learning. This research directly links to the teacher responses, allowing the teacher to think about appropriate instructional decisions before the lesson.
- The CTS clarification of the learning targets better prepares observers to identify evidence that will indicate student learning.

Show Slide 38. Invite participants to return to the Lesson Study Reflection (Handout 7.4.2) and think about the work just completed on the instructional sequence. Ask the group how this work would prepare a team for the observation phase of the lesson study cycle. How could the CTS process they just completed support or enhance that aspect of their own lesson study work? Ask them to record thoughts or insights on the handout.

Ask participants to share some of their reflections with the group. Draw out the following ideas (Slide 39):

- CTS helps group members anticipate student responses based on research. Surfacing these ideas prior to teaching the lesson better equips observers for what they may encounter.
- CTS helps group members stay rooted in the content and the intended learning goals. With a deeper understanding of the content, observers are able to more effectively assess student learning during the observations.

After Teaching the Research Lesson

Share that in this lesson study group, each observer agreed to focus on one table group of students during the preassessment and subsequent table discussion. Indicate that for purposes of their learning experience in this session, participants will look at four samples of student work to simulate the initial portion of the lesson study observation done by the Pleasant Hills team. Distribute the four samples of student writing collected during the lesson (Handout 7.4.6). Ask participants to follow the prompt on Step 6 on Handout 7.4.4 and Slide 40 to examine these samples and record their thoughts on the extent to which they demonstrate understanding of the intended learning targets. (Remind participants that in an actual lesson study there would be additional observation data and other evidence of learning collected during the lesson, for example, students "working" in addition to looking at "student work.")

Show Slide 41. Ask table groups to have a post-observation discussion focused on analyzing particular pieces of evidence. (Point to Step 7 of the CTS–Lesson Study Scaffold.) Remind them to talk from the evidence, both what they found in CTS readings and their research of the lesson. During their debriefing have participants respond to the following prompts from Step 7 on page 6 on Handout 7.4.4 and take notes on the points raised:

- Describe the extent to which the learning goals were met based on the evidence gathered during the observation.
- Identify what key activities, strategies, or portions of the instructional sequence may have contributed to the students achieving or falling short of the learning goals.

Show Slide 42 and invite participants to once again step outside of the simulation and return to their earlier reflections. Ask them to think about the debriefing phase of the lesson study cycle. How could the CTS process support or enhance the ability of group members to reflect on their observations and assess the impact of the lesson on students' learning? Ask participants to share a few reflections with the group. Draw out the following ideas (Slide 43):

- CTS provides a research-based rationale to focus learning objectives and thereby direct activities. Having this foundation directs debriefings to "what the research says" rather than just "what I think."
- CTS connects observers to research on student thinking as well as a map to how student ideas develop over time. These findings provide an important grounding for discussion of evidence of learning and where students are in the continuum.
- CTS provides access to resources on many topics. Often a lesson observation surfaces unanticipated discussion. With access to CTS resources, a team can explore those topics raised by the student and better understand how they may have related to the intended focus of the research lesson.

Indicate that the results of the debriefing session provide the lesson study group evidence to guide modifications needed to improve the quality of the lesson. Show Slide 44. Step 8 in the scaffold reminds teams to identify needed modifications or revisions to the lesson based on evidence of student learning. Point out that making a written record of the rationale for changes is important in helping teams stay anchored in an evidence-based approach.

Show Slide 45. Explain that usually lesson study groups are structured so that another member of the team goes on to teach the modified lesson, as described in Step 9 of the scaffold. After Step 9,

Facilitator Note

If the structure of your lesson study group doesn't allow such a cycle to take place, simply skip Step 9 and proceed to Step 10 to document what the group learned throughout their lesson study. Indicate that for the purpose of the simulation in this session, they will move forward to Step 10.

the group repeats Step 6 with a new member of the team teaching the lesson. Afterwards, the group returns to Step 7 for a second debriefing and then continues through the remaining steps in the scaffold.

Show Slide 46 and share that once the team completes the lesson study cycle, they document their work in a report. The report synthesizes what was learned through the process and its impact on students as well as additional insights applicable to instructional practice. The report provides an important reflection point for the group and also creates a written record of their work that can be returned to in the future. Share the Pleasant Hills research report on Handout 7.4.8 that shows what insights the team gained and how they will share their findings.

Reflections on CTS Influence on Lesson Study

Indicate that throughout the session they paused to reflect on how CTS supported the process of lesson study at each stage—designing, observing, and debriefing the research lesson. At the start of the session, they thought about lesson study as more than just a "process" or "protocol," but rather a complex professional learning strategy comprised of three interdependent strands. Show Slide 47. Draw their attention to the strand of lesson study called habits of mind, discussed earlier, including the need to maintain safe environment for discourse, a research stance, and self-efficacy and responsibility for the process. Ask participants to reflect on their experience with the CTS–Lesson Study Scaffold and consider how the scaffold supported these strands—focusing on the two strands: habits of mind and big ideas. Ask them to record their thoughts on Handout 7.4.2: Lesson Study Reflections.

After a few minutes, invite participants to share some of their reflections with the group. Draw out the following ideas (Slide 48):

- CTS strengthened the lesson study group's ability to remain grounded in the big ideas the lesson study process is intended to support by focusing on content, student learning, and instruction.
- CTS supports the group's efforts to take a research stance by talking from evidence and referring to the research of the lesson done by the group.
- CTS allows lesson study members to take greater personal responsibility by giving them access to research-based resources that can support their learning and their practice.

> **Facilitator Note**
>
> You may also wish to make this template available electronically so that your participants can fill it out on their computers.

Link these remarks with previous reflections from throughout the day to highlight the "value added" of using CTS for lesson study.

Indicate that templates that support lesson study groups to use the same processes they did in this session are included in their handout packet. (Have them pull out Handout 7.4.7: CTS–Lesson Study Template.) They can use this to guide their own lesson study groups through the ten-step CTS–Lesson Study process. Review the cover sheet of the template and highlight the suggested times indicated for each step to review what it would take for them to complete the process with their own lesson study groups.

Next Steps and Personal Reflections

Ask participants to think about their current or upcoming lesson study activities. Invite them to take a few minutes to write down how they intend to use CTS to support their work. (If participants are all engaged in a lesson study, consider having this closing be a time for them to take a few minutes and identify the CTS guides that they can use with their own group [refer them back to Handout 7.4.5] so that when they return to their workplaces, they are ready to share their CTS work in an immediately applicable way.)

Show Slide 49. Before closing, invite participants to write a "Note to Self" memo. Ask them to write a memo in response to the following stem: "Note to self: I will improve my lesson study work by . . ." Ask them to use the note as a reminder of two or three things they will do based on what they learned in this session. Give participants a few minutes to write down their thoughts and then invite them to find a partner and exchange their ideas. If time permits, ask for a few report outs.

Show Slide 50. Summarize the day by reminding everyone that using CTS–Lesson Study can enhance the quality of their research lesson, their classroom observations, and their post-lesson debriefing. Moreover, CTS also supports the other strands of lesson study—big ideas and habits of mind—that move this strategy beyond a mechanical set of procedures and toward professional discourse focused on improving instruction and student learning. Tell them that in addition to the resources used today, there are additional tools and resources that can be found on the CTS Web site at www.curriculumtopicstudy.org.

Final Facilitator Notes

This session uses the work of an existing lesson study team to simulate the entire lesson study cycle. An alternative strategy for existing lesson study teams would be to conduct this session in real time over the course of the lesson study cycle. Rather than using the work of the "Pleasant Hills Lesson Study Group," the facilitator would use the blank CTS–Lesson Study Template (Handout 7.4.7) to apply the design to a current lesson study activity. Over multiple sessions, the lesson study team would complete the template using their instructional materials and selected lesson to complete the CTS–Lesson Study process. This approach works well with teams already familiar with lesson study, but less familiar with CTS. Suggested time allocations for completing the CTS–Lesson Study process in this way are noted on the CTS–Lesson Study Template.

CTS Professional Development Strategy Application 5: CTS Action Research

DESCRIPTION OF ACTION RESEARCH AS A CTS PROFESSIONAL DEVELOPMENT STRATEGY

CTS action research is a specific type of classroom-based research that combines CTS and an examination of students' thinking. It is a type of teacher research that relies less on rigid data collection protocols and experimental designs and more on methods that involve observation, exploration, and reflection on teaching and learning in an individual's own classroom. The CTS action research can be conducted by individual teachers in their classrooms or as a collaborative action research project conducted by small groups of teachers. The strategy has been used within a variety of professional development programs, especially those with the goals of deepening teacher reflection on practice and on learning to inquire into one's own practice. For example, the Maine Mathematics and Science Alliance used CTS action research as the professional development capstone for a three-year math-science partnership (MSP) grant called Science Content, Conceptual Change, and Collaboration (SC4). The SC4 program targeted conceptual change teaching in physical science for elementary teachers. The culmination of the three years of professional development was a physical science CTS action research project. (*Note:* To see a collection of CTS action research monographs from the SC4 project, visit the CTS Web site at www.curriculumtopic study.org.)

Action research has been defined as "an investigation conducted by the person or the people empowered to take action concerning their own actions, for the purpose of improving their future actions" (Sagor, 2005, p. 4). First introduced in the 1940s by Kurt Lewin as a means to encourage researchers and community-based practitioners to collaborate to solve societal problems, action research has evolved into many forms, yet has maintained its core focus on taking actions to address problems and learning from those actions to inform new action. All forms of action research involve some type of ongoing process of systematic study in which teachers examine teaching and learning by posing questions, collecting and analyzing data, drawing conclusions from the evidence gathered and analyzed, and reflecting on what was learned through the action research process. One of the benefits of teacher action research as a teacher learning strategy is that teachers have ownership in the process and are committed to promoting changes in practice indicated by their findings (Loucks-Horsley et al., 2010).

CTS action research differs from other types of action research because it focuses specifically on understanding student thinking related to important concepts or skills in the standards. The literature search that precedes the action research comes from doing a CTS utilizing the national standards and research on learning. The CTS provides a lens through which teachers understand the importance of the standards-based content and how it is made accessible to students through effective instruction and a coherent curriculum. The data collected and analyzed from their own students and context is looped back to the CTS findings in order to inform action steps that will be taken as a result of the study.

GOALS OF A CTS ACTION RESEARCH PROJECT

The goals of the project are

- to investigate questions about students' ideas of direct and immediate interest to a teacher,
- to deepen teachers' understanding of commonly held student ideas and implications for curriculum and instruction, and
- to generate and share CTS findings and new knowledge with colleagues.

AUDIENCE

CTS action research is appropriate for teachers at all levels of the teacher professional continuum. It can be used with preservice teachers in a methods course or experienced teachers seeking new ways to grow professionally, as well as teacher leaders looking for ways to generate and share new knowledge about teaching and learning. It is important for action research participants to have had previous experience with CTS, including knowing how to carry out a full topic study. (To develop initial experience with curriculum topic studies, see the suggestions and designs in Chapters 4 and 5 of this *Leader's Guide.*)

STRUCTURE OF CTS ACTION RESEARCH

The CTS action research process consists of ten steps in five phases as described in Table 7.3. A description of each of these steps is also provided on Handout 7.5.1: CTS Action Research—10-Step Planning Guide in the Chapter 7, action research folder.

Table 7.3   The CTS Action Research Process

Phases	*Steps*
Defining purpose and questions	1. Define your purpose 2. Formulate three research questions 3. Formulate overarching question and title
Literature search	4. Conduct CTS background research
Collecting evidence	5. Identify and collect data 6. Organize and display data
Making sense of evidence	7. Analyze data 8. Connect findings to actions
Reflection and resources	9. Reflection 10. References and artifacts

CTS ACTION RESEARCH PRODUCTS

CTS action research culminates in a product that can be shared with others. This product can take the form of a poster session, an informal report, or a published paper. The CD-ROM at the back of this *Leader's Guide* includes photographs of CTS action research posters shared at the NSTA Teacher Research Day at the 2008 NSTA Conference on Science Education in Boston. Handout 7.5.4: Making Invisible Student Ideas About Evaporation Visible is an example of a monograph written by a teacher who experienced CTS action research for the first time.

GUIDELINES FOR INTRODUCING AND LEADING CTS ACTION RESEARCH

Materials Needed by Facilitator

CTS Parent Book

- *Science Curriculum Topic Study: Bridging the Gap Between Standards and Practice* (Keeley, 2005).
- If participants do not have their own CTS parent book, make copies of the CTS guide "States of Matter" (p. 173 in the CTS parent book) if you are using this topic as your example in the action research introductory session.

Resource Books

- Access to all the CTS resource books used with Sections I–VI of the CTS guides

Assessment Probes

- See examples of CTS-developed probes that can be used with CTS topics on Handout 7.5.5. To find or develop probes on topics other than the one used in this workshop, see "Additional Resources for CTS Action Research" in the Chapter 7 folder under Action Research on the CD-ROM. This lists sources of probes and other resources. You may also develop your own assessment probes using the process in Chapter 4 of the CTS parent book or through the designing assessment probes session in Chapter 6, pages 204–215 of this *Leader's Guide.*

Local Standards

- Access to state standards, district frameworks, or curriculum materials guides

Application 7.5 CTS Action Research PowerPoint Presentation

- PowerPoint slides for Introducing CTS Action Research (in the Chapter 7 folder on the CD-ROM)

Supplies and Equipment

- Computer and LCD projector to show PowerPoint presentation
- Flip chart easel, pads, and markers
- Paper for participants to take notes
- Sticky notes for notes and marking charts
- Optional: Annenberg Private Universe Videos—see "Additional Resources for CTS Action Research" in the Chapter 7 folder under Action Research on the CD-ROM.

Wall Chart

No wall charts are required, but you may want to post group norms to refer to during the action research sessions.

Facilitator Preparation

- Review PowerPoint slides for Action Research (in Chapter 7 folder on the CD-ROM). Insert your own graphics and additional information as desired. (Optional: Print out copies of the PowerPoint slides for participants.)
- Prepare an evaluation form to collect feedback from participants.

Materials Needed for Participants

(*Note:* All handouts and the summary notes can be found in the Chapter 7 folder under Action Research on the CD-ROM.)

- Handout 7.5.1: CTS Action Research—10-Step Planning Guide
- Handout 7.5.2: CTS Action Research Posters or Monographs
- Handout 7.5.3: Guidelines for Student Interviews
- Handout 7.5.4: Monograph: Making Invisible Student Ideas About Evaporation Visible
- Handout 7.5.5: Probes and CTS Crosswalk
- Application 7.5: CTS Action Research PowerPoint slides (optional)
- Tri-fold poster boards (optional)

Time

CTS action research is a professional development strategy conducted over several weeks or months. The time frame for conducting action research varies. The following is a suggested breakdown of times needed to introduce action research and carry out an action research project. Times will vary according to your project and meeting times with teachers.

- Introduction to CTS Action Research—half day
- Preresearch planning time—half day
- CTS background research—half day
- Classroom research—allow one to three months for teachers to implement their research projects
- Midpoint check in—half day
- Monograph or poster preparation—one to two days
- Optional symposium—one day

DIRECTIONS FOR INTRODUCING CTS ACTION RESEARCH (HALF-DAY SESSION—3 HOURS)

Welcome participants with Slide 1. Show Slide 2. Ask participants what they know about action research and whether any of them have had previous experience doing action research. Give them a few minutes to discuss the questions on the slide in small groups and then ask for a few volunteers to share their initial thoughts (5 minutes).

Show Slides 3 and 4 and using the points on the slides, explain what CTS action research is, how it differs from other forms of action research, and how it is different from formal educational research. Explain how action research findings are usually not generalized across classrooms, like scientific research. However, because CTS action research includes a component on examining commonly held ideas students have about science that are in the cognitive research literature, it is highly likely that other teachers will find their students have ideas similar to the ones uncovered by the teacher action researcher. Show Slide 5 and explain that the purpose of this introductory session will be to introduce them to the CTS action research process and prepare them to start their own action research study. Poll the group to find out how familiar they are with CTS. Explain how there are different processes for conducting action research. Explain that the process they will be using is unique in its use of CTS as an integral part in conducting the literature search that precedes the classroom research (5 minutes).

Refer participants to Handout 7.5.1: CTS Action Research—10-Step Planning Guide. Explain that in this introduction they will walk through the ten steps on the guide, practice some of the steps, and examine a monograph prepared by a first-time CTS action researcher to see how this teacher's action research culminated in a CTS action research product.

Show Slide 6 and explain that they will all use the same example in this introductory session to practice the steps of CTS action research. Describe how the first step involves identifying a problem situated in a teacher's instructional practice. Share the scenario on Slide 6 as an example of a learning problem identified by a middle school teacher.

Give participants a couple minutes to think about and respond to the questions on Slide 6 as if they were the teacher interested in researching this problem. Ask the group to share a few responses (5 minutes).

Show Slide 7. Explain that the three types of questions on the slide will guide their research. As a group, generate a sample question for each of the three types. Point out that the first question guides the CTS; the second question guides the data collection; and the third question guides analysis and informs actions to be taken.

Facilitator Note

If you are working with a group in which all members are addressing the same content and grade levels, you may substitute for this example an actual example of your own that addresses the content and grade level of the teachers with whom you are working. If you substitute the example, be sure to make changes to the rest of the slides so they fit the example you selected.

Show Slide 8 and mention that these standard or boilerplate questions can be used by any teachers by inserting the topic or concept they wish to research. Show examples of each of the three questions using the concept of evaporation. Ask if there are any questions about the process before going to the next slide. Remind them they will be looking later at an example of a teacher's action research study that shows the three questions (5 minutes).

Show Slide 9. Give participants an opportunity to generate possible titles for their research project based on the question they come up with (5 minutes). Show them the example on Slide 9 of a title a teacher action researcher came up with to research her students' understanding of evaporation.

Show Slide 10. Explain how CTS is used to do a literature search, which is an important step to ground the action researcher in the content, standards, and research on learning before conducting research with the students. Ask which CTS guide would best address the learning problem for these questions. Allow a few people to suggest guides and explain that the best guide for addressing this content is "States of Matter" (although "Water Cycle" could be used as well). Have participants turn to the CTS guide, "States of Matter," on page 173 of the CTS parent book or to the copies of this guide you provided.

Explain that they will use the CTS process to look for information related to the learning problem only: students' ideas about what happens during a change in state from liquid to a gas. They are not doing a complete study of the topic, but rather using this topic to extract the information they need for their research. Using the prompts on Slide 10 ask the groups to divide up the readings for the six

CTS sections among different people at the table—this could be done by interest or just by counting off one to six. Each person will read a section looking only for text that relates to the concept of evaporation and related ideas about liquids and gases. Since the scenario for this example of an action research project is based on a middle school teacher's desire to learn more about her students' ideas related to evaporation, participants will only study the Grades 3–5 (or K–4) and 6–8 (or 5–8) readings from Sections II and III. Point out that it is important to go back to the previous grade span in order to identify precursor ideas that students may have missed. Section V will provide an opportunity to examine the ideas across the K–12 continuum of learning.

Have each group read their sections, and when they are done, go through each section reporting and recording the key findings on a sheet of chart paper to later share with the whole group (30 minutes).

When groups are finished, have each group share their study results on states of matter and how their findings relate to understanding the concept of evaporation. Briefly discuss with the group how these findings are important to the action research (e.g., they clarify the content and learning goals, give a sense of what to anticipate from the students, may reveal gaps in the curriculum, etc.). In essence, the CTS serves as the literature search in a CTS action research project (15 minutes).

Show Slide 11. Explain that this is the point in the action research process where they will begin to plan their research in the classroom. The first step is to identify the data they will need and ways to collect it. Refer to the example of the probe on the slide and explain that there are other probes related to the topics of states of matter and the water cycle listed on Handout 7.5.5. Ask participants to discuss in pairs how they might use one of these probes or a similar one to collect data. Share the source of the probe (Keeley et al., 2007) with participants and explain there are four volumes in this series. These probes were developed using the CTS process so that they are connected to ideas in the standards and the research on student learning. Chapter 4 in the CTS parent book also describes a process for developing their own probes that may be useful when they undertake their research. Provide an opportunity for the group to brainstorm other ways of collecting data (10 minutes).

Show Slide 12 and explain that these are some of the other types of data teachers can collect from their students; refer them to Handout 7.5.3: Guidelines for Student Interviews. Explain that student interviews are often used in action research. A structured interview is one in which a sequence of questions or "probes," sometimes accompanied by a set of tasks that students are asked to complete, is carefully chosen beforehand. Interviews can be audiotaped or videotaped and then transcribed for later analysis. The researcher should take limited notes during the interview; it's important that the note taking does not distract the student or disrupt the interview process. Right after the interview, while it is fresh in their minds, transcribe the important parts of the interview that they might want to use.

Because interviews with every student in a class can be time consuming, action researchers often select a sample of students to interview. For example, selecting a broad range of students that represents a cross-section of the diversity in the class can provide a sample portrait of how different students think about ideas in science.

Ask if anyone has conducted a student interview either formally or informally for the purpose of examining student thinking and ideas. Ask them to describe what it was like. Ask for a few responses. Give everyone a few minutes to look over the guidelines for student interviews on Handout 7.5.3 and ask any questions. To practice, have the group generate a list of questions they might ask a student about evaporation. Remind them to refer to their findings from doing the CTS readings. Ask for a report out of possible questions and write them on the chart paper at the front of the room. Ask everyone to find a partner and practice a mock student interview with one person being the interviewer and the other acting as the student. Encourage the interviewer to use the guidelines on the handout and the

> **Facilitator Note**
>
> You might consider preceding this session with the session on designing CTS assessment probes in Chapter 6 of this *Leader's Guide*, pages 204–215; if participants have previously experienced this CTS application, remind them about the assessment design process.

Facilitator Note

If you do not have access to the videotapes, the series is shown free on streaming video at www.learner.org (search for Private Universe in Science). Sessions 1, 2, and 5 have excellent examples of student interviews. You can also search for and see Private Universe video clips by *Benchmark* learning goal or *Atlas* strand at http://hsdvl.org.

questions he or she generated. Give each pair five minutes to practice; then ask them to switch roles and give five more minutes. Debrief the experience and discuss what was challenging about the interviewing process and how they might use it. As an optional extension or replacement for the mock interviews, consider showing the interviews between students and an experienced interviewer in the Private Universe series (see list of suggested resources on the CD-ROM). If you show the videotape, ask participants to watch for evidence of the guidelines as the interviewer talks with the student. (30 minutes)

Show Slide 13 and refer to Steps 6 and 7 on Handout 7.5.1. Explain that this is where they will use the data they collect, organize it for analysis, and make inferences and conclusions related to their findings. Ask if there are any questions. Show Slide 14 and refer to Steps 8, 9, and 10 on Handout 7.5.1. Explain the steps, discuss possible artifacts to include, and ask if there are any questions. Explain they will see an example of these steps in a real action research project in a few minutes. Answer other questions about the ten steps. (15 minutes)

Show Slide 15. Explain that after completing an action research project, it will be time to think about sharing your research with others. Discuss the importance of publicly sharing your research. Refer to the Handout 7.5.2: CTS Action Research Posters or Monographs. Explain that this handout provides the structure for organizing your writing or poster display. Point out the photos on Slides 16, 17, and 18 as examples of ways teachers displayed their research. Slide 16 is a middle school project on students' ideas about conservation set in the context of dissolving substances. Slide 17 is an elementary project on students' ideas about magnetism. Slide 18 is a project that examined students' ideas about evaporation. These displays have been used at conferences, symposia, and even displayed in teachers' staff rooms and district professional development centers for others to examine and learn from. Often copies of the written monograph accompany the poster for teachers to take away. Provide time for questions about the products. (15 minutes)

Show Slide 19. Ask everyone to examine Handout 7.5.4: Making Invisible Student Ideas About Evaporation Visible. Explain that the teacher who conducted this research was a first-time action researcher. With a partner, have pairs read through the monograph, stopping after each section to discuss how it connects to each of the ten steps in the CTS action research process described in Handout 7.5.1 (or the short, summarized version in Handout 7.5.2). Give them time to discuss the teacher's findings as a whole and how they think the project contributed to the teacher's learning as well as the value of sharing it with colleagues. Finally, show Slide 18 (again) to see how the teacher prepared a poster of her findings as well as the monograph. (30 minutes)

Show Slides 20 and 21. Answer any questions and invite any thoughts participants might like to share. Include information on next steps that may be specific to your work with the teachers, such as what the expectations are for them to do an action research project, support, honoraria, and any other information that may be specific to the context you are using this CTS strategy in. End with having each participant reflect on what they have learned about action research and ask for five to six people to publicly share their thoughts by filling in the stem "I'm looking forward to _____." If they have ideas already about what they want to research, encourage them to include it in their response. (10 minutes)

Planning Meeting (Follow-Up to the Half-Day Introductory Session)

After the introductory session, plan a follow-up half-day session as a planning meeting in which participants can begin to plan their CTS action research study. This can be held during the second half of the day, following the introductory session, or convened again after teachers have had time to think about what they want to research. At the planning meeting, try to set up a table of resources for teachers to use. (See annotated resource list on the CD-ROM.) Consider using a tuning protocol, which is a process for providing feedback, such as the one developed by the Coalition of Essential Schools and available at www.essentialschools.org/cs/resources/view/ces_res/54.

The follow-up meeting should be an informal opportunity for the teachers to interact with each other, collaborate with others if they are doing a joint research project, and get feedback from you, the leader, as well as other "critical friends" who can help them with their project. Areas for discussion and input include identifying data collection and analysis techniques, suggested resources, and so forth. Build in time at the end for sharing action research plans and for people to pair up with one other person to get feedback on their plans.

At the end of the planning meeting, everyone who will do an action research project should have Steps 1 through 3 completed following the steps on the CTS Classroom Action Research—10 Step Planning Guide. Steps 1 through 3 involve establishing the purpose, formulating three questions to guide research, and formulating an overarching question and title. They should also be thinking about what they need to complete Steps 4 through 6 (i.e., do the background research, collect data, and organize and display data). If your participants are going to produce posters of their research, at this time you may want to provide them with tri-fold poster boards so they can start to plan how they will share findings.

CTS Background Research

After teachers have completed Steps 1 through 3, they will do a CTS as part of Step 4. A half-day follow-up session can be set aside for this if teachers do not have their own CTS resources; or alert teachers that they will need to set aside about three hours to complete Step 4 before conducting research with their own students.

Conducting the Classroom Research

Provide at least a month and preferably no more than three months for teachers to carry out their action research project. Depending on your context for working with the teachers, you may want to designate periodic check-in times, including a midpoint meeting where teachers can report on their progress thus far and get feedback from others.

Concluding and Sharing the Research

Teachers should plan on one to two days to analyze their findings and prepare their monographs and poster boards. You may consider having teachers present their monographs and poster boards at a symposium that can be held as part of your project or consider having teachers present at the NSTA Teacher Research Day in conjunction with the NSTA National Conferences. (See CTS Web site for more information on this opportunity and a link to the SC4 MSP Project in which teachers presented their CTS Action Research at the 2008 NSTA National Conference. The CTS Web site also includes a link to the collection of monographs prepared by the teacher researchers who participated in CTS Action Research.)

CTS Professional Development Strategy
Application 6: CTS Seminars

DESCRIPTION OF SEMINARS AS A CTS
PROFESSIONAL DEVELOPMENT STRATEGY

CTS seminars are content-focused seminars in which small groups of teachers engage in discussion and dialogue of readings from a CTS. The seminar provides a structured format for teachers to discuss their CTS readings, which are read previous to attending the seminar. It is a way for teaching professionals to raise questions and discuss important content-related ideas, rejuvenate themselves, and come together as a learning community.

CTS seminars enable teachers to engage in intellectual text-based conversations that build the learning capacity of a group. These are not freewheeling discussions; they incorporate a type of Socratic seminar structure that supports inquiry into a curricular topic. The close examination and discussion of the findings that result from a topic study not only contribute to teachers' content and pedagogical knowledge of the concepts and skills in the topics they teach, but also contribute to building the knowledge base of the group and their schools. Through the seminar format, a CTS learning community focuses on developing and strengthening the skills and dispositions of analytical professional reading, use of common professional language, careful listening, citing evidence from text, disagreeing respectfully, being open-minded about new ideas, and connecting theory from the text of standards and research to practice.

The adult learning skills teachers draw upon during a CTS seminar parallel the K–12 critical skills teachers want students to acquire and use when discussing science expository text. These skills include oral discourse, use of academic language, respectful discussion, healthy skepticism, critical thinking and reasoning, constructing meaning from reading, and applying learning from text to real-life situations.

GOALS OF A CTS SEMINAR

The goals of a CTS seminar are to

- provide a structured opportunity for teachers to have content-focused intellectual discussions about topics they teach,
- deepen teachers' content knowledge and understanding of pedagogical implications that surface from doing a curriculum topic study, and
- surface awareness of one's own ideas and ideas of others as they relate to the topic studied.

AUDIENCE

It is suggested that a CTS seminar include no fewer than five participants and no more than twelve. Readings, questions, and conversations will differ according to the audience's grade level, school context, familiarity with CTS, and content backgrounds of the teachers.

KEY COMPONENTS OF A CTS SEMINAR

CTS Text Readings

Choose a topic that is relevant to the group attending the seminar. CTS seminar discussions revolve around the text-based readings from CTS. Topics and their associated text readings are chosen based on the needs of the group to learn more about a particular science curricular topic. The CTS text readings anchor the discussion in the important content of the topic and provide an opportunity for teachers to cite and use findings from CTS to discuss implications for teaching and learning. Supplementary readings, which may include content readings, primary research articles from journals, or technical professional literature, offer an opportunity for teachers to collaboratively tackle more difficult readings than they might read on their own.

Questions

Once the topic is selected, the seminar leader prepares questions to facilitate discussion of the CTS findings from the text resources. Even though you might not get to all the questions during a seminar, it is important to plan questions ahead of time to help participants collaboratively construct a deeper understanding of the text readings. Use the questions on pages 37–39 of the CTS parent book to focus on specific sections of the CTS guide or develop your own. It is also important to include a few open-ended, thought-provoking questions that teachers may not be able to answer quickly from their text notes, but may need to think about before responding. Questions should be clear and succinct and not open to a wide array of interpretations.

Seminar Leader

The CTS seminar leader has several tasks. These tasks include the following:

Preparation: The seminar leader is responsible for making sure participants have books or copies of the readings and know what the assigned readings are, developing the questions for the seminar, and planning for after-seminar follow-up where necessary (Handouts 7.6.1, 7.6.2, and 7.6.3 can be used to prepare to lead CTS seminars).

Participation: The primary role of the seminar leader during the seminar is to listen, think, and question. Seminar leaders participate to some extent, but it is important that their voices are not the ones most frequently heard in the room. Their role is to help participants get deeply into the text readings and elicit and surface ideas for all to engage with.

Maintaining an engaging, respectful, professional environment: The seminar leader keeps participants engaged, makes sure all ideas are listened to and participants are not interrupted or ignored, keeps the group on task and away from tangential storytelling, and encourages the skills of group dialogue and discussion. Seminar leaders should continually encourage participants to explain where their ideas come from so that they learn to develop the skill of citing CTS evidence during the discussion. The seminar leader also establishes and posts norms that all participants agree to during the seminars. If the seminar leader decides to provide a written summary after the seminar has taken place, then the seminar leader will also take notes during the seminar.

Reflection

Reflection is a critical component that links the knowledge gained from CTS findings, the discussion among participants, and thoughts about what could be done as a result to improve teaching and learning. Every seminar builds in time for shared reflections at the close of the seminar to examine how the group met its group goal(s). Reflection should also include time to examine the extent to which teachers' personal goals were met.

STRUCTURE OF A CTS SEMINAR

The CTS seminar structure includes three parts: (1) pre-seminar activities, (2) seminar questions and discussion, and (3) post-seminar activities.

Pre-Seminar Activities

The pre-seminar activities take place prior to and right before the seminar. Prior to the seminar, participants read the CTS sections, make notes in preparation for the discussion, and choose a personal learning goal related to the topic. Activities that happen right before the seminar begins include establishing or reviewing the seminar norms and setting two group goals for the seminar—one content and one process goal. Individuals may also set their own personal goals related to these.

The Seminar

The seminar itself includes three phases: the opening, the core, and the closing. The opening should begin with a question that requires participants to refer to their CTS text readings in order to set the stage for practicing "text-based" discussions. It should be broad enough to allow everyone to have an entry point into the CTS readings and stimulate conversation. The core forms the bulk of the seminar. Core questions focus on specific aspects of the CTS readings and are designed to get participants to delve deeply into the CTS findings. The closing question helps participants connect the group's discussion to their own practice. The closing is an opportunity for participants to contextualize and personalize their CTS findings.

Post-Seminar Reflection

The post-seminar reflection happens at the very end of the seminar. Participants consider the content and process goal(s) they established for the group and reflect on how well the seminar helped them reach the goal(s). Participants also reflect on their individual goals. Attention to this part of the seminar structure is critical, as reflection is often shortchanged for the sake of time. It is through reflection that participants can improve upon the social and intellectual skills that are part of the seminar experience.

GUIDELINES FOR LEADING CTS SEMINARS

Materials Needed by Seminar Leader

CTS Parent Book

Science Curriculum Topic Study: Bridging the Gap Between Standards and Practice (Keeley, 2005)

Resources

- All CTS resource books or copies made of selected readings
- Discussion questions prepared in advance

- Handout 7.6.1 Seminar Norms
- Handout 7.6.2 Seminar Leader Checklist
- Handout 7.6.3 Resource Checklist

Materials Needed by Participants

- CTS guide with selected seminar readings identified
- CTS selected resource books or copies made of selected readings
- Notes made by participants on readings to bring to the seminar

Time

Seminars range from 1 to 1.5 hours, depending on topics, lengths of readings, and number of questions selected. They are usually held after school or during the school day.

Participants' Responsibility

Participants have three main tasks in any CTS seminar: (1) preparation—by doing the readings in advance and highlighting, annotating, or taking notes; (2) participation—by active listening and thinking, speaking, and referring to the text and not their own personal opinions and stories; and (3) respect—acknowledging differences, honoring the process, and being aware of "air time" so that everyone has a chance to participate.

> **Facilitator Note**
>
> The best way to introduce teachers to this strategy is to model a CTS seminar using the following directions. Because it is a seminar, there are no PowerPoint slides and few handouts. The materials come primarily from the CTS book and CTS readings. The questions used depend on what direction the discussion takes.

DIRECTIONS FOR LEADING THE SEMINAR

Welcome (5–10 minutes)

Welcome participants, review group norms if necessary (refer to Handout 7.6.1), and generate two goals on chart paper, one for content and one for group process. For example, two group goals for a middle school CTS seminar on conservation of matter might include the following:

1. Examine how we can embed conservation-of-matter concepts into our existing science units (content goal).

2. Improve our ability to practice wait time after someone speaks (process goal).

Post the goals on a chart for all participants to see. Also provide time for participants to share one personal goal that describes what they hope to gain from the seminar with a partner or triad. If the seminar group is small, have participants go around and briefly share their goals with the group.

Opening Question (5–10 minutes)

The opening question is broad and allows everyone an entry point into the conversation. Try to encourage all participants to respond to the opening question. Example: Why is conservation of matter such an important concept to teach in middle school science?

The Core (30–45 minutes)

The seminar leader poses the prepared questions in an order that fits with the flow of the conversation and may pose additional questions as needed. The seminar leader guides the conversation and at times may interject with probes and extenders such as the following:

"Can you tell us more about that?"

"Can you show us where you found that in the CTS resources?"

"Would anyone like to add anything?"

During the core discussion, the seminar leader keeps the group focused on the findings that emerge from the text and redirects when personal stories or opinions draw the discussion away from the CTS readings. When the discussion of one question appears to be exhausted, the seminar leader raises a new question. The seminar leader makes sure participants see the link between the discussion questions and the CTS text. Here are some example questions a seminar leader might use during the core for a seminar focused on the topic of conservation of matter:

1. What does the reading reveal about what is important to teach in order for students to be able to understand and use the principle of conservation of matter?

2. What misconceptions do we need to be aware of that may pose barriers to students' learning?

3. What other matter-related ideas impact students' understanding of conservation of matter?

4. What kinds of contexts and instructional approaches would be useful for us to consider for developing a deep understanding of conservation-of-matter ideas?

5. What does the *Atlas* map reveal about prerequisite ideas we need to pay attention to? Does it appear that we have any gaps in our district curriculum that may affect our students' understanding?

The Closing (15–20 minutes)

The closing allows participants to connect their discussion about the readings to their own practice. The closing may consist of one or two questions posed by the seminar leader that relate directly to the participants' context and the nature of the discussion during the seminar. Seminar leaders may use a question developed prior to the seminar or may pose a new one based on major points from the readings that participants were most interested in discussing. Examples using the conservation of matter topic might include the following:

- Considering what we just discussed, how well do the CTS findings describe the way our students currently learn about conservation of matter?
- What are you thinking about doing differently, and why, based on what we discussed today?

The Reflection (15–20 minutes)

The seminar leader transitions from closing to reflection by referring participants to the chart that lists group goals. Ask participants to comment on the extent to which they felt the seminar met the group's goals. Also provide time for reflection on their own personal goals, asking if people would like to share what they learned from the seminar that helped them meet those goals.

Next Steps (5 minutes)

If the seminar is a continuing series, take a few minutes for the group to identify the next topic, date, location, and seminar leader.

Optional

Seminar leaders or a designee may wish to write a summary of the seminar to share with participants or disseminate at the next seminar meeting.

CTS Professional Development Strategy Application 7: CTS Mentoring and Coaching

DESCRIPTION OF MENTORING AND COACHING AS A CTS PROFESSIONAL DEVELOPMENT STRATEGY

Mentoring and coaching are teacher-to-teacher professional development strategies that provide one-on-one opportunities for teachers to improve their practice with the help of an experienced colleague. Mentoring usually involves an experienced teacher working with a new or novice teacher. Dunne and Villani (2007) identified four of the most important roles played by mentors. These are a collegial guide, orienting new teachers to the culture of a school; a consultant, working with new teachers to resolve difficulties and adopt strategies for addressing problems; a seasoned teacher, offering professional insights based on extensive experience in the classroom; and a coach, leading new teachers through collaborative inquiry into practice and to expand their instructional repertoire. The coaching role can involve two experienced teachers working together, with one teacher having more expertise in a certain area that can be used to guide the less experienced teacher. Instructional coaches are called upon to lead groups of other teachers as they examine student work, to organize demonstration lessons, to assist with curriculum planning and take other actions that help teachers reach high performance and use of effective practice. For example, sometimes coaches are used to support implementation of new instructional materials. They might demonstrate use of the materials in the classroom, observe and provide feedback to teachers new to the materials, and meet with the new users to reflect on student work derived from the new instructional materials.

Over the years, different forms of coaching used by mentors with new teachers as well as by coaches with their colleagues have emerged with different purposes and correspondingly different techniques. Many of these traditional forms incorporate a supervisory model focused on observations and use a pre-conference/observation/post-conference cycle. More recently, coaching has shifted to focus less on a supervisory model and more on collaborative learning and provision of support to teachers who are making changes in their instructional practice at all stages of their careers (Loucks-Horsley et al., 2010).

CTS mentoring and coaching differs from other forms of mentoring and coaching because of its specific focus on coaching and mentoring around the science content and science content-specific pedagogy. In some districts, new teachers are assigned mentors who are not in their content area. These experienced mentors can help them acclimate to the school and provide general instructional techniques, but are limited in how they can help with science content or support a new teacher of science in effectively planning lessons that target key ideas in science. Likewise, general coaches don't know the science or science-specific pedagogy to help their colleagues implement new instructional materials or approaches that are specific to the discipline of science. CTS mentoring and coaching addresses the subject-specific needs of science teachers. It is not a "model" of mentoring or coaching, but rather is intended to be used with existing mentoring models to add the content specificity mentors and coaches need to support science teachers to improve their practice.

The CTS mentoring and coaching strategy uses a special form of coaching called content-focused coaching, in which the tools of CTS are used to support individuals or groups of teachers in designing, implementing, and reflecting on standards and developing research-based lessons.

GOALS OF CTS MENTORING AND COACHING

The goals of CTS mentoring and coaching are

- to increase the capacity of mentors and coaches to address standards and research on learning in their work,
- to familiarize new teachers and experienced teachers who are new to teaching science or are using new practices in science with the CTS tools and resources, and
- to provide "experts at teachers' fingertips" when they have questions about science content and pedagogy.

AUDIENCE

CTS mentoring and coaching is designed for K–12 mentors and coaches who have been through (or are currently engaged in) professional training to become a coach or mentor and are now interested in using CTS to enhance the focus on content in their mentoring and coaching work. It is also designed for people who provide training for mentors and coaches who are interested in embedding the CTS tools into their professional development programs.

STRUCTURE OF CTS MENTORING AND COACHING

The CTS mentoring and coaching strategy is guided by the CTS Instructional Coaching Framework for Mentors and Coaches shown in Figure 7.3. It consists of

- identifying the topic of a lesson and the content one needs to know to teach it (CTS topic guide and Section I);
- examining the key ideas and concepts identified by the standards that are learning targets to be aligned with the lesson (CTS Sections III and VI);
- examining instructional contexts, effective strategies, relevant phenomena, connections to other ideas, and a coherent sequence of learning (CTS Sections II and V);
- learning about the research on learning in order to understand the commonly held ideas students bring to the lesson, research-identified difficulties students might have, and developmental considerations (CTS Section IV);
- examining one's own beliefs about teaching and learning;
- knowledge of one's own students' prior experiences, learning styles, preconceptions, and learning difficulties; and
- using all of the above to collaboratively plan a lesson.

The CTS Instructional Coaching Framework is designed to be embedded into the existing structure for mentoring or coaching used in any school or district.

> **Facilitator Note**
>
> The CTS instructional coaching model was originally used with mentors and instructional coaches who were part of the National Science Foundation–funded Northern New England Co-Mentoring Network. (For more information see the Web site at www.nnecn.org.) Mentors involved in this program applied CTS throughout their mentoring work and taught the teachers with whom they worked to consult CTS when they have questions about content and pedagogical content knowledge.

Figure 7.3 CTS Instructional Coaching Framework for Mentors and Coaches

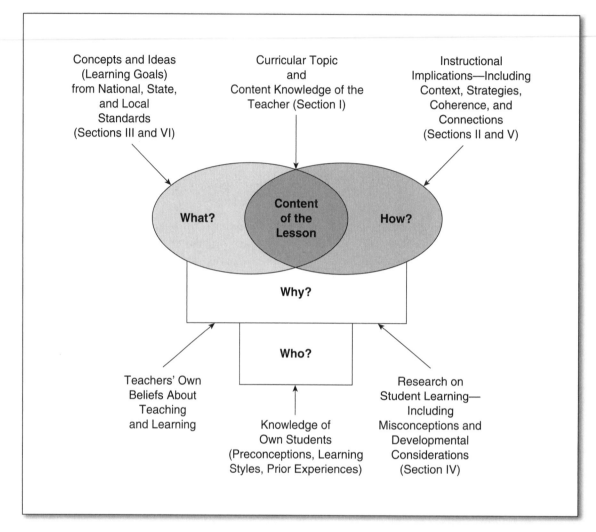

GUIDELINES FOR INTRODUCING CTS
MENTORING AND COACHING TO MENTORS AND COACHES

Below we provide two different suggestions for using CTS with mentors and coaches.

Introducing the Use of CTS in a Mentoring and Coaching
Context Including the Use of the Instructional Coaching Framework

Facilitators use this session to introduce CTS mentoring and coaching to groups of mentors or coaches, including introducing the CTS Instructional Coaching Framework.

Time

Depending on participants' familiarity with CTS, this introductory session can take anywhere from 2 to 2.5 hours.

Materials Needed

CTS Parent Book

- *Science Curriculum Topic Study: Bridging the Gap Between Standards and Practice* (Keeley, 2005), or copies of the CTS guide, "Physical Properties and Change," page 170 in the CTS parent book

Resource Books

- Copies of all readings from resource books for the CTS guide "Physical Properties and Change"

Handouts

- Handout 7. 7.1 CTS Instructional Coaching Framework for Mentors and Coaches
- Handout 7.7.2 Coaching Scenario
- Handout 7.7.3 Instructional Coaching Questions

Directions for Introducing CTS Coaching and Mentoring

Show Slide 1 and welcome participants. Show Slide 2 and review the goals for this session.

Show Slide 3 and explain that there are a variety of forms of coaching and different labels for them. This session will focus on the last one—content coaching.

Show Slide 4 and explain that CTS content coaching is designed to improve teaching and promote student learning by having a coach or mentor and a teacher seeking assistance work collaboratively using the tools of CTS. Review the characteristics of CTS content coaching listed on the slide. Ask if anyone has used a similar coaching model. Share examples (if any). Ask participants why content coaching is needed in science as opposed to more generic forms of coaching. Accept responses and add to the discussion as needed, drawing on points raised under the introduction to this application, "Description of Mentoring and Coaching as a CTS Professional Development Strategy."

Show Slide 5 and explain that this session is going to address both coaching and mentoring. While there are differences between the two, there are also similarities. CTS content coaching and mentoring are designed to address the overlapping similarities. Give participants a few minutes at their tables to discuss what they think are the similarities and differences between the two.

Show Slide 6 and point out the similarities in black. Then point out the differences in red. Explain how CTS mentoring and coaching focus only on teaching and learning in the context of science content and pedagogy. Because this is a common focus, this session will combine mentoring and coaching with a focus on what mentors and instructional coaches do in the coaching cycle.

Show Slide 7. Explain that mentoring is part of coaching and how both mentoring and coaching are part of a larger goal to build professional learning communities that use research and tools like CTS to continuously improve practice.

Show Slide 8 and explain that the CTS mentoring and coaching approach is designed to be embedded within an existing mentoring or coaching model in schools and districts. Point out how the red bullets are unique to the way CTS is used in mentoring and coaching. For example, mentors and coaches have access to a "tool kit" of content-specific resources used in CTS.

Show Slide 9 and explain that although this session will focus just on using the instructional planning tools, mentors and coaches who are familiar with CTS and the designs in this book can also lead other professional learning strategies that build community among their colleagues. Inclusion of these strategies can be thought of as a type of enhanced mentoring that they might want to consider.

Facilitator Note

For activities for introducing the CTS parent book and the CTS resource books shown on Slides 11 and 12, turn to Chapter 4, Module A1, in this *Leader's Guide*.

Show Slide 10 and reinforce the idea that "content talk" is at the core of the approach we use in CTS.

Show Slides 11 and 12 to remind participants of the core resources used in the CTS mentoring and coaching approach. Hold up a copy of each book if there are some participants who may not be familiar with them or invite participants to take a few minutes looking through the books.

Show Slide 13 and point out why CTS is used for mentoring and coaching. Ask participants who have had previous experience with CTS why they think it is so useful in working with new and novice teachers.

Show Slide 14 and explain that a major focus of CTS coaching is on instructional planning. An instructional framework was developed to guide coaches and mentors in using CTS-informed instructional planning in their collaborative relationships. It shows how the sections of CTS are used along with the teachers' knowledge of their own students to plan effective lessons. Point out the framework on the slide or have everyone refer to Handout 7.7.1: CTS Instructional Coaching Framework for Mentors and Coaches and review the parts of the framework and the sections of CTS that relate to each, including the following:

1. *Content of the Lesson*: Identifying the topic of a lesson and the content one needs to know to teach it (CTS guide and Section I)

2. *What Content to Teach*: An examination of the key ideas and concepts identified by the standards that are learning targets to be aligned with the lesson (CTS Sections III and VI)

3. *How to Teach Content Effectively*: Examining instructional contexts, effective strategies, relevant phenomena, connections to other ideas, and a coherent sequence of learning (CTS Sections II and V)

4. *Why Students Have Different Ideas or Learning Difficulties*: Learning about the research on learning in order to understand the commonly held ideas students bring to the lesson, research-identified difficulties students might have, and developmental considerations (CTS Section IV)

5. *Why Our Own Beliefs Can Affect How We Teach*: Examining one's own beliefs about teaching and learning (drawing on our own knowledge)

6. *Knowing Who Our Students Are as Learners*: Recognition of diversity and knowledge of one's own students' prior experiences, learning styles, preconceptions, and learning difficulties (drawing on our own knowledge and/or gathering additional data)

7. Using all of the above to collaboratively plan a lesson

Show Slide 15 and explain that CTS and the CTS Instructional Coaching Framework help mentors and coaches do these things more effectively. Pass out Handout 7.7.3: Instructional Coaching Questions and explain that these questions are based on the framework and guide coaches and their colleagues through CTS-focused instructional planning. Give participants time to look at and review the questions, making connections back to the framework.

Show Slide 16 and give participants five minutes to pause and make sense of what they have learned thus far. Ask them to summarize what the framework and instructional coaching questions provide and identify questions that need to be addressed before they practice using the framework. After five minutes, take time to answer questions.

Show Slide 17 and pass out Handout 7.7.2: Coaching Scenario. Set the stage by reviewing the scenario and asking participants to work in pairs to use sections of CTS to plan for the lesson in the scenario. Ask one to pretend to be the coach and the other the "coached." Encourage the coaches to turn back to Handout 7.7.3: Instructional Coaching Questions to guide their work. Tell them they do not have to come up with a finished lesson, but rather engage in the steps and consult CTS readings to gather enough information to describe what a lesson might involve. For example, when they turn to the questions related to the content of the lesson, they will look at CTS Section III or VI and ask, "What does that suggest about the content of the lesson?" (Allow 45–60 minutes for this exercise.)

Go around and listen in on the groups. If people are stuck, help them find the right CTS section and reading or answer questions. Remind everyone to use the questions on Handout 7.7.3 to stay on task.

Show Slide 18 and have pairs discuss the question prompts. When groups are ready, have four to five pairs volunteer to briefly share their planning, citing how CTS informed their lesson planning and what they found helpful.

Show Slide 19. End the session by having participants reflect on how they might use CTS in their mentoring and coaching. Share any next steps such as how everyone can get access to all the CTS resources so they can use the framework and the instructional coaching questions in their buildings. Ask everyone to think of how they will introduce some aspect of the CTS tools to the teachers with whom they are working. Allow time for writing and then ask for volunteers to share some of their ideas.

Ask everyone to provide written feedback on the session.

Optional Design for Introducing CTS Mentoring and Coaching: Mentor-Coach Use Scenarios

Facilitators who are working with mentors and coaches who have had prior professional development in coaching and mentoring (or are experiencing it in their session) and have had experience using CTS may want to incorporate the mentor-coach use scenarios as a way to practice using CTS to guide and inform their mentoring and coaching sessions. This session can be used prior to introducing the CTS Instructional Coaching Framework for Mentors and Coaches.

Materials Needed

- Handout 7.7.4: Mentor-Coach Use Scenarios
- CTS parent book and resource books to do the readings on Handout 7.7.4

Directions for Using the Scenarios

If you are using this activity as part of a mentoring or coach training program, identify where you will use the scenarios. Describe how the scenarios will help mentors and coaches see how CTS can be used in a mentoring or coaching situation and provide an opportunity for them to practice it in a mock scenario.

Distribute Handout 7.7.4: Mentor-Coach Use Scenarios. Ask the participants to pair up and assign a use scenario to each pair of teachers or have them select their own.

Ask pairs to complete the sections of CTS needed to address their scenario and then engage in a mock coaching or mentoring conversation based on their CTS findings. (Allow 30–40 minutes for this exercise.)

Have pairs reflect on the value of exploring the CTS resources and discuss in the larger group how CTS can be used in their work.

After this smaller "taste" of CTS, you may want to follow up by using the full design for introducing CTS mentoring and coaching in this section.

A FINAL WORD

This chapter, and in fact the whole book, has many "moving parts." Detailed background information is included to share as much as we could about our own experiences with CTS. Scripts are provided for many of our designs so that facilitators will have as much guidance as we had when we field tested the CTS designs and applications. Hundreds of handouts, examples, and facilitator resources are included on the CD-ROM. We encourage you to use any and all of these materials to the extent that they work for you and to adapt them to your own style, interest, and audience. Our ultimate goal is not that the particular materials are used exactly as we have suggested, but rather that leaders have many, varied designs, tools, and resources to help teachers develop and use the specialized professional knowledge of science content and pedagogy. Through the many activities in this book, we think that is quite possible. Please let us know what you discover as you use CTS and this guide and share any new ideas and designs with us by visiting the Web site, www.curriculumtopicstudy.org.

References

Ajewole, G. A. (1991). Effects of discovery and expository instructional methods on the attitude of students to biology. *Journal of Research in Science Teaching, 28,* 401–409.

Allen, J. (2007). *Inside words: Tools for teaching academic vocabulary, grades 4–12.* Portland, ME: Stenhouse.

American Association for the Advancement of Science. (1989). *Science for all Americans.* New York: Oxford University Press.

American Association for the Advancement of Science. (1993). *Benchmarks for science literacy.* New York: Oxford University Press.

American Association for the Advancement of Science. (2001). *Designs for science literacy.* New York: Oxford University Press.

American Association for the Advancement of Science. (2001–2007). *Atlas of science literacy* (Vols. 1–2). New York: Oxford University Press.

American Association for the Advancement of Science. (2008). *Benchmarks for science literacy online.* Available at http://www.project2061.org/publications/bsl/online.

Anderson, J. R. (1995). *Learning and memory.* New York: Wiley.

Atkin, J. M., & Karplus, R. (1962). Discovery or invention? *The Science Teacher, 29*(2), 121–143.

Beane, J. (Ed.). (1995). *Toward a coherent curriculum.* Alexandria, VA: Association for Supervision and Curriculum Development.

Birman, B., Desimone, L., Garet, M., & Porter, A. (2000). Designing professional development that works. *Educational Leadership, 57*(8), 28–33.

Bransford, J., Brown, A., & Cocking, R. (2000). *How people learn.* Washington, DC: National Academies Press.

Brooks, A., & Driver, R. (with Hind, D.). (1989). *Progression in science: The development of pupils' understanding of physical characteristics of air across the age range 5–16 years.* Leeds, UK: Centre for Studies in Science and Mathematics Education, University of Leeds.

Brown, C., Smith, M., & Stein, M. (1996, April). *Linking teacher support to enhanced classroom instruction.* Paper presented at the annual meeting of American Educational Research Association, New York.

Bybee, R. (1997). *Achieving scientific literacy.* Portsmouth, NH: Heinemann.

Cohen, D. K., & Hill, H. C. (1998). *Instructional policy and classroom performance: The mathematics reform in California (RR-39).* Philadelphia: Consortium for Policy Research in Education.

Cohen, D. K., & Hill, H. (2000). Instructional policy and classroom performance: The mathematics reform in California. *Teachers College Record, 102*(2), 294–343.

Danielson, C. (1996). *Enhancing professional practice: A framework for teaching.* Alexandria, VA: Association for Supervision and Curriculum Development.

Darling-Hammond, L. (2000). Teacher quality and student achievement: A review of state policy evidence. *Education Policy Archives, 8.* Retrieved July 11, 2009, from http://epaa.asu.edu/epaa/v8n1.

De Posada, J. M. (1999). The presentation of metallic bonding in high school science textbooks during three decades: Science educational reforms and substantive changes of tendencies. *Science Education, 83*(4), 423–447.

Donovan, M. S., & Bransford, J. D. (Eds.). (2005). *How students learn: Science in the classroom.* Washington, DC: National Academies Press.

Driver, R., Squires, A., Rushworth, P., & Wood-Robinson, V. (1994). *Making sense of secondary science: Research into children's ideas.* London: Routledge-Falmer.

DuFour, R., & Eaker, R. (1998). *Professional learning communities at work.* Bloomington, IN: National Educational Service.

DuFour, R., Eaker, R., & DuFour, R. (Eds.). (2005). *On common ground: The power of professional learning communities.* Bloomington, IN: National Educational Service.

Dunne, K., & Villani, S. (2007). *Mentoring new teachers through collaborative coaching: Linking teacher and student learning.* San Francisco: WestEd.

Eaker, R., DuFour, R., & Burnette, R. (2002). *Getting started: Reculturing schools to become professional learning communities.* Bloomington, IN: National Educational Service.

Erickson, L. (1998). *Concept-based curriculum and instruction.* Thousand Oaks, CA: Corwin.

Feger, S., & Arruda, E. (2008). *Professional learning communities: Key themes from the literature.* Providence, RI: Brown University. Retrieved July 11, 2009, from www.alliance.brown.edu/db/ea_catalog.php.

Frayer, D. A., Frederick, W. C., & Klausmeier, H. J. (1969). A schema for testing the level of concept mastery. *Technical Report No. 16.* Madison: University of Wisconsin Research and Development Center for Cognitive Learning.

Garmston, R. J., & Wellman, B. M. (2008). *The adaptive school: A sourcebook for developing collaborative groups* (2nd ed. with CD-ROM). Norwood, MA: Christopher-Gordon.

Garnett, P. J., & Treagust, D. F. (1990). Implications of research of students' understanding of electrochemistry for improving science curricula and classroom practice. *International Journal of Science Education, 12,* 147–156.

Goldhaber, D. D., & Brewer, D. J. (2000). Does teacher certification matter? High school teacher certification status and student achievement. *Educational Evaluation and Policy Analysis, 22*(2), 129–146.

Goldsworthy, A. (1997). *Making sense of primary science investigations.* Hatfield, England: Association for Science Education.

Gould, A., Willard, C., & Pompea, S. (2000). *Real reasons for seasons.* Berkeley, CA: Lawrence Hall of Science.

Harvard-Smithsonian Center for Astrophysics. (2003). *Essential science for teachers: Life science.* Washington, DC: Annenberg Media. Available at http://www.learner.org/courses/essential/life.

Hazen, R. M., & Trefil, J. (1991). *Science matters: Achieving scientific literacy.* New York: Anchor Books.

Hazen, R. M., & Trefil, J. (2009). *Science matters: Achieving scientific literacy* (rev. ed.). New York: Anchor Books.

Hord, S. M. (1997). *Professional learning communities: Communities of continuous inquiry and improvement.* Austin, TX: Southwest Educational Development Laboratory.

Hord, S. M., & Sommers, W. A. (2008). *Leading professional learning communities: Voices from research and practice.* Thousand Oaks, CA: Corwin.

Jacobs, H. H. (2004). *Getting results with curriculum mapping.* Alexandria, VA: Association for Supervision and Curriculum Development.

Jolly, A. (2007). *Building Professional Learning Communities.* Retrieved July 28, 2009, from http://www.edweek.org/chat/transcript_11_19_2007.html.

Keeley, P. (2005). *Science curriculum topic study: Bridging the gap between standards and practice.* Thousand Oaks, CA: Corwin.

Keeley, P., Eberle, F., & Dorsey, C. (2008). *Uncovering student ideas in science, volume 3: Another 25 formative assessment probes.* Arlington, VA: NSTA Press.

Keeley, P., Eberle, F., & Farrin, L. (2005). *Uncovering student ideas in science, volume 1: 25 formative assessment probes.* Arlington, VA: NSTA Press.

Keeley, P., Eberle, F., & Tugel, J. (2007). *Uncovering student ideas in science, volume 2: 25 more formative assessment probes.* Arlington, VA: NSTA Press.

Keeley, P., & Rose, C. M. (2006). *Mathematics curriculum topic study: Bridging the gap between standards and practice.* Thousand Oaks, CA: Corwin.

Keeley, P., & Tugel, J. (2009). *Uncovering student ideas in science, volume 4: 25 new formative assessment probes.* Arlington, VA: NSTA Press.

Kennedy, M. (1999). Form and substance in mathematics and science professional development. *NISE Brief, 3*(2), 1–7.

Lee, V. E., Smith, J. B., & Croninger, R. G. (1995). *Another look at high school restructuring: Issues in restructuring schools.* Madison: Center on Organization and Restructuring of Schools, School of Education, University of Wisconsin–Madison.

Lipton, L., & Wellman, B. (2004). *Pathways to understanding: Patterns and practices in the learning-focused classroom* (3rd ed). Sherman, CT: MiraVia.

Loucks-Horsley, S., Hewson, P. W., Love, N., & Stiles, K. E. (1998). *Designing professional development for teachers of science and mathematics.* Thousand Oaks, CA: Corwin.

Loucks-Horsley, S., Stiles, K. E., Mundry, S., Love, N., & Hewson, P. (2010). *Designing professional development for teachers of science and mathematics* (3rd ed.). Thousand Oaks, CA: Corwin.

Love, N., Stiles, K. E., Mundry, S., & DiRanna, K. (2008). *The data coach's guide to improving learning for all students: Unleashing the power of collaborative inquiry.* Thousand Oaks, CA: Corwin.

Maine Department of Education. (2007). *Maine learning results: Parameters for essential instruction.* Augusta, ME: Author.

Maine Mathematics and Science Alliance. (n.d.). *PRISMS: Phenomena and representations for the instruction of science in middle schools.* Retrieved July 11, 2009, from http://www.prisms.mmsa.org.

Marks, H. M., Louis, K. S., & Printy, S. M. (2000). The capacity for organizational learning: Implications for pedagogical quality and student achievement. In K. Leithwood (Ed.), *Understanding schools as intelligent systems* (pp. 239–265). Stamford, CT: JAI Press.

Marzano, R., Pickering, D., & Pollock, J. (2001). *Classroom instruction that works: Research-based strategies for increasing student achievement.* Alexandria, VA: Association for Supervision and Curriculum Development.

McLaughlin, M. W., & Talbert, J. (2001). *Professional communities and the work of high school teaching.* Chicago: University of Chicago Press.

McNeill, K. L., & Krajcik, J. (2008). *Inquiry and scientific explanations: Helping students use evidence and reasoning.* Retrieved July 11, 2009, from http://learningcenter.nsta.org/product_detailaspx?id=10.2505/9781933531267.11.

Michaels, S., Shouse, A. W., & Schweingruber, H. A. (2008). *Ready, set, science! Putting research to work in K–8 classrooms.* Washington, DC: National Academies Press.

Monk, D. H. (1994). Subject area preparation of secondary mathematics and science teachers and student achievement. *Economics of Education Review, 13,* 125–145.

Mundry, S., & Stiles, K. E. (Eds.). (2009). *Professional learning communities for science teaching: Lessons from research and practice.* Arlington, VA: NSTA Press.

Murphy, C. U., & Lick, D. W. (2001). *Whole-faculty study groups: Creating student-based professional development.* Thousand Oaks, CA: Corwin.

National Council of Teachers of Mathematics. (2000). *Principles and standards for school mathematics.* Reston, VA: Author.

National Council of Teachers of Mathematics. (2006). *Curriculum focal points: For prekindergarten through grade 8 mathematics.* Reston, VA: Author.

National Research Council. (1996). *National science education standards.* Washington, DC: National Academies Press.

National Research Council. (2001). *Classroom assessment and the National Science Education Standards.* Washington, DC: National Academies Press.

National Research Council. (2007). *Taking science to school: Learning and teaching science in grades K–8.* Committee on Science Learning Kindergarten through Eighth Grade; Richard A. Duschl, Heidi A. Schweingruber, and Andrew W. Shouse, Editors; Board on Science Education, Center for Education, Division of Behavioral and Social Sciences and Education. Washington, DC: National Academies Press.

National Science Teachers Association. (2007). *Principles of professionalism for science educators.* Arlington, VA: Author.

National Staff Development Council. (2001). *Standards for staff development.* Oxford, OH: Author.

Northwest Regional Educational Laboratory. (2008). *Lesson study facilitator's guide.* Retrieved June 26, 2009, from http://www.nwrel.org/lessonstudy/ls-guide.

Posner, G. J., Strike, K. A., Hewson, P. W., & Gertzog, W. A. (1982). Accommodation of a scientific conception: Toward a theory of conceptual change. *Science Education, 66*(2), 211–227.

Programme for International Student Assessment. (2007, April). *PISA 2006 science competencies for tomorrow's world.* Paris: Organisation for Economic Co-Operation and Development.

Roseman, J. E., Kesidou, S., Stern, L., & Caldwell, A. (1999). Heavy books light on learning: Not one middle grades science text rated satisfactory by AAAS's Project 2061. *Science Books & Films, 35*(6). Retrieved July 27, 2009, from http://www.project2061.org/about/press/pr990928 .htm.

Saginor, N. (2008). *Diagnostic classroom observation: Moving beyond best practice.* Thousand Oaks, CA: Corwin.

Sagor, R. (2005). *The action research guidebook: A four-step process for educators and school teams.* Thousand Oaks, CA: Corwin.

Sanger, M. J., & Greenbowe, T. J. (1997). Common student misconceptions in electrochemistry: Galvanic, electrolytic, and concentration cells. *Journal of Research in Science Teaching, 34*, 377–398.

Schmoker, M. (2004). Tipping point: From feckless reform to substantive instructional improvement. *Phi Delta Kappan, 85*(6), 424–432.

Shulman, L. S. (1986). Those who understand: Knowledge growth in teaching. *Educational Researcher, 15*(2), 4–14.

Shymansky, J. A., Youre, L. D., & Good, R. (1991). Elementary school teachers' beliefs about and perceptions of elementary school science, science reading, science textbooks, and supportive instructional factors. *Journal of Research in Science Teaching, 28*, 437–454.

Sparks, D. (2002). *Designing powerful professional development for teachers and principals.* Oxford, OH: National Staff Development Council.

Strube, P., & Lynch, P. P. (1984). Some influences on the modern science text: Alternative science writing. *European Journal of Science Education, 6*(4), 321–338.

Sutherland, L. M., Meriweather, A., Rucker, S., Sarratt, P., Hines-Hale, Y., Moje, E. B., et al. (2006). "More emphasis" on scientific explanation: Developing conceptual understanding while developing scientific literacy. In R. E. Yager (Ed.), *Exemplary science in grades 5–8: Standards-based success stories.* Arlington, VA: NSTA Press.

Van den Akker, J. (1998). The science curriculum: Between ideals and outcomes. In B. J. Fraser & K. G. Tobin (Eds.), *International handbook of science education* (pp. 421–447). Dordrecht, Boston: Kluwer Academic.

Vescio, V., Ross, D., & Adams, A. (2008). A review of research on the impact of professional learning communities on teaching practice and student learning. *Teaching and Teacher Education, 24*(1), 80–91.

Wei, R. C., Darling-Hammond, L., Andree, A., Richardson, N., & Orphanos, S. (2009). *Professional learning in the learning profession: A status report on teacher development in the United States and abroad.* Dallas, TX: National Staff Development Council.

Weiss, I. R., Gellatly, G. B., Montgomery, D. L., Ridgeway, C. J., Templeton, C. D., & Whittington, D. (1999). *Executive summary of the local systemic change through teacher enhancement year four cross-site report.* Chapel Hill, NC: Horizon Research.

Weiss, I. R., Pasley, J. D., Smith, P. S., Banilower, E. R., & Heck, D. J. (2003). *Looking inside the classroom: A study of K–12 mathematics and science education in the United States.* Chapel Hill, NC: Horizon Research.

Wiggins, G., & McTighe, J. (2005). *Understanding by design* (2nd ed.). Alexandria, VA: Association for Supervision and Curriculum Development.

Wiley, D., & Yoon, B. (1995). Teacher reports on opportunity to learn: Analyses of the 1993 California learning assessment systems. *Educational Evaluation and Policy Analysis, 17*, 355–370.

Index

Note: A "t" after a page number indicates a relevant table included in the page range, "f" indicates a figure, and "n" indicates a Facilitator Note.

CORWIN

A SAGE Company

The Corwin logo—a raven striding across an open book—represents the union of courage and learning. Corwin is committed to improving education for all learners by publishing books and other professional development resources for those serving the field of PreK–12 education. By providing practical, hands-on materials, Corwin continues to carry out the promise of its motto: **"Helping Educators Do Their Work Better."**

National Science Teachers Association

The National Science Teachers Association is the largest professional organization in the world promoting excellence and innovation in science teaching and learning for all. NSTA's membership includes more than 55,000 science teachers, science supervisors, administrators, scientists, business and industry representatives, and others involved in science education.

Maine
MATHEMATICS
and SCIENCE Alliance

- **Vision**: The Maine Mathematics and Science Alliance endeavors to enhance science, technology, engineering, and mathematics education to elevate student aspirations and achievement, so all students will meet or exceed state and national standards.
- **Mission**: The MMSA will provide and conduct research, development, and implementation strategies, and form active partnerships to support excellence in STEM curriculum, instruction, and assessment in schools and districts for educators at all stages of their careers.
- **Beliefs**: All activities of the Maine Mathematics and Science Alliance (MMSA) are based on five core beliefs. They are:
  - Strong mathematics and science content knowledge and the skills of inquiry and problem solving.
  - Data-informed planning and decision-making.
  - Research based instructional practice and professional development.
  - Equity of opportunity.
  - Rigorous alignment with state and national standards.

WestEd, a national nonpartisan, nonprofit research, development, and service agency, works with education and other communities to promote excellence, achieve equity, and improve learning for children, youth, and adults. WestEd has 16 offices nationwide, from Washington and Boston to Arizona and California. Its corporate headquarters are in San Francisco. More information about WestEd is available at WestEd.org.